Electronic Amplifiers and Circuit Design

Electronic Amplifiers and Circuit Design

Contributors

Yuanyuan Li, Wenke Lu et al.

www.aurisreference.com

Electronic Amplifiers and Circuit Design

Contributors: Yuanyuan Li, Wenke Lu et al.

Published by Auris Reference Limited
www.aurisreference.com

United Kingdom

Copyright 2016
Printed in 2017 for Sale in the Indian Subcontinent

The information in this book has been obtained from highly regarded resources. The copyrights for individual articles remain with the authors, as indicated. All chapters are distributed under the terms of the Creative Commons Attribution License, which permit unrestricted use, distribution, and reproduction in any medium, provided the original author and source are credited.

Notice

Contributors, whose names have been given on the book cover, are not associated with the Publisher. The editors and the Publisher have attempted to trace the copyright holders of all material reproduced in this publication and apologise to copyright holders if permission has not been obtained. If any copyright holder has not been acknowledged, please write to us so we may rectify.

Reasonable efforts have been made to publish reliable data. The views articulated in the chapters are those of the individual contributors, and not necessarily those of the editors or the Publisher. Editors and/or the Publisher are not responsible for the accuracy of the information in the published chapters or consequences from their use. The Publisher accepts no responsibility for any damage or grievance to individual(s) or property arising out of the use of any material(s), instruction(s), methods or thoughts in the book.

Electronic Amplifiers and Circuit Design

ISBN: 978-1-78154-918-6

British Library Cataloguing in Publication Data
A CIP record for this book is available from the British Library

Printed in the United Kingdom

Exclusively distributed by CBS Publishers & Distributors Pvt. Ltd.

Sales & Distribution Rights only for India, Pakistan, Bangladesh, Sri Lanka, Nepal and Bhutan. This book is not to be sold outside these territories.

Contents

List of Abbreviations vii
List of Contributors ix
Preface xv

Chapter 1   **Circuit Design of Surface Acoustic Wave Based Micro Force Sensor** 1
Yuanyuan Li, Wenke Lu, Changchun Zhu, Qinghong Liu, Haoxin Zhang, and Chenchao Tang

Chapter 2   **Switching Optimization for Class-G Audio Amplifiers with Two Power Supplies** 21
Patrice Russo, Firas Yengui, Gael Pillonnet, Sophie Taupin, Nacer Abouchi

Chapter 3   **Photonic Integrated Semiconductor Optical Amplifier Switch Circuits** 37
R. Stabile and K.A. Williams

Chapter 4   **A Digital Auto-Zeroing Circuit to Reduce Offset in Sub-Threshold Sense Amplifiers** 73
Peter Beshay, Joseph F. Ryan and Benton H. Calhoun

Chapter 5   **Evolvable Metaheuristics on Circuit Design** 89
Felipe Padilla, Aurora Torres, Julio Ponce, María Dolores Torres, Sylvie Ratté and Eunice Ponce-de-León

Chapter 6   **Design of a Switched Capacitor Negative Feedback Circuit for a Very Low Level DC Current Amplifier** 115
Hiroki Higa, Naoki Nakamura

Chapter 7   **The Analysis of the Performance of Multi-Beam Forming in Memory Nonlinear Power Amplifier** 135
Huiyong Li, Xun Li and Chen Wei

Chapter 8   **A High Dynamic Range Ultralow Current-Mode Amplifier With Pico-Ampere Sensitivity For Biosensor Applications** 149
Lei Zhang, Zhiping Yu, Xiangqing He

Chapter 9	A 12-Bit Track and Hold Amplifier for Giga-Sample Applications 163
	Francesco Cannone and Gianfranco Avitabile
Chapter 10	Symbolic Analysis of Analog Circuits Containing Voltage Mirrors and Current Mirrors.. 181
	E. Tlelo-Cuautle, C. Sánchez-López, E. Martínez-Romero, Sheldon X.-D. Tan
Chapter 11	Design and Stability Analysis of CFOA-based Amplifier Circuits Using Bode Criterion 195
	Ivailo Pandiev
Chapter 12	Equivalent Circuit Models for Optical Amplifiers............................ 223
	Jau-Ji Jou and Cheng-Kuang Liu
Chapter 13	The Switched Mode Power Amplifiers... 249
	Elisa Cipriani, Paolo Colantonio, Franco Giannini and Rocco Giofrè
	Citations 291
	Index... 293

List of Abbreviations

PCVD	Ambient-pressure chemical vapor deposition
BEOL	Back-end-of-line
BTI	Bias Temperature Instability
CMP	Chemical-mechanical-planarization
CAD	Computer aided design
CM	Current mirror
CFOAs	Current-feedback operational amplifiers
DAZ	Digital auto-zeroing
EBL	Electron beam photolithography
ENOB	Equivalent number of bits
EOT	Equivalent oxide thickness
EDFAs	Erbium-doped fiber amplifiers
EAs	Evolutionary algorithms
EHW	Evolvable hardware
FP	Fabry-Perot
FTFN	Four terminal floating nullor
GBW	Gain–bandwidth–product
GA	Genetic algorithm
GFETs	Graphene field-effect transistors
HT	Harmonic Tuning terminations
IRNC	Input–referred–noise–current
IRO	Input–referred–offset
IBS	Integrated biosensing system
IC	Integrated circuits
IDT	interdigital transducers
LDs	Laser diodes
MEMS	microelectromechanical systems
NCP	Negative Charge Pump
PS	Pattern Search
PEAQ	Perceptual Evaluation of Audio Quality
PA	Power amplifier
PCB	Printed circuit board
RF	Radio frequency
S/H	Sample & hold
SOAs	Semiconductor optical amplifiers
SAs	Sense amplifiers
SOLT	Short-open-load-through
SDR	Software defined radio
SR	Software radio
SSPAs	Solid state power amplifiers
SFDR	Spurious-free dynamic range

SAW	surface acoustic wave
SC	Switched capacitor
SCF	Switched capacitor filter
SCNF	Switched capacitor negative feedback circuit
SEF	Switched emitter follower
THD	Total Harmonic Distortion
THAs	Track and hold amplifiers
THA	Track-and-hold amplifier
TW	Traveling wave
VLSI	Very large scale integration
VF	Voltage follower
VM	Voltage mirror
VFOAs	Voltage-feedback operational amplifiers
ZVDS	Zero Voltage Derivative Switching
ZVS	Zero Voltage Switching

List of Contributors

Yuanyuan Li
College of Information Science and Technology, Donghua University, Shanghai 201620, China
College of Electronic and Electrical Engineering, Shanghai University of Engineering Science, Shanghai 201620, China

Wenke Lu
College of Information Science and Technology, Donghua University, Shanghai 201620, China

Changchun Zhu
College of Electronics and Information Engineering, Xi'an Jiaotong University, Xi'an 710049, China

Qinghong Liu
Xi'an Leitong Science & Technology Co. Ltd., Xi'an 710049, China

Haoxin Zhang
Xi'an Leitong Science & Technology Co. Ltd., Xi'an 710049, China

Chenchao Tang
College of Electronic and Electrical Engineering, Shanghai University of Engineering Science, Shanghai 201620, China

Patrice Russo
Lyon Institute of Nanotechnology (INL-UMR5270), University of Lyon, Lyon, France
ST Microelectronics, Grenoble, France

Firas Yengui
Lyon Institute of Nanotechnology (INL-UMR5270), University of Lyon, Lyon, France

Gael Pillonnet
Lyon Institute of Nanotechnology (INL-UMR5270), University of Lyon, Lyon, France
ST Microelectronics, Grenoble, France

Sophie Taupin
ST Microelectronics, Grenoble, France

Nacer Abouchi
Lyon Institute of Nanotechnology (INL-UMR5270), University of Lyon, Lyon, France

R. Stabile
Eindhoven University of Technology The Netherlands

K.A. Williams
Eindhoven University of Technology The Netherlands

Peter Beshay
The Charles L. Brown Department of Electrical and Computer Engineering, University of Virginia, Charlottesville, VA 22904, USA

Joseph F. Ryan
Intel Corporation, Hillsboro, OR 97124, USA

Benton H. Calhoun
The Charles L. Brown Department of Electrical and Computer Engineering, University of Virginia, Charlottesville, VA 22904, USA

Felipe Padilla
Aguascalientes University, México
École de Technologie Supérieure Canada

Aurora Torres
Aguascalientes University, México

Julio Ponce
Aguascalientes University, México

María Dolores Torres
Aguascalientes University, México

Sylvie Ratté
École de Technologie Supérieure Canada

Eunice Ponce-de-León
Aguascalientes University, México

Hiroki Higa
Faculty of Engineering, University of the Ryukyus, Okinawa, Japan

Naoki Nakamura
Faculty of Engineering, University of the Ryukyus, Okinawa, Japan

Huiyong Li
School of Electronic Engineering, University of Electronic Science and Technology of China

Xun Li
School of Electronic Engineering, University of Electronic Science and Technology of China

Chen Wei
School of Electronic Engineering, University of Electronic Science and Technology of China

Lei Zhang
Graduated with honors from the Department of Electronic Engineering in Tsinghua University, Beijing, China

Zhiping Yu
Institute of Microelectronics, Tsinghua University, Beijing 100084, People's Republic of China

Xiangqing He
BS degree from Tsinghua Univer- sity, Beijing, China

Francesco Cannone
Politecnico di Bari, Bari, Italy

Gianfranco Avitabile
Politecnico di Bari, Bari, Italy

E. Tlelo-Cuautle
INAOE, Tonantzintla, Mexico
C. Sánchez-López
UAT, Apizaco, Tlaxcala, Mexico

E. Martínez-Romero
IMSE-CSIC, Sevilla, Spain
Sheldon X.-D. Tan
University of California, Riverside, CA, USA

Ivailo Pandiev
Department of Electronics, Faculty of Electronic Engineering and Technologies, Technical University – Sofia, Sofia, 1797, Bulgaria

Jau-Ji Jou
National Kaohsiung University of Applied Sciences Taiwan

Cheng-Kuang Liu
National Taiwan University of Science and Technology Taiwan

Elisa Cipriani
University of Roma Tor Vergata Italy

Paolo Colantonio
University of Roma Tor Vergata Italy

Franco Giannini
University of Roma Tor Vergata Italy

Rocco Giofrè
University of Roma Tor Vergata Italy

Mansour Barari
Electrical Engineering Department, Islamic Azad University, South Tehran Branch, Tehran 11365/4435, Iran

Hamid Reza Karimi
Department of Engineering, Faculty of Engineering and Science, University of Agder, 4898 Grimstad, Norway

Farhad Razaghian
Electrical Engineering Department, Islamic Azad University, South Tehran Branch, Tehran 11365/4435, Iran

Timur Ibrayev
School of Engineering, Nazarbayev University, Astana, Republic of Kazakhstan

Irina Fedorova
School of Engineering, Nazarbayev University, Astana, Republic of Kazakhstan

Akshay Kumar Maan
Griffith University, Brisbane, Australia
Enview R & D Labs, Trivandrum, India

Alex Pappachen James
School of Engineering, Nazarbayev University, Astana, Republic of Kazakhstan

Hongming Lyu
Institute of Microelectronics, Tsinghua University, Beijing, 100084, China
Department of Electrical & Computer Engineering, University of California, San Diego, La Jolla, CA 92093, USA9

Qi Lu
Institute of Microelectronics, Tsinghua University, Beijing, 100084, China

Yilin Huang
Institute of Microelectronics, Tsinghua University, Beijing, 100084, China

Teng Ma
Shenyang National Laboratory for Materials Science, Institute of Metal Research, Chinese Academy of Sciences, Shenyang, 110016, China

Jinyu Zhang
Institute of Microelectronics, Tsinghua University, Beijing, 100084, China

Xiaoming Wu
Institute of Microelectronics, Tsinghua University, Beijing, 100084, China

ZhipingYu
Institute of Microelectronics, Tsinghua University, Beijing, 100084, China

Wencai Ren
Shenyang National Laboratory for Materials Science, Institute of Metal Research, Chinese Academy of Sciences, Shenyang, 110016, China.

Hui-Ming Cheng
Shenyang National Laboratory for Materials Science, Institute of Metal Research, Chinese Academy of Sciences, Shenyang, 110016, China

Huaqiang Wu
Institute of Microelectronics, Tsinghua University, Beijing, 100084, China

Tsinghua National Laboratory for Information Science and Technology (TNList), Beijing, 100084, China

He Qian
Institute of Microelectronics, Tsinghua University, Beijing, 100084, China
Tsinghua National Laboratory for Information Science and Technology (TNList), Beijing, 100084, China

Preface

An amplifier is an electronic device that increases the voltage, current, or power of a signal. Amplifiers are used in wireless communications and broadcasting, and in audio equipment of all kinds. Electronic circuit design comprises the analysis and synthesis of electronic circuits. The text *Electronic Amplifiers and Circuit Design* presents the basic principles of amplifiers and circuit design, basic per-stage building blocks, and feedback. A surface acoustic wave (SAW) based micro force sensor is designed in first chapter, which is based on the theories of wavelet transform, SAW detection, and pierce oscillator circuits. Second chapter presents a system-level method to decrease the power consumption of integrated audio Class-G amplifiers for mobile phones by using the same implementation of the level detector, but by changing the parameters of the switching algorithm. Third chapter addresses the engineering of semiconductor optical amplifier (SOA) gates for high-connectivity integrated photonic switching circuits. In fourth chapter, we propose a circuit that reduces the sense amp offset using an auto-zeroing scheme with automatic temperature, voltage, and aging tracking. Evolvable metaheuristics on circuit design have been discussed in fifth chapter. Sixth chapter describes the design of a switched capacitor negative feedback circuit for a very low level dc current amplifier. In seventh chapter, a form of the multi-beamforming signal and a nonlinear model with memory for power amplifier (PA) are given. In eighth chapter, a novel ULCA aiming at the application of signal pre-amplification in the integrated biosensing system has been proposed and verified by using SMIC 0.18 µm CMOS technology. Ninth chapter presents a track-and-hold amplifier (THA), based on the switched emitter follower topology, suitable for emerging receiver's architectures and data acquisition systems. Tenth chapter focuses on symbolic analysis of analog circuits containing voltage mirrors and current mirrors. In eleventh chapter, the frequency stability of small-signal high-speed amplifier circuits using Bode criterion has been analyzed theoretically. In twelfth chapter, using a new circuit model for EDFAs, the static and dynamic characteristics of EDFAs can be analyzed conveniently through the aid of a SPICE simulator. The switched mode power amplifiers have been presented in last chapter.

Chapter 1

CIRCUIT DESIGN OF SURFACE ACOUSTIC WAVE BASED MICRO FORCE SENSOR

Yuanyuan Li,[1,2] Wenke Lu,[1] Changchun Zhu,[3] Qinghong Liu,[4] Haoxin Zhang,[4] and Chenchao Tang[2]

[1]College of Information Science and Technology, Donghua University, Shanghai 201620, China

[2]College of Electronic and Electrical Engineering, Shanghai University of Engineering Science, Shanghai 201620, China

[3]College of Electronics and Information Engineering, Xi'an Jiaotong University, Xi'an 710049, China

[4]Xi'an Leitong Science & Technology Co. Ltd., Xi'an 710049, China

ABSTRACT

Pressure sensors are commonly used in industrial production and mechanical system. However, resistance strain, piezoresistive sensor, and ceramic capacitive pressure sensors possess limitations, especially in micro force measurement. A surface acoustic wave (SAW) based micro force sensor is designed in this paper, which is based on the theories of wavelet transform, SAW detection, and pierce oscillator circuits. Using lithium niobate as the basal material, a mathematical model is established to analyze the frequency, and a peripheral circuit is designed to measure the micro force. The SAW based micro force sensor is tested to show the reasonable design of detection circuit and the stability of frequency and amplitude.

INTRODUCTION

Wavelet transform finds its application in many disciplines and fields such as in image processing, water-sound, earthquake detection, biomedicine, mechanical vibration, pronunciation recognition, communication, chemical industry, and torrent analysis [1]. SAW devices are early examples of microelectromechanical systems (MEMS) because of the coupling needed between the electrical and mechanical properties as discussed by Ballantine et al. [2]. The method of implementing wavelet transform with SAW devices has been first proposed by Peng et al. [3–5]. The wavelet transform device of SAW can benefit from the excellent properties of the SAW devices, namely, passive, small size, low cost, excellent temperature stability, high reliability, and high reproducibility, which overcomes the complicated algorithms and high power for VLSI [6, 7], and big size and low reproducibility for optical devices [8]. Wave propagation along the surface allows the sensitivity of the wavelet transform device of SAW to change in the external environment and the development of these sensors for applications such as gas detection, changes in fluid viscosity, determination of stiffness constants of mechanical vibration, and detection of the onset of ice formation on aerospace structures [9–11].

Nowadays, micromanipulation has performed to design either mobile micro robots or a precise positioning device under the control of mechanical systems. Various tools for manipulating micro parts and assembling micro systems have been developed and integrated. Semiconductor strain gauges are preferred when small forces have been measured. M. Jungwirth has described the micromechanical precision pressure sensor but in delay lines. A silicon based micro force sensor has been developed with larger electromechanical coupling coefficient k2. The three-axis micro force sensor has been designed by Jungwirth et al. with the problem of measurement uncertainty. The dual axis micro force sensor for robotic manipulations needs to use strain gauges [12–15].

This paper proposes to use substrate materials of small electromechanical coupling coefficient k2 (128° Y shear $LiNbO_3$) in the manufacture to design the wavelet transform device of SAW based micro force sensor, which has high accuracy and sensitivity testing precision. In addition, a mathematical model is established to analyze the frequency of our sensor and a peripheral detection circuit is designed. Within the scope of effective measurement, the SAW based micro force sensor possesses good linearity, consistency, and repeatability in performance. Besides, it uses the piezoelectric properties and the temperature stability of the crystal, as well as the frequency signal instead of the conventional pressure sensor with voltage signal, which makes the signal processing of this device more digital and possesses more stable performances.

The succeeding sections are organized as follows. Section 2 introduces the fundamental principles of designing interdigital transducers (IDT) for wavelet transform device of SAW. In Section 3, the peripheral detection circuit is explained. In addition, the linear regression model is provided in Section 4. Finally Section 5 delivers the testing and analysis of SAW based micro force sensor which is used for measuring various micro force.

DESIGN IDT FOR WAVELET TRANSFORM DEVICE OF SAW

The wavelet function is

$$\psi_s(t) = \frac{1}{\sqrt{s}}\psi\left(\frac{t}{s}\right), \quad (1)$$

where s denotes the scale of wavelet function.

The wavelet transform of signal f(t) is

$$\begin{aligned} WT_s(\tau) &= f(t) * \psi_s(t) \\ &= \int_R f(t) \frac{1}{\sqrt{s}} \psi\left(\frac{\tau-t}{s}\right) dt \\ &= \frac{1}{\sqrt{s}} \int_R f(t) \psi\left(\frac{\tau-t}{s}\right) dt. \end{aligned} \quad (2)$$

When $\psi_s(t)$ is a Morlet wavelet function, formula (1) is converted into [3–5]

$$\psi_s(t) = \frac{1}{\sqrt{s}} e^{-(1/2)(t/s)^2} e^{j2\pi(f_0/s)t} = P_s(t) e^{j2\pi(f_0/s)t}, \quad (3)$$

where $P_s(t)$ is the wavelet-envelope function, $P_s(t) = (1/\sqrt{s})e^{-1/2(t/s)^2}$, and f_0/s is the center frequency. When, s=2k, k is a random number from $-\infty$ to $+\infty$. The wavelet function shown in formula (3) is converted into the Morlet dyadic wavelet function [4, 5]. The microwave communication equipment of this paper design needs a single-scale wavelet transform processor of the center frequency f_0, which can be rewritten as

$$f_0 = \frac{v_s}{\lambda} = \frac{v_s}{2(a+b)}, \quad (4)$$

where $\lambda = 2(a+b)$ is the wavelength, v_s is the speed of the SAW, and a and b are the width and the interval of IDT.

The delay line in wavelet transform device of SAW designed in this paper exhibits the basic structure in Figure1 and two acoustic electric transducers on the piezoelectric characteristic substrate material polishing surface, named input IDT and output IDT.

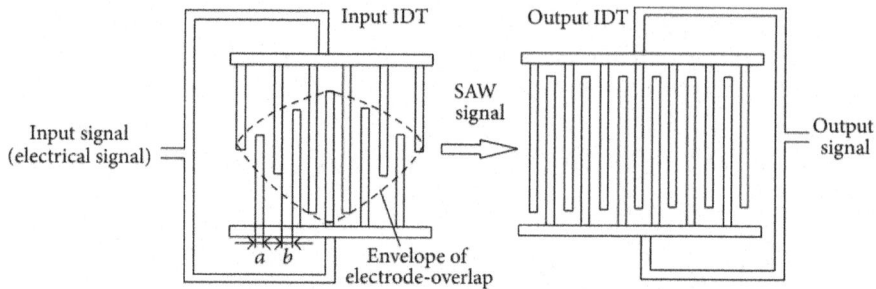

Figure 1: The transducer substrate of wavelet transform device of SAW.

As the Morlet wavelet transform of SAW devices is based on formula (3), once all the parameters in (4) are relatively fixed, its center frequency of each scale should be corresponding with the only device. The device has been designed and produced in this paper that is shown in Figure 2.

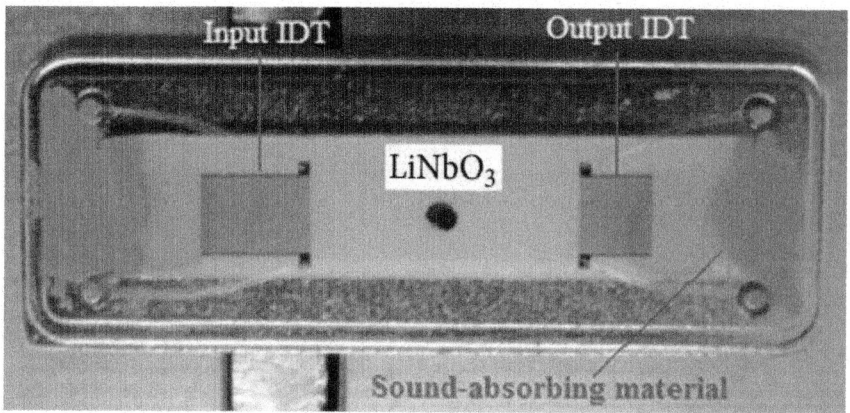

Figure 2: The fabricated single-scale wavelet transform device of SAW.

DESIGN OF THE PERIPHERAL DETECTION CIRCUIT

The Principle of Peripheral Detection Circuit

After making the fabricated device, the peripheral detection circuit needs to be designed to get F_m and f which are the pressure and the output frequency of the SAW based micro force sensor. Figure 3 shows the general structure of the SAW micro force sensor with IDTs.

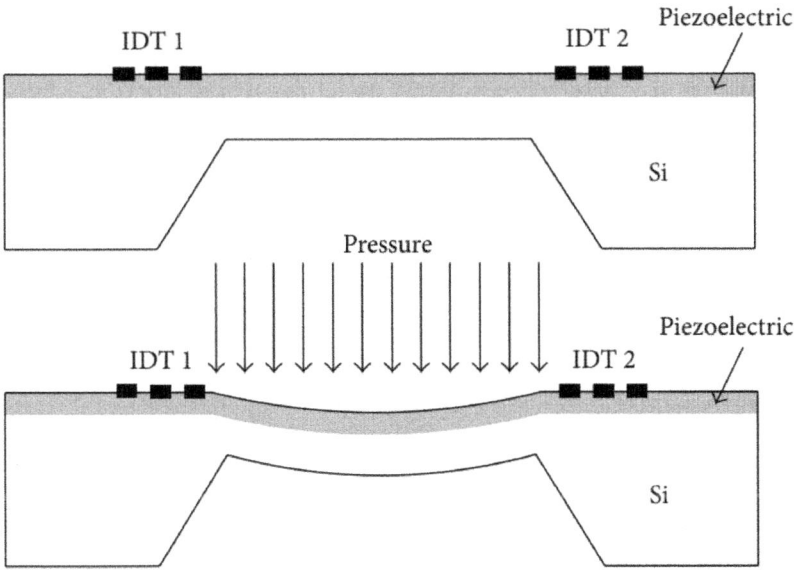

Figure 3: General structure of the SAW micro force sensor with two IDTs.

If the SAW resonator density of the piezoelectric substrate is m, the unit is m/g², the dielectric constant is ε, the electrical conductivity is σ, the elastic parameter is c, the unit is n/m², the environment temperature is t, and the pressure is p [16], v_s based on the theory of surface disturbance is given as

$$v_s = v(m, \varepsilon, \sigma, c, t, p). \tag{5}$$

After derivation, (5) can be changed to

$$\frac{\Delta v_s}{v_s} = \frac{1}{v_s}\left(\frac{\partial v}{\partial m}\Delta m + \frac{\partial v}{\partial \varepsilon}\Delta \varepsilon + \frac{\partial v}{\partial \sigma}\Delta \sigma \right.$$
$$\left. + \frac{\partial v}{\partial c}\Delta c + \frac{\partial v}{\partial t}\Delta t + \frac{\partial v}{\partial p}\Delta p\right). \tag{6}$$

Equation (6) has been used to calculate the relationship between the propagation speed of available and the SAW oscillator frequency which is equal to the following equation:

The change in the oscillator frequency and the measurement of the parameters can be obtained by using (7), which is the working principle of the SAW micro force sensor circuit.

The Actual Oscillator Circuit

The mixing frequency circuit is divided into two oscillator circuits, one for the reference oscillator circuit with a fixed frequency, and the other for the detector circuit. The frequency difference can be received from the mixing circuit. Figure 4 shows the circuit framework.

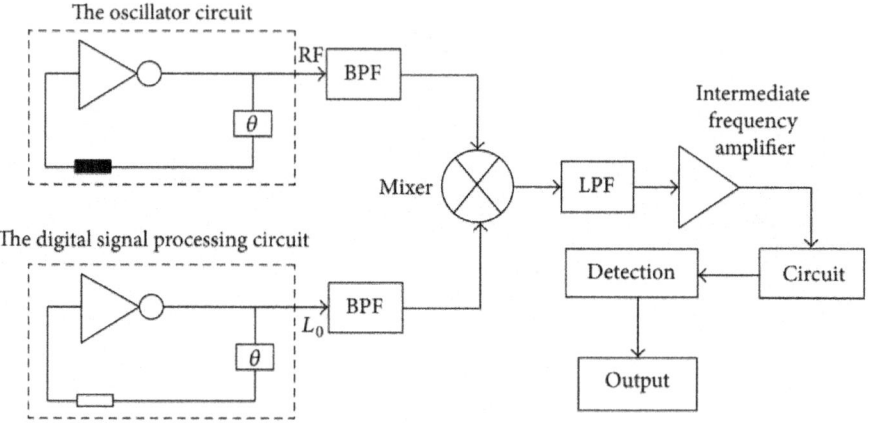

Figure 4: The circuit framework of fixed frequency circuit.

The oscillator circuit with fixed frequency in the dual-channel structure has an output signal, which is used as the stable signal source during measurement. The digital signal processing circuit is composed of a filter, a mixer, an amplifier, and a shaping circuit.

For a surface acoustic generator as the sensitive element of the SAW micro force sensor, the frequency stability directly affects the resolution of the testing precision. The oscillator frequency has depended on the conditions of the feedback loop phase. Therefore, the improvement of the SAW oscillator frequency stability also improves the performance of the micro force sensor.

The Pierce oscillator circuit has a better stability compared with Colpitts and Clapp circuits [17]. Figure 5shows the oscillator circuit of the SAW based micro force sensor based on the principle of Pierce circuit and the combination with the design requirements.

Circuit Design of Surface Acoustic Wave Based Micro Force Sensor

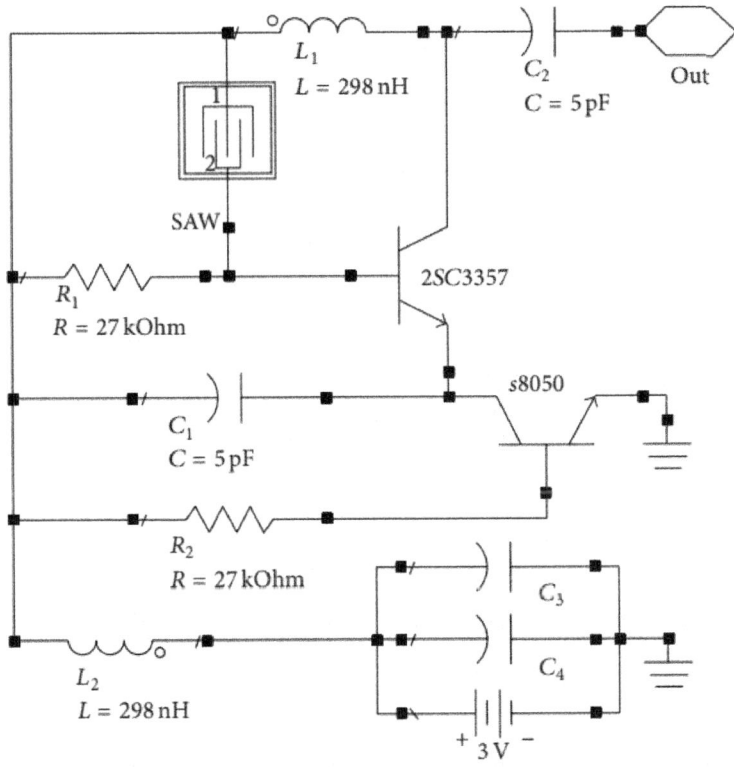

Figure 5: The Pierce circuit of the SAW based micro force sensor.

The high-frequency triode 2SC3357 affects the resonance amplifier in the circuit by ensuring the normal startup of the oscillator circuit. R_1 is the DC bias resistance of this triode, and adjusting the value changes with DC working points. R_2 is the DC bias resistance of the triode s8050 that changes the DC working points and transforms the bias current of the circuit. This triode functions as the current source and stabilizes the circuit working state. The SAW resonator affects the frequency selection in the feedback loop by inducing the SAW oscillator frequency stability and improving the antijamming ability.

To ensure that the feedback loop circuit phase is in balance, the inductance L has been added to the Pierce circuit to eliminate the effect of the stability of the DC voltage source from the oscillator circuit. The LC parallel resonant circuit also affects the DC bypass circuit. The feedback coefficient F is related to C_1 and C_2 and is given as

$$F = \frac{C_1}{C_2}, \qquad (8)$$

where C_3, C_4, and L_2 are the decoupling devices of the DC source that can eliminate the effect of the oscillator circuit from the LC filter function.

Determine the Parameters

The circuit layout, selection of components, and calculated parameters are important. The work frequency of the micro force sensor is 50 MHz; the center frequency of the LC parallel resonant circuit should be

$$f_0 = \frac{1}{2\pi\sqrt{LC}}. \tag{9}$$

To ensure that the amplitude of oscillator circuit has better characteristics than the initial conditions, C_2 should be smaller. In addition, the designed oscillator frequency is given; the capacitance is selected as 5.1 pF. From (10), L can be obtained as

$$L = \frac{1}{(2\pi f_0)^2 C}. \tag{10}$$

Because parasitic parameters in practical circuits affect their performances, L must be adjustable. By changing the number of turns of the air-core coil, the coil inductance can be fine-tuned, and the frequency of the LC parallel resonant circuit will be in accordance with the working frequency of the SAW based micro force sensor. The air-core coil inductance has been calculated by the following equation:

$$L = \frac{0.01 \times D \times N^2}{(L_N/D) + 0.44}, \tag{11}$$

where L is the coil inductance (in μH), D is the diameter of the coil (in cm), N is the number of coil turns, and L_N is the coil length (in cm).

An oscillator frequency source should have good stability, phase noise, and high Q value [18]. Figure 6 shows the equivalent circuit, with C_2 expressed in picofarads.

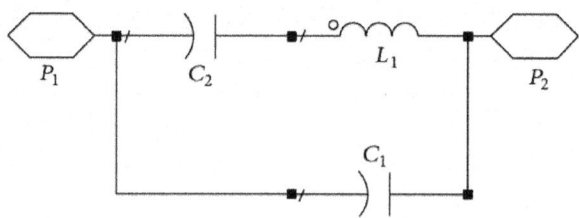

Figure 6: The equivalent circuit of the SAW device.

The SAW resonator can work in series or parallel resonant frequencies upon operating in the feedback loop. The transistor 2SC3357 and s8050 provide the DC bias current in the oscillator circuit design based on the requirements.

THE LINEAR REGRESSION MODEL OF THE SAW BASED MICRO FORCE SENSOR

Establishment of the Linear Regression Model

After getting the pressure of SAW sensor (F_m) and the output frequency (f) from the actual oscillator circuit, the fitting function should be established. The relationship between the pressure of this sensor (F_m) and the output frequency (f) is given as follows:

$$F_m = k_0 + k_1 \Delta f + k_2 \Delta f^2 + k_3 \Delta f^3 + \cdots + k_n \Delta f^n \cdots . \qquad (12)$$

If F_i, f_i, i=1, 2, …,n, and n=5, the method of least squares can be used to solve this function is given as

$$F_m = k_0 + k_1 \Delta f + k_2 \Delta f^2 + k_3 \Delta f^3 \\ + k_4 \Delta f^4 + k_5 \Delta f^5 + k_6 \Delta f^6. \qquad (13)$$

The regression coefficients $k_0, k_1, k_2, k_3, k_4, k_5$, and k_6 can be calculated through formula (13).

Calculate the Input and Output Variable Regression

The sum of the variance (Q) between the dependent and independent variables is given as [19]

$$Q(k_0 - k_6) = \sum_{i=1}^{N} \Delta_i^2 = \sum_{i=1}^{N} [\Delta f_i - F_i]^2, \qquad (14)$$

where N is the sampling point of the independent variables. Therefore, the least squares estimation calculates the minimum value of , which implies that $\hat{k}_0, \hat{k}_1, \hat{k}_2, \hat{k}_3, \hat{k}_4, \hat{k}_5,$ and $\hat{k}_6$ must suit the following equation:

$$= \sum_{i=1}^{N} [\Delta f_i - \Gamma_i]^2 = \min_{\hat{k}_0 - \hat{k}_6} \sum_{i=1}^{N} [\Delta f_i - \Gamma_i]^2, \qquad (15)$$

where $\hat{k}_0, \hat{k}_1, \hat{k}_2, \hat{k}_3, \hat{k}_4, \hat{k}_5,$ and $\hat{k}_6$ are the respective least squares estimations of $k_0, k_1, k_2, k_3, k_4, k_5,$ and k_6

When

$$\left.\frac{\partial Q(k_0,k_1,k_2,k_3,k_4,k_5)}{\partial k_0}\right|_{k_0=\hat{k}_0} = 0,$$

$$\left.\frac{\partial Q(k_0,k_1,k_2,k_3,k_4,k_5)}{\partial k_1}\right|_{k_0=\hat{k}_1} = 0,$$

$$\left.\frac{\partial Q(k_0,k_1,k_2,k_3,k_4,k_5)}{\partial k_2}\right|_{k_0=\hat{k}_2} = 0,$$

$$\left.\frac{\partial Q(k_0,k_1,k_2,k_3,k_4,k_5)}{\partial k_3}\right|_{k_0=\hat{k}_3} = 0,$$

$$\left.\frac{\partial Q(k_0,k_1,k_2,k_3,k_4,k_5)}{\partial k_4}\right|_{k_0=\hat{k}_4} = 0,$$

$$\left.\frac{\partial Q(k_0,k_1,k_2,k_3,k_4,k_5)}{\partial k_5}\right|_{k_0=\hat{k}_5} = 0,$$

$$\left.\frac{\partial Q(k_0,k_1,k_2,k_3,k_4,k_5)}{\partial k_6}\right|_{k_0=\hat{k}_6} = 0. \tag{16}$$

The equation (17) can be derived by formula (15) and (16)

$$\sum_{i=1}^{N}\left[\left(\hat{k}_0 + \hat{k}_1\Delta f_i + \hat{k}_2\Delta f_i^2 + \hat{k}_3\Delta f_i^3 \right.\right.$$
$$\left.\left. + \hat{k}_4\Delta f_i^4 + \hat{k}_5\Delta f_i^5 + \hat{k}_6\Delta f_i^6\right) - F_i\right] = 0,$$

$$\sum_{i=1}^{N}\left[\left(\hat{k}_0 + \hat{k}_1\Delta f_i + \hat{k}_2\Delta f_i^2 + \hat{k}_3\Delta f_i^3 \right.\right.$$
$$\left.\left. + \hat{k}_4\Delta f_i^4 + \hat{k}_5\Delta f_i^5 + \hat{k}_6\Delta f_i^6\right) - F_i\right]\Delta f_i = 0,$$

$$\sum_{i=1}^{N}\left[\left(\hat{k}_0 + \hat{k}_1\Delta f_i + \hat{k}_2\Delta f_i^2 + \hat{k}_3\Delta f_i^3 \right.\right.$$
$$\left.\left. + \hat{k}_4\Delta f_i^4 + \hat{k}_5\Delta f_i^5 + \hat{k}_6\Delta f_i^6\right) - F_i\right]\Delta f_i^2 = 0,$$

$$\sum_{i=1}^{N}\left[\left(\hat{k}_0 + \hat{k}_1\Delta f_i + \hat{k}_2\Delta f_i^2 + \hat{k}_3\Delta f_i^3 \right.\right.$$
$$\left.\left. + \hat{k}_4\Delta f_i^4 + \hat{k}_5\Delta f_i^5 + \hat{k}_6\Delta f_i^6\right) - F_i\right]\Delta f_i^3 = 0,$$

$$\sum_{i=1}^{N}\left[\left(\hat{k}_0 + \hat{k}_1\Delta f_i + \hat{k}_2\Delta f_i^2 + \hat{k}_3\Delta f_i^3 \right.\right.$$
$$\left.\left. + \hat{k}_4\Delta f_i^4 + \hat{k}_5\Delta f_i^5 + \hat{k}_6\Delta f_i^6\right) - F_i\right]\Delta f_i^4 = 0,$$

$$\sum_{i=1}^{N}\left[\left(\hat{k}_0 + \hat{k}_1\Delta f_i + \hat{k}_2\Delta f_i^2 + \hat{k}_3\Delta f_i^3 \right.\right.$$
$$\left.\left. + \hat{k}_4\Delta f_i^4 + \hat{k}_5\Delta f_i^5 + \hat{k}_6\Delta f_i^6\right) - F_i\right]\Delta f_i^5 = 0,$$

$$\sum_{i=1}^{N}\left[\left(\hat{k}_0 + \hat{k}_1\Delta f_i + \hat{k}_2\Delta f_i^2 + \hat{k}_3\Delta f_i^3 \right.\right.$$
$$\left.\left. + \hat{k}_4\Delta f_i^4 + \hat{k}_5\Delta f_i^5 + \hat{k}_6\Delta f_i^6\right) - F_i\right]\Delta f_i^6 = 0. \tag{17}$$

Equation (18) can be derived from formulas (14) and (17). Consider

$$N\hat{k}_0 + \hat{k}_1 \sum_{i=1}^{N} \Delta f + \hat{k}_2 \sum_{i=1}^{N} \Delta f^2 + \hat{k}_3 \sum_{i=1}^{N} \Delta f^3$$

$$+ \hat{k}_4 \sum_{i=1}^{N} \Delta f^4 + \hat{k}_5 \sum_{i=1}^{N} \Delta f^5 + \hat{k}_6 \sum_{i=1}^{N} \Delta f^6 = \sum_{i=1}^{N} F_i ,$$

$$\hat{k}_0 \sum_{i=1}^{N} \Delta f + \hat{k}_1 \sum_{i=1}^{N} \Delta f^2 + \hat{k}_2 \sum_{i=1}^{N} \Delta f^3 + \hat{k}_3 \sum_{i=1}^{N} \Delta f^4$$

$$+ \hat{k}_4 \sum_{i=1}^{N} \Delta f^5 + \hat{k}_5 \sum_{i=1}^{N} \Delta f^6 + \hat{k}_6 \sum_{i=1}^{N} \Delta f^7 = \sum_{i=1}^{N} F_i \Delta f_i ,$$

$$\hat{k}_0 \sum_{i=1}^{N} \Delta f^2 + \hat{k}_1 \sum_{i=1}^{N} \Delta f^3 + \hat{k}_2 \sum_{i=1}^{N} \Delta f^4 + \hat{k}_3 \sum_{i=1}^{N} \Delta f^5$$

$$+ \hat{k}_4 \sum_{i=1}^{N} \Delta f^6 + \hat{k}_5 \sum_{i=1}^{N} \Delta f^7 + \hat{k}_6 \sum_{i=1}^{N} \Delta f^8 = \sum_{i=1}^{N} F_i \Delta f_i^2 ,$$

$$\hat{k}_0 \sum_{i=1}^{N} \Delta f^3 + \hat{k}_1 \sum_{i=1}^{N} \Delta f^4 + \hat{k}_2 \sum_{i=1}^{N} \Delta f^5 + \hat{k}_3 \sum_{i=1}^{N} \Delta f^6$$

$$+ \hat{k}_4 \sum_{i=1}^{N} \Delta f^7 + \hat{k}_5 \sum_{i=1}^{N} \Delta f^8 + \hat{k}_6 \sum_{i=1}^{N} \Delta f^9 = \sum_{i=1}^{N} F_i \Delta f_i^3 ,$$

$$\hat{k}_0 \sum_{i=1}^{N} \Delta f^4 + \hat{k}_1 \sum_{i=1}^{N} \Delta f^5 + \hat{k}_2 \sum_{i=1}^{N} \Delta f^6 + \hat{k}_3 \sum_{i=1}^{N} \Delta f^7$$

$$+ \hat{k}_4 \sum_{i=1}^{N} \Delta f^8 + \hat{k}_5 \sum_{i=1}^{N} \Delta f^9 + \hat{k}_6 \sum_{i=1}^{N} \Delta f^{10} = \sum_{i=1}^{N} F_i \Delta f_i^4 ,$$

$$\hat{k}_0 \sum_{i=1}^{N} \Delta f^5 + \hat{k}_1 \sum_{i=1}^{N} \Delta f^6 + \hat{k}_2 \sum_{i=1}^{N} \Delta f^7 + \hat{k}_3 \sum_{i=1}^{N} \Delta f^8$$

$$+ \hat{k}_4 \sum_{i=1}^{N} \Delta f^9 + \hat{k}_5 \sum_{i=1}^{N} \Delta f^{10} + \hat{k}_6 \sum_{i=1}^{N} \Delta f^{11} = \sum_{i=1}^{N} F_i \Delta f_i^5 ,$$

$$\hat{k}_0 \sum_{i=1}^{N} \Delta f^6 + \hat{k}_1 \sum_{i=1}^{N} \Delta f^7 + \hat{k}_2 \sum_{i=1}^{N} \Delta f^8 + \hat{k}_3 \sum_{i=1}^{N} \Delta f^9$$

$$+ \hat{k}_4 \sum_{i=1}^{N} \Delta f^{10} + \hat{k}_5 \sum_{i=1}^{N} \Delta f^{11} + \hat{k}_6 \sum_{i=1}^{N} \Delta f^{12} = \sum_{i=1}^{N} F_i \Delta f_i^6 . \quad (18)$$

Equation (18) can be written as

$$N\hat{k}_0 + A\hat{k}_1 + B\hat{k}_2 + C\hat{k}_3 + D\hat{k}_4 + E\hat{k}_5 + F\hat{k}_6 = M,$$

$$A\hat{k}_0 + B\hat{k}_1 + C\hat{k}_2 + D\hat{k}_3 + E\hat{k}_4 + F\hat{k}_5 + G\hat{k}_6 = P,$$

$$B\hat{k}_0 + C\hat{k}_1 + D\hat{k}_2 + E\hat{k}_3 + F\hat{k}_4 + G\hat{k}_5 + H\hat{k}_6 = Q,$$

$$C\hat{k}_0 + D\hat{k}_1 + E\hat{k}_2 + F\hat{k}_3 + G\hat{k}_4 + H\hat{k}_5 + I\hat{k}_6 = R,$$

$$D\hat{k}_0 + E\hat{k}_1 + F\hat{k}_2 + G\hat{k}_3 + H\hat{k}_4 + I\hat{k}_5 + J\hat{k}_6 = S,$$

$$E\hat{k}_0 + F\hat{k}_1 + G\hat{k}_2 + H\hat{k}_3 + I\hat{k}_4 + J\hat{k}_5 + K\hat{k}_6 = T,$$

$$F\hat{k}_0 + G\hat{k}_1 + H\hat{k}_2 + I\hat{k}_3 + J\hat{k}_4 + K\hat{k}_5 + L\hat{k}_6 = U, \qquad (19)$$

where

$$A = \sum_{i=1}^{N} \Delta f_i, \qquad B = \sum_{i=1}^{N} \Delta f_i^2,$$

$$C = \sum_{i=1}^{N} \Delta f_i^3, \qquad D = \sum_{i=1}^{N} \Delta f_i^4,$$

$$E = \sum_{i=1}^{N} \Delta f_i^5, \qquad F = \sum_{i=1}^{N} \Delta f_i^6,$$

$$G = \sum_{i=1}^{N} \Delta f_i^7,$$

$$H = \sum_{i=1}^{N} \Delta f_i^8, \qquad I = \sum_{i=1}^{N} \Delta f_i^9,$$

$$J = \sum_{i=1}^{N} \Delta f_i^{10}, \qquad K = \sum_{i=1}^{N} \Delta f_i^{11},$$

$$L = \sum_{i=1}^{N} \Delta f_i^{12},$$

$$H = \sum_{i=1}^{N} \Delta f_i^8, \qquad I = \sum_{i=1}^{N} \Delta f_i^9,$$

$$J = \sum_{i=1}^{N} \Delta f_i^{10}, \qquad K = \sum_{i=1}^{N} \Delta f_i^{11},$$

$$L = \sum_{i=1}^{N} \Delta f_i^{12},$$

$$M = \sum_{i=1}^{N} F_i, \qquad P = \sum_{i=1}^{N} F_i \Delta f_i^1,$$

$$Q = \sum_{i=1}^{N} F_i \Delta f_i^2, \qquad R = \sum_{i=1}^{N} F_i \Delta f_i^3,$$

$$S = \sum_{i=1}^{N} F_i \Delta f_i^4, \qquad T = \sum_{i=1}^{N} F_i \Delta f_i^5,$$

$$U = \sum_{i=1}^{N} F_i \Delta f_i^6. \tag{20}$$

Coefficients k_0 to k_6 in (13) can be obtained, and the input and output variable regression of the SAW based micro force sensor can be calculated by solving (18).

TESTING AND ANALYSIS OF THE SAW BASED MICRO FORCE SENSOR

Actual Circuit Detection Results

The output frequency of the SAW based micro force sensor is conducted by using the network analyzer equipment E5061A (Figure 7). Force-measuring elements employ a cantilever beam loaded with 0–20 kPa pressure and add 2 kPa to this beam at each time [20].

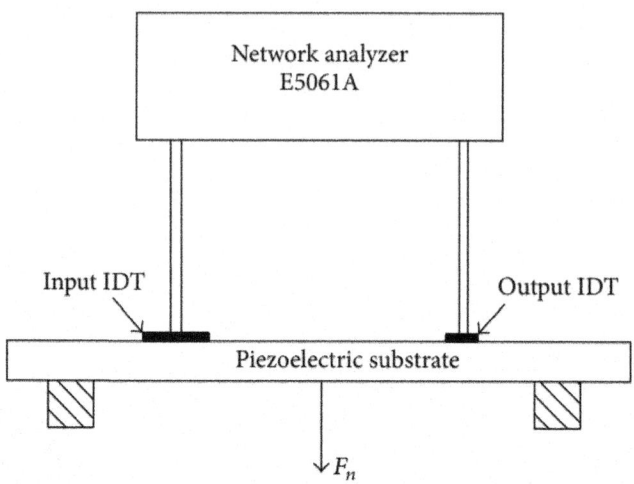

Figure 7: Schematic of the SAW based micro force sensor.

Figure 8 shows the test schematic. The SAW based micro force sensor has been designed with a single channel. For the inductive components, the electromagnetic field interference is more sensitive; two inductors have been placed at 90° to reduce interference between these components. Figure 9 shows the actual circuit. In the process of real production, the device location and the connection between our devices are more important.

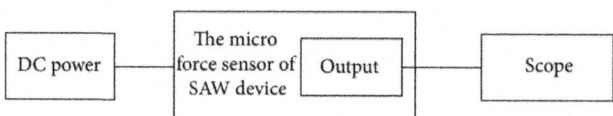

Figure 8: Test schematic diagram.

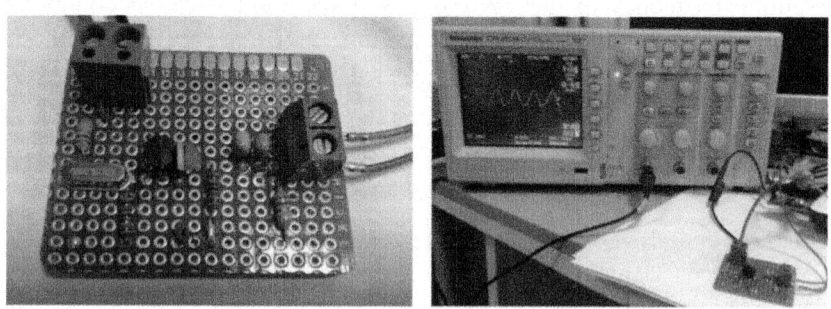

Figure 9: The actual circuit testing diagram.

The Fitting Curve for the Measured Frequency

Figure 10 shows the actual test waveform diagram [21].

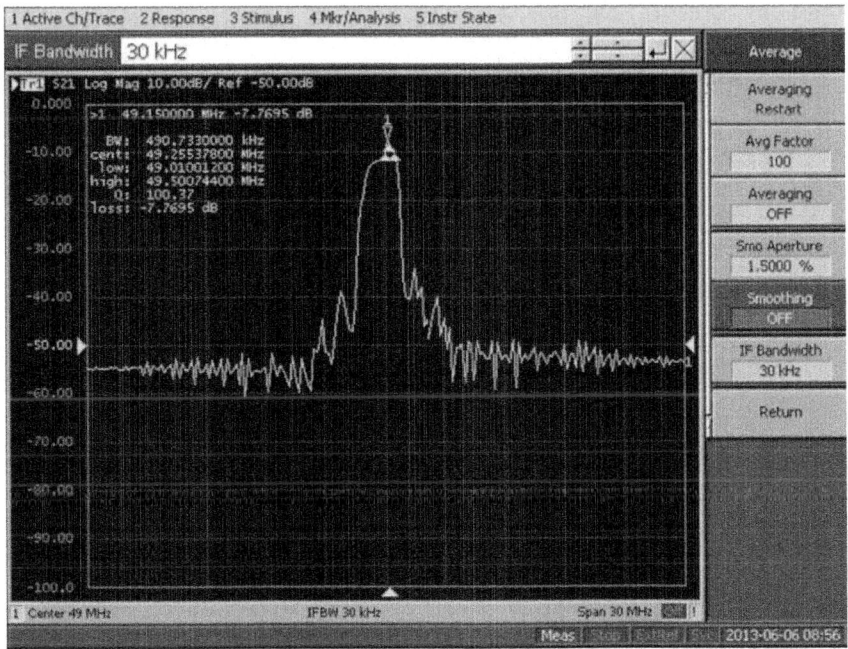

Figure 10: The actual test waveform diagram.

To eliminate the system instability, the experimental data has been averaged in Table 1.

Table 1: Difference in frequency data for various pressures.

F_m	(kPa)									
	0	2	4	6	8	10	12	14	16	18	20
f (MHz)	49.255389	49.255391	49.255378	49.255393	49.255434	49.255459	49.255474	49.255481	49.255447	49.255486	49.255497
f (MHz)	49.255321	49.255312	49.255298	49.255296	49.255338	49.255343	49.255357	49.255359	49.255319	49.255361	49.255363
Δf (Hz)	68	79	80	97	96	116	117	122	128	125	134

Frequency experiment data samples have been generated from (18) according to the different micro pressure reading, and the estimated regression coefficients $\hat{k}_0, \hat{k}_1, \hat{k}_2, \hat{k}_3, \hat{k}_4, \hat{k}_5,$ and $\hat{k}_6$ can be calculated. The method of least squares can reduce the error due to measurement inaccuracies caused by the SAW device based micro force sensor. Through the minimization of squared errors of our actual data, the least squares method has the ability to find the best matching function data. Additionally, the fitting curve of the experimental data shows that the circuit design is reasonable and the device exhibits good linearity (Figure 11). The pressure F_m is proportional to the output frequency f, which has proved that the SAW detection and pierce oscillator circuit have correct logic function.

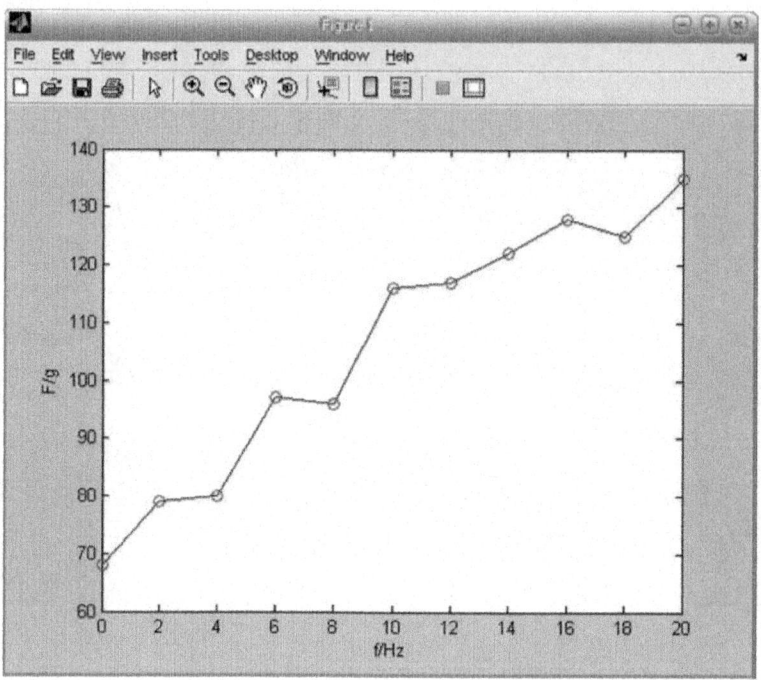

Figure 11: The experimental data fitting curve.

Based on (13), the available in different micro force sensor output difference frequency transfer function is

$$F_m(\Delta f) = 3.1441 \times 10^{-1} + 5.9057 \times 10^{-2} \Delta f_i$$
$$- 2.3012 \times 10^{-4} \Delta f_i^2 + 5.4127 \times 10^{-7} \Delta f_i^3$$
$$- 6.0364 \times 10^{-10} \Delta f_i^4$$
$$+ 3.0200 \times 10^{-13} \Delta f_i^5 - 5.0540 \times 10^{-17} \Delta f_i^6. \qquad (21)$$

Because of the load impedance having affected the amplitude output of the micro force sensor, the impedance of the oscilloscope probe during testing is set at 1 mΩ to stabilize the oscillator amplitude. The working circuit voltage and current are 3V and 10mA. Using the difference in frequency data for various pressures in Table1, the Oscilloscope measurement result has been calculated in Table 2.

Table 2: Oscilloscope measurement results.

Measurement parameter	Average value	Minimum value	Maximum value
Voltage output (mV)	320	314	326
Frequency output (MHz)	49.2553965	49.255296	49.255497

CONCLUSION

This paper has addressed three problems in designing the SAW based micro force sensor, namely, the envelope of IDT, the variable regression, and the fitting curve analysis for the measured frequency, which are according to the wavelet transform method, SAW detection, and pierce oscillator circuits.

The paper has also proposed the mixing frequency circuit for the reference oscillator and the detector circuit. The changes between the oscillator frequency and the measurement of the parameters can be obtained through the pierce oscillator circuits, and the difference in frequency data for various pressures also can be established by the linear regression model. By affecting the production of amplitude of impedance and parasitic capacitance, the output frequency of 50 MHz still has stabilized features and performance. The parameters of frequency, amplitude, frequency stability, and amplitude stability have been measured during the device response and the sensor simulation.

According to the Morlet wavelet transform, SAW detection, and pierce oscillator circuits, the experimental results have confirmed that the SAW based micro force sensor can implement high reproducibility. When the input transducer of our device has been designed with the envelope of the conducting strips, our device performs well linearity. In addition, the device uses the piezoelectric properties and the temperature stability of the crystal, as well as the frequency signal instead of the conventional pressure sensor with voltage signal, which has made the signal processing of this device more digital and possessed more stable performances. The SAW based micro force type can be fabricated and has high performance in either mobile micro robots or a precise positioning device under the control of mechanical system.

Conflict of Interests

The authors declare that there is no conflict of interests regarding the publication of this paper.

Acknowledgment

This work was supported by the National Natural Science Foundation of China (Grant no. 61274078), the Research Innovation and Project of the Shanghai Municipal Education Commission (Grant no. 13ZZ049), the Doctoral Scientific Fund Project of the Ministry of Education of China (Grant no. 20120075110006), and the Foundation of Shanghai University of Engineering Science (Grant no. nhky-2013-10).

REFERENCES

1. I. Daubechies, "The wavelet transform, time-frequency localization and signal analysis," IEEE Transactions on Information Theory, vol. 36, no. 5, pp. 961–1005, 1990.
2. D. S. Ballantine, R. M. White, S. J. Martin, et al., Acoustic Wave Sensor: Theory, Design, and Physico-Chemical Applications, Academic Press, New York, NY, USA, 1997.
3. Y. H. Peng, Wavelet Transform and Its Engineering Application, Science Press, Beijing, China, 1999, (Chinese).
4. W. K. Lu, C. C. Zhu, J. H. Liu, and Q. Liu, "Implementing wavelet transform with SAW elements,"Science in China E: Technological Sciences, vol. 46, no. 6, pp. 627–638, 2003.
5. W.-K. Lu, C.-C. Zhu, J.-H. Liu, and P.-Y. Wei, "Study on implementation of surface-acoustic-wave type of the wavelet-transformation and reconstruction element," Acta Electronica Sinica, vol. 30, no. 8, pp. 1156–1159, 2002 (Chinese).
6. K. Andra, C. Chakrabarti, and T. Acharya, "A VLSI architecture for lifting-based forward and inverse wavelet transform," IEEE Transactions on Signal Processing, vol. 50, no. 4, pp. 966–977, 2002.
7. K. K. Parhi and T. Nishitani, "VLSI architecture for discrete wavelet transforms," IEEE Transactions on Very Large Scale Integration (VLSI) Systems, vol. 1, no. 2, pp. 191–202, 1993.
8. X. Chen, X. Zhang, K. Chen, and Q. Li, "Optical wavelet-matched filtering with bacteriorhodopsin films," Applied Optics, vol. 36, no. 32, pp. 8413–8416, 1997.
9. M. von Schickfus, R. Stanzel, T. Kammereck, D. Weiskat, W. Dittrich, and H. Fuchs, "Improving the SAW gas sensor: device, electronics and sensor layer," Sensors and Actuators B: Chemical, vol. 19, no. 1–3, pp. 443–447, 1994.

10. M. K. Tan, L. Y. Yeo, and J. R. Friend, "Rapid fluid flow and mixing induced in microchannels using surface acoustic waves," Europhysics Letters, vol. 87, no. 4, pp. 537–563, 2009.
11. W. C. Wilson, D. C. Malocha, N. Y. Kozlovski, et al., "Orthogonal frequency coded SAW sensors for aerospace SHM applications," Sensors Journal, vol. 9, no. 11, pp. 1546–1556, 2009.
12. M. Jungwirth, H. Scherr, and R. Weigel, "Micromechanical precision pressure sensor incorporating SAW delay lines," Acta Mechanica, vol. 158, no. 3-4, pp. 227–252, 2002.
13. S. Muntwyler, F. Beyeler, and B. J. Nelson, "Three-axis micro-force sensor with sub-micro-Newton measurement uncertainty and tunable force range," Journal of Micromechanics and Microengineering, vol. 20, no. 2, pp. 3165–3170, 2010.
14. Y. C. Kim, Y. S. Ihn, H. Moon et al., "Low cost dual axis micro force sensor for robotic manipulations,"Microsystem Technologies, vol. 17, no. 5–7, pp. 1197–1205, 2011.
15. K. J. Singh, O. Elmazria, F. Sarry et al., "Enhanced sensitivity of SAW-based Pirani vacuum pressure sensor," IEEE Sensors Journal, vol. 11, no. 6, pp. 1458–1464, 2011.
16. C. B. Wen and C. C. Zhu, "Time synchronous dyadic wavelet processor array using surface acoustic wave devices," Smart Materials and Structures, vol. 15, no. 4, pp. 939–945, 2006.
17. W. K. Lu, C. C. Zhu, J. F. Zhang, C. Shi, and X. Z. Lü, "Study of small size wavelet transform processor and wavelet inverse-transform processor using SAW devices," Measurement: Journal of the International Measurement Confederation, vol. 44, no. 5, pp. 994–999, 2011.
18. Y. Kang, Design of Surface Acoustic Wave Devices and Its Application Oscillator Circuit, Chang'an University, 2011.
19. X. Z. Lü, Interfacial Stress Sensor for Artificial Skin Application, Donghua University, 2012.
20. Y. Y. Li, W. K. Lu, C. C. Zhu, et al., "Acoustic electric generation for morlet wavelet transform of surface acoustic wave device," Research Journal of Applied Sciences, Engineering and Technology, vol. 5, no. 4, pp. 1203–1207, 2013.
21. Y. Y. Li, W. K. Lu, and C. C. Zhu, "Pspice equivalent circuit model for implementation of surface acoustic wave filter," Journal of Donghua University, vol. 29, no. 2, pp. 148–152, 2012.

Chapter 2

SWITCHING OPTIMIZATION FOR CLASS-G AUDIO AMPLIFIERS WITH TWO POWER SUPPLIES

Patrice Russo[1,2], Firas Yengui[1], Gael Pillonnet[1,2], Sophie Taupin[2], Nacer Abouchi[1]

[1]Lyon Institute of Nanotechnology (INL-UMR5270), University of Lyon, Lyon, France

[2]ST Microelectronics, Grenoble, France

ABSTRACT

This paper presents a system-level method to decrease the power consumption of integrated audio Class-G amplifiers for mobile phones by using the same implementation of the level detector, but by changing the parameters of the switching algorithm. This method uses an optimization based on a simplified model simulation to quickly find the best power supply switching strategy in order to decrease the losses of the internal Class-AB amplifier. Using a few relevant equations of Class-G on the electrical level and by reducing the number of calculation points, this model can dramatically reduce the calculation time to allow power consumption evaluation in realistic case conditions compared to the currently available tools. This simplified model also evaluates the audio quality reproduction thanks to a psycho-acoustic method. The model has been validated by comparing model results and practical measurements on two industrial circuits. This proposed model is used by an optimizer based on a genetic algorithm associated with a pattern search algorithm to find the best power supply switching strategy for the internal Class-AB amplifier. The optimization results improve life-time performance by saving at least 25% in power consumption for typical use-case (1 mW) compared to the industrial circuit studied and without losses in audio quality.

INTRODUCTION

The battery-powered systems, such as mobile phone, PDA and MP3, integrate more and more complex and power-consuming functions: large screen, GPS and audio applications, etc. The IC designers are faced with two main challenges: higher integration and lower power consumption to reduce PCB area and increase battery life. In this work, we focus on a headphone application for cell phones. Indeed, this application uses a large part of the total power consumption of the mobile phone when the consumer is listening the music (when the screen doesn't work and no other application is running). In the first generation of headphone amplifiers, Class-AB topology was preferred because of its low relative complexity and good audio performance compared to other topologies. However this solution suffers from a limited efficiency given by:

$$\eta_{AB} = \frac{\pi}{4} \times \frac{V_{OUT}}{V_{DD}} \qquad (1)$$

where, V_{DD} is the power supply of the Class-AB amplifier and V_{OUT} the RMS audio output voltage. In (1), the quiescent current is considered as negligible. Due to the low RMS voltages of a standard audio signal and the power required by the headphone, the efficiency is only a few percent because of the low ratio between the output signal level and the power supply (V_{OUT}/V_{DD}), and also the quiescent current.

Other solutions could be used for headphone applications, such as switching (Class-D) or hybrid (Class-G, H and K) amplifiers. Recent work proposed Class-D amplifiers for headphone applications [1,2] but they suffer from a higher static current consumption with a DC coupling capacitor and unpredictable electromagnetic interference. The Class-K amplifier also suffers from high consumption [3]. Class-G topology has been proposed to reduce the V_{OUT}/V_{DD} ratio, see (1), by powering the linear amplifier with a dynamic power supply. Reducing the current consumption by a factor of 3 (at average output power) compared to the Class-AB, means there are now alternative solutions in headphone applications [4-8]. **Figure 1** shows a Class-G block diagram and an example of power supply variation with a real audio signal. An integrated Class-G amplifier is therefore composed of a dynamic power supply converter associated with a switching power supply algorithm and a linear amplifier.

Present Class-G amplifiers [6-8] use two different power supply rails to decrease the conducted power

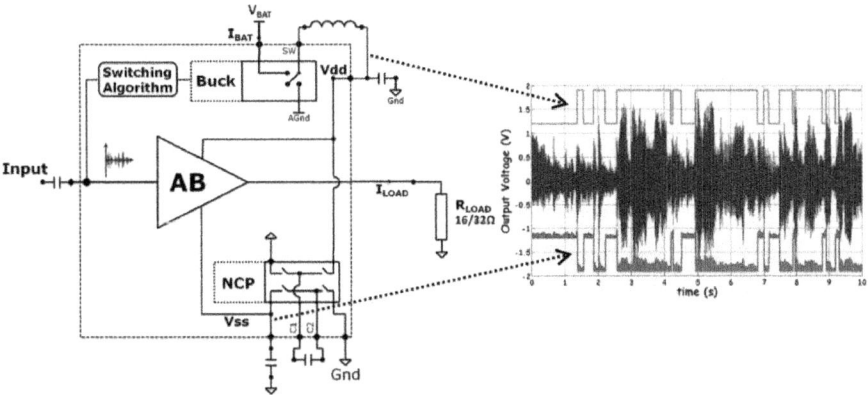

Figure 1. Class-G amplifier with two supplies.

losses $(V_{DD} - V_{OUT})/I_{LOAD}$ in the linear amplifier, and only a few papers introduce hybrid amplifiers with more than two power supplies [9-11]. Class-G exhibits higher efficiency at low output power where the audio amplifier is widely used. Indeed, average listening power is ten times less than the maximal output power. In addition to their improved efficiency, Class-G amplifiers exhibit a comparable audio quality to classic Class-AB solution [6-8].

Due to the complexity of multi-level Class-G electrical implementation, which could increase the global power consumption, the multi-levels (also called Class-H) are not yet used in industrial applications.

To find the optimal algorithm of power switching, a thorough analysis of amplifier efficiency for many different cases has to be undertaken. Indeed, [10] and [12] present an optimization made by experts in this domain and suggest optimal parameters for the switching algorithm in amplifiers with two or three power supplies. [4], [5] and [13] focus more on optimizing the electrical implementation by proposing solutions on the improvement of consumption and harmonic distortion (i.e. one component of the audio quality).

The objective of this work is therefore to optimize the power consumption that could be achieved with a ClassG with two power supplies for headphone applications by finding the optimal value of several parameters of the level detector. Section 2 is devoted to the modeling of the amplifier. Then, the optimization issues are described in Section 3. The last section discusses model validation and optimization results based on an existing Class-G amplifier.

MODELIZATION

Objectives

Present Class-G amplifiers contain more than a few thousand transistors and therefore require several weeks of simulation with a sinusoidal signal of a few milliseconds. In order to reduce this calculation time and enable the simulation of longer test signals such as music, a fast and accurate model is suggested. As the simulation time is strictly linked to the level of abstraction, as indicated on **Figure 2**, behavioral modeling enables a good compromise between time and precision of the simulations.

This model must allow post-processing on audio signals (interpolation, and cutting up of sound tracks, etc.), and enable us to evaluate the sound reproduction quality of the audio signal (notably using the PEAQ method, explained in Subsection 2.5). The parameters of our model must also be optimized via different search algorithms (cf. Section 3). We therefore chose a Matlab model using the version R2007a, rendering all these actions possible within the same interface.

Power Supply Switching Algorithm

The synoptic diagram of the power supply switching algorithm for a class G2 amplifier is presented in **Figure 3** and was implemented in our Matlab model.

Depending on the input voltage, the buck converter provides two different power supplies as shown in **Figure 1**. If the output voltage exceeds the upper threshold $\alpha \times |V_{ss}|$, then the higher supply is applied with a given rise time (**Figure 4**). Then, when the input audio signal falls under the lower threshold $\beta \times |V_{ss}|$, the lower supply is selected after a time, called the decay time. A parameter called the attack time was included in the model. This parameter is a delay between the time when the music is above to $\alpha \times |V_{ss}|$, and the time when the lower supply starts to move to the upper supply. The fall time depends on the discharge of the capacitor of the Negative Charge Pump (NCP) and the buck converter. All these parameters are taken into account in our proposed model.

Switching Optimization for Class-G Audio Amplifiers with Two Power... 25

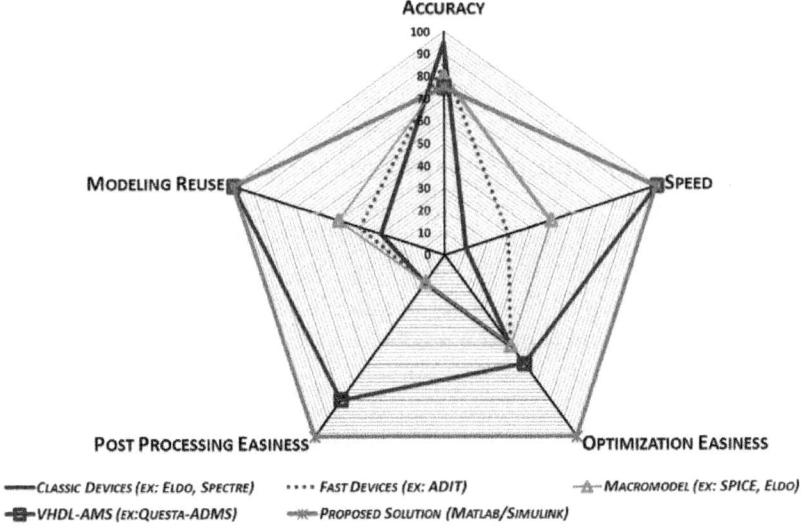

Figure 2. Modeling comparison.

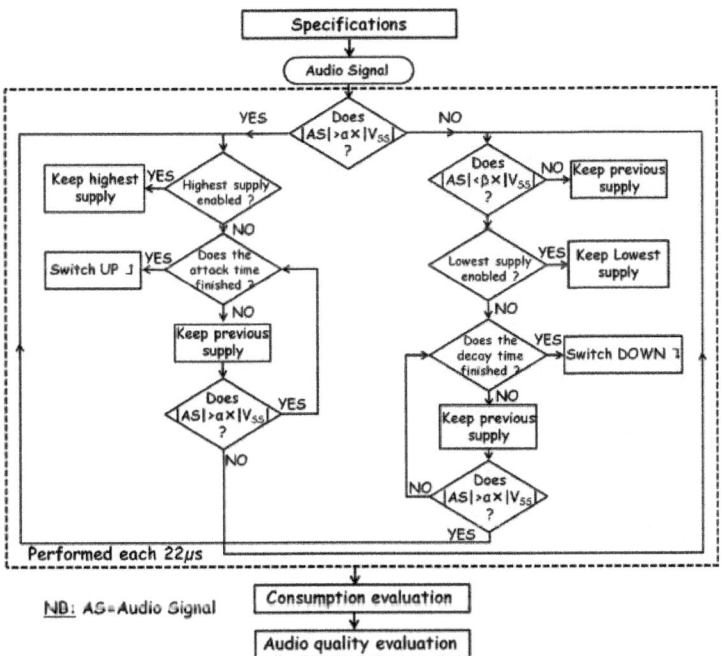

Figure 3. Algorithm synoptic.

Modeling of the Power Converters and the Linear Amplifier

In order to only study the leverage of the switching algorithm, only the transient parameters that influence the consumption and the audio quality are finely modeled. Therefore, the linear amplifier is modeled as having a fixed gain, and its linearity, its noise and immunity to the power supply are considered ideal. Regarding the power converters, the buck is modeled with an ideal line and load transient, but its efficiency is 80% (like the current available buck) [6]. The NCP is also considered ideal (no losses R_{ON} in MOS switches) but its equivalent resistor (R_{EQ} = 5 Ω) is modeled since it's a contributor to the clipping of the signal in transient analysis.

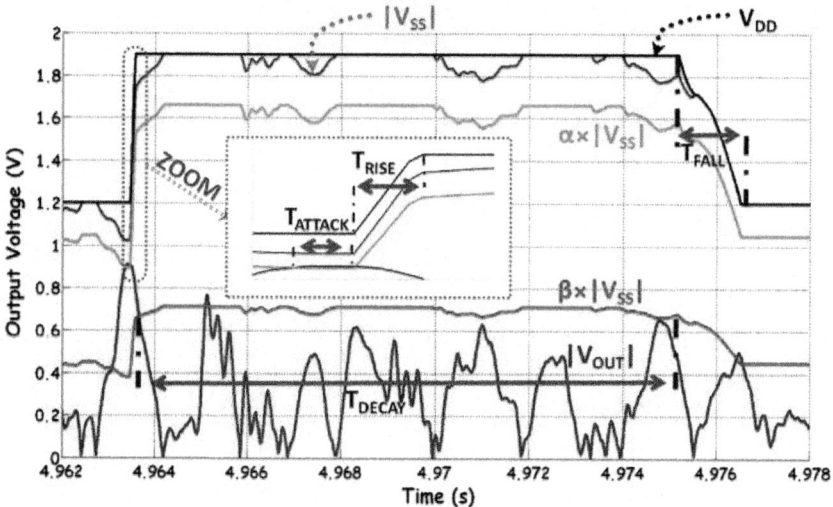

Figure 4. Switching algorithm.

The buck converter and the NCP can be linearized and modeled by simplified equations while keeping a good accuracy to predict the total current consumption. The equation of the consumption, in a two mono-channel configuration can be expressed by:

$$I_{BAT} = \frac{V_{DD}}{\eta_{BUCK} \times V_{BAT}}(I_{Q\_BUCK} + 2I_{LOAD}) + I_{Q\_BAT} \quad (2)$$

$$I_{LOAD} = \frac{V_{OUT}}{R_{LOAD}} \quad (3)$$

where, I_{BAT} is the supply current, I_{LOAD} the load current, I_{Q_BUCK} the quiescent current of the buck converter, η_{BUCK} the buck converter efficiency, and I_{Q_BAT} the quiescent current on the battery.

The effects of the switch on the DC/DC converter (f_{SW} = 750 kHz) are negligible for the prediction of the consumption of the amplifier. This averaged behavior implies a reduction in the number of calculation points. In our model, the time step is 22 µs which corresponds to the inverse of the sampling rate of our audio test signal (44.1 kHz), instead of a few ns in electrical simulators.

Choice of Input Signals

Using a realistic audio test signal instead of a pure sine wave is essential because the behavior of Class-G amplifiers is quite different. Indeed, for a sine wave at 1 kHz with a crest factor of 3 dB (the ratio between V_{PEAK} and V_{RMS}), there is no switch down for a decay time greater than 100 µs (when the output power leads the power supply to switch). However, the audio signal leads the amplifier to switch up and down (**Figure 3**). The behavior of our model is thus working in realistic conditions, contrary to a simulator at transistor level which would require too long a calculation time if a real audio wave was used. The study takes into account that the consumer can listen to several kinds of music (jazz, rap and techno, etc.), which have a typical crest factor between 5 and 20 [14]. Based on this principle, three signals with different crest factors were chosen. Choosing three test signals allow a faster simulation and a best convergence in optimization, while providing a representative sample of the main cases in order to optimize our model. In addition to these test signals (n°1, 2 and 3), two other test signals are used in section 4 to prove the robustness of the optimization (n°4 and 5). **Table 1** summarizes the input signals used in this paper.

Figure 5 shows the current consumption for the previous test signals, and then shows the actual amplifier switch for an output power above 3 mW for a crest factor under 10 dB. Typical uses of the audio amplifier for headphone required low output power (until 100 µW and 1 mW) which means that the current amplifier is oversized. This figure also highlights the fact that the switching of the power supply of Class-G amplifier is dependent on the crest factor.

Table 1. Signal used for test.

Signal	Type	CF (dB)	Artist/Title	Length (s)
N°1	Sine Wave (f = 1 KHz)	3	-	0.05
N°2	Audio Track	14	Janis Joplin/Me and Bobby Mc Gee	10
N°3	Audio Track	7	David Guetta	10
N°4	Audio Track	17	Red Hot Chili Peppers/Under the Bridge	10
N°5	Audio Track	13	Diana Krall/Bye Bye Blackbird	10

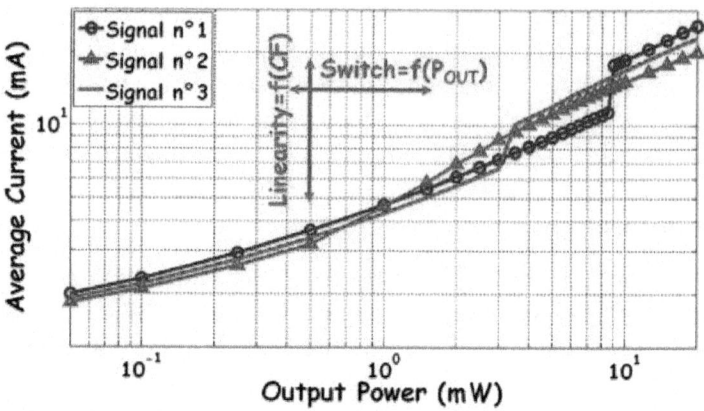

Figure 5. Current consumption for different test signals.

Evaluation of Audio Quality

Audio quality is as important as the reduction in power consumption. The model makes it possible to use the objective method to evaluate the audio quality, because Total Harmonic Distortion (THD) could only be used with a pure sine wave. For an objective method, the Perceptual Evaluation of Audio Quality (PEAQ) was chosen. This standard uses a number of psycho-acoustic measurements which are combined to give a measure of the quality difference between the reference and post-processing signal [15]. The method then returns to an ODG (Objective Difference Grade) value between 0 and –4 which reflects the impairment heard by human ears. The value 0 means than the difference is imperceptible and value –4 means than the difference is very annoying for the listener. To complete this method, listening tests (subjective method) were performed.

OPTIMIZATION APPROACH

The aim of this section is to present our optimization approach to find the best strategy to switch the two power supplies of the Class-G audio amplifier. This is done in order to reach an optimal current consumption without losing audio quality, and at the same time independent of the input signals. This approach consisted in solving, numerically, design problems while respecting the limitation due to circuit constraints and specifications. Two factors require special attention when such a problem is analyzed, namely, the problem formulation and the search algorithm.

Problem Formulation

Design Variable

Ten parameters are defined: battery voltage, load, two power supply voltages, and six switching parameters (rise, fall, attack and decay times, α and β). We will optimize five design variables (degrees of freedom): V_{DD_LOW}, decay time, attack time, α and β. Moreover the range of V_{DD_LOW} is limited by the IC design constraint topology and the output power selected is 1mW, even if other output power is considered.

Objective Function

The problem consists in minimizing supply current consumption for all three selected audio input signals. Despite the multi-objective approach, we study as a monoobjective problem using the aggregation approach [16]. It is one of the most often used methods for generation of Pareto optimal solutions. Our optimization algorithms allow us to minimize the objective function expressed as:

$$f = \sum_{i=1}^{3} wi \times Ii \text{ with } wi > 0 \text{ and } \sum_{i=1}^{3} wi = 1 \quad (4)$$

where, Ii represents the supply current consumption for each input signal, i Î [1, 3] the number of the objective and wi the weighting coefficient. In our case, wi = 1 because no preference between each objective is made.

Constraints

Constraints are conditions that must be satisfied in order to find a feasible design. Inequality constraints are used: each of the three ODGs has to be above -0.5 with the three input signals.

Optimization Algorithm

Once the problem has been formulated, we must choose the best optimization algorithm that allows us to minimize the objective function under the constraints. The Genetic Algorithm (GA) is one of the most popular and robust algorithms. It is based on natural genetic and natural selection mechanisms and some fundamental ideas are borrowed from genetics in order to artificially construct an optimization procedure. The GA acts over a population of potential solutions, applying intensification (crossover) and diversification (mutation) operators to explore the problem space. The fittest individuals are selected and give birth to a new population in the hope of improving the solution quality. More details on the mechanism of GAs can be found in [17]. GA is useful for a global search solution. However it is very slow and poor in a localized search. The direct search algorithm, Pattern Search (PS), on the contrary, is often able to find local optima for constrained optimization problems, but it cannot guarantee that the solution is the global optimum of the problem. It ensures computational robustness when it starts from a feasible initial solution [18]. By combining GA with PS, an algorithm referred to as the GA-PS hybrid algorithm is formulated in this paper. In other words, the GA looks at the whole solution space to obtain a quasi-optimal solution. Then, the PS is used to increase the quality and speed of convergence to the optimal solution.

Cascade Simulation-Based Optimization

To optimize the Class-G amplifiers for different types of input signal simultaneously, we use a multiple simulation-based optimization to find the optimal solution with respect to the three audio signals. **Figure 6** presents the concept of our approach.

For each iteration of the optimization loop, a simulation of our model, presented in Section 2, was performed for each audio input signal to find the performance such as current consumption I_{BAT} and the quality factor ODG.

RESULTS

Model Validation

An existing circuit [6] was modeled with our proposed method (Section 2). Several measurements on [6] have been done to find the input parameters of the simplified electrical equations given in Equation (3) and to find the switching algorithm (α, β, etc).

In **Figure 7**, the measurement test bench is presented and allows us to compare the current found by the measurement of [6] and our proposed model. The setup configuration is a 47 Ω load, 3.6 V power supply and signal n°2. The error is less than 5% over all the output power range. Other input signals gave the same results. The comparison was also made with other existing Class-G [6] and it showed that the error is less than 10% with the same conditions. These results confirm that the Class-G amplifier model gives a reliable current consumption and can be used by the optimizer.

Optimizer Algorithm Comparison

For our application, we compared three algorithm optimizations to show the effectiveness of the proposed hybrid GA-PS algorithm. GA-PS is compared to GA and another hybrid algorithm GA-SQP, which is based on a GA coupled with the local search algorithm: Sequential Quadratic Programming (SQP). The optimization algorithms used in this study are part of the MATLAB optimization toolbox. In **Table 2**, we compared the output power of 1mW with signal 1, 2 and 3. The objective function is the mean of the consumption for these three input signals. The best result is obtained using a GA-PS optimization, in terms of minimizing the objective function while keeping an acceptable time of optimization. We therefore used this solution in order to reduce the consumption of our amplifier.

Indeed, over 100 000 simulations would be necessary with an exhaustive search. Moreover, extrapolating the results of a GA-PS optimization with a model using a transistor and macro-model would require 134 years to

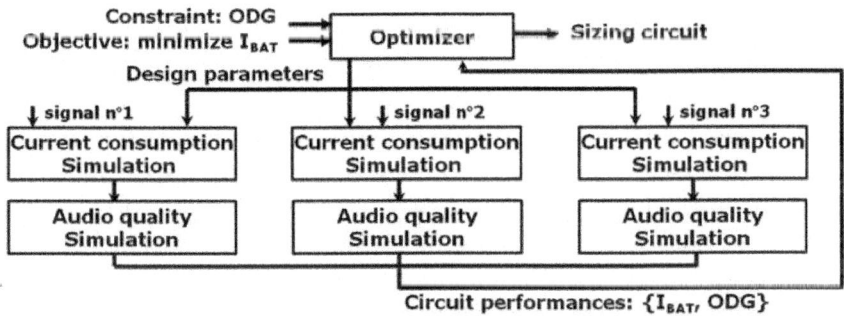

Figure 6. Cascade simulation-based optimization.

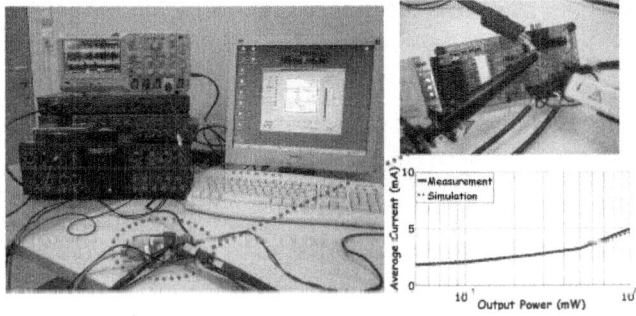

Figure 7. Comparison between [6] and our model.

Table 2. Algorithm comparison.

	Hybrid GA-SQP	Hybrid GA-PS
Number of Iterations	576	554
Objective function (mA)	2.6	2.41
Simulation Time (min)	47	45
Results	☹	☺

find an optimal solution, since a single simulation with an audio signal during ten seconds takes three months.

Optimization Results

At the moment, the minimum supply voltage of a ClassG amplifier is 1.2 V [6]. However, because of the real power needed in general, we made an optimization allowing the constraint V_{DD_LOW} to be above 700 mV (which is the actual limit for the power amplifier) in order to see the leverage of optimization by tuning V_{DD_LOW} while keeping the other constraints as before. The test signal used is signal 4. It is used to perform the optimization in order to prove the robustness of the proposed optimization. The results shown in **Figure 8** presents the gain obtained compared to [6] from 20 µW until 20 mW and highlight the need to lower the supply voltage of the power amplifier, since this reduction reduces the consumption for low power without degrading the consumption at high power. Moreover, the algorithm used in this optimization always respects the condition for the audio quality. **Figure 7** also shows that the gains in consumption start to reduce after 2 mW for $V_{DD\_LOW} > 1.2$ V and 5 mW for $V_{DD\_LOW} > 0.7$ V. This result is explained by the facts that highest is the output power, lower are the switches of the amplifiers until the blocking of the upper supply.

We can note that for a Class-G amplifier with two power supplies, the reduction of V_{DD_LOW} under 700 mV is not justified. Indeed, the optimization performed did not show a reduction in consumption.

This conclusion leads us to perform the optimization with constraints V_{DD_LOW} above 700 mV, as shown in

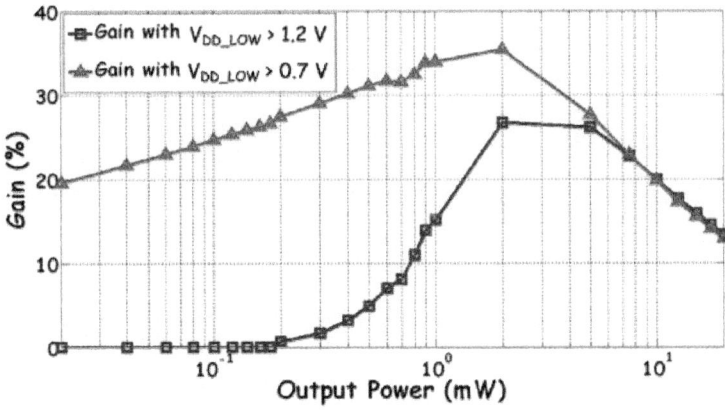

Figure 8. Current consumption vs. V_{DD_LOW} for signal n°4.

Table 3. Here, we present the results for one music used for the optimization (signal n°2) and two signals not used for the optimization (signal n°4 and 5), to prove the robustness of the proposed optimization.

It can be noted that the results presented in **Table 3** do not reduce the audio quality compared to the initial configuration of [6], which means that the ODG is always above −0.5. The results found by the optimizer are summarized in **Table 4** and compared to two industrial circuits [6,7]. These parameters are not directly found in the datasheet but are obtained using reverse engineering on their test board. This table shows that the parameters of current industrial circuits are oversized. The threshold voltage can be placed closer to the supply voltage, the decay time has to be reduced and the lowest power supply should be minimized even if the highest power supply is at 1.9 V. However, like the industrial circuits [6,7], the attack time is not reliable to gain in consumption without deteriorate the audio quality. But this parameter has to be tried for Class-G with more than two power supplies in order to see if the results could be better. Indeed, all the parameters present in **Table 4** can't be applied for Class-G with more than two power supplies.

Figure 9 shows the gain in the current consumption from 20 µW to 5mW between the initial configuration from [6] and the optimization. 18% at 100 µW and at least 25% at 1 mW are saved with an optimal configuration of the switching power supply algorithm. Even the two signals not used for the optimization give a gain in con

Table 3. Comparison for different class-G amplifiers.

Input Signals	Results for	Current Consumption (mA)		
		0.1 mW	0.5 mW	1 mW
Signal n°2	Previous Work [6]	2.1	3.2	4.61
	This Work	1.6	2.35	3.28
Signal n°4	Previous Work [6]	2.11	3.41	4.84
	This Work	1.62	2.37	3.21
Signal n°5	Previous Work [6]	2.11	3.22	4.68
	This Work	1.61	2.36	3.24

Table 4. Value of the optimized parameters.

Parameters	[6]	[7]	This Work
V_{DD_LOW} (V)	1.2	1.3	0.7
α	$7/8 \times |V_{ss}|$	$5/8 \times |V_{ss}|$	$7.2/8 \times |V_{ss}|$
β	$3/8 \times |V_{ss}|$	$3/8 \times |V_{ss}|$	$6.5/8 \times |V_{ss}|$
Decay Time (ms)	130	4.5	0.1
Attack Time (s)	0	0	0

Figure 9. Gain in current consumption because of optimization.

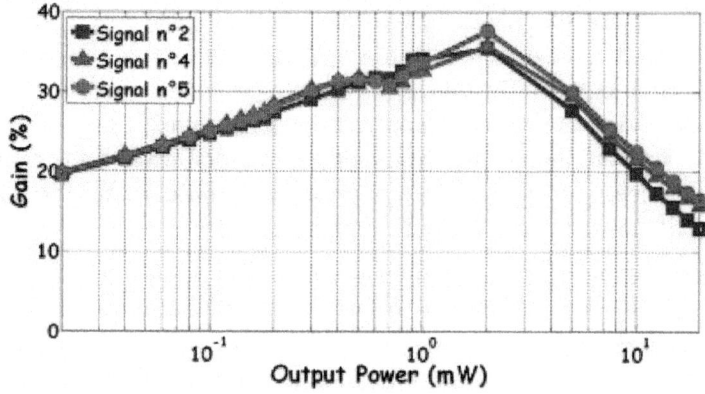

sumption at low and nominal power. Like previously, for high power, they are few switch of the amplifier and the gain start to decrease after 2 mW. It is well to remember that the same level detector than [6] is used (based on logic gates and comparator which are few consuming).

CONCLUSION

In this paper, an original equation-based model has been introduced and associated with a hybrid optimization algorithm in order to reduce the current consumption of audio Class-G amplifiers by choosing the best parameters of the power supply switching algorithm. The proposed model has been validated. It also saves simulation time to predict the power consumption and keeps audio quality with various input signals. The optimizer coupled to this model allows us to find the best power supply switching strategy for a Class-G amplifier with two supplies by giving the optimal value of the parameters (V_{DD_LOW}, α, β, decay time and attack time). At least 25% of power consumption can be saved by optimizing the switching algorithm compared to an existing Class-G circuit with the same electrical implementation. In addition, the model is robust for operating conditions since the optimization was done for multiple input signals without loss of audio quality thanks to the PEAQ method. In future work, this approach will be used with Class-G amplifiers with more than two-power supplies, in order to optimize on all the range of power.

Acknowledgements

The authors gratefully acknowledge the standard linear division of ST Microelectronics for their valuable technical help.

REFERENCES

1. K. Kang, J. Roh, Y. Choi, H. Roh, H. Nam and S. Lee, "Class-D Audio Amplifier Using 1-bit Fourth-Order Delta-Sigma Modulation," IEEE Transaction on Circuits and Systems II, Vol. 55, No. 8, 2008, pp. 728-732. doi:10.1109/TCSII.2008.922457
2. G. Pillonnet, N. Abouchi, R. Cellier and A. Nagari, "A 0.01% THD, 70 dB PSRR Single Ended Class D Using Variable Hysteresis Control For Headphone Amplifiers," IEEE International Symposium on Circuits and Systems, Taipei, 24-27 May 2009, pp. 1181-1184. doi:10.1109/NEWCAS.2010.5603759
3. E. Sturtzer, G. Pillonnet, A. Huffenus, N. Abouchi, F. Goutti and V. Rabary, "Improved Class-K Amplifier for Headset Applications," 8th IEEE International NEWCAS Conference, Montreal, 20-23 June 2010, pp. 185-188. doi:10.1109/NEWCAS.2010.5603759
4. A. Lollio, G. Bollati and R. Castello, "A Class-G Headphone Amplifier in 65 nm CMOS Technology," IEEE Journal of Solid State Circuits, Vol. 45, No. 12, 2010, pp. 2530-2542.doi:10.1109/JSSC.2010.2076450

5. A. Downey and G. Wierzba, "A Class-G/FB Audio Amplifier," IEEE Transactions on Consumer Electronics, Vol. 53, No. 4, 2007, pp. 1537-1545. doi:10.1109/TCE.2007.4429249
6. Datasheet ST-M TS4621. http://www.st.com
7. Datasheet TI TPA6140. http://www.ti.com
8. Datasheet NS LM48824. http://www.national.com
9. R. Bortoni, S. N. Filho and R. Seara, "Analysis, Design and Assessment of Class A, B, AB, G and H Audio Power Amplifier Output Stage Based on Matlab® Software," 110th Audio Engineering Society Conferences, Amsterdam, 12-15 May 2001
10. F. H. Raab, "Average Efficiency of Class-G Power Amplifier," IEEE transaction on Consumer Electronics, Vol. 32, No. 2, 1986, pp. 145-150. doi:10.1109/TCE.1986.290146
11. E. Mendenhall, "Computer Aided Design and Analysis of Class B and Class H Power Amplifier Output Stage," Audio Engineering Society 101st Convention, 1996.
12. T. Sampei, S. Ohashi, Y. Ohta and S. Inoue, "Highest Efficiency and Super Quality Audio Amplifier Using MOS Power FETS in Class G Operation," IEEE Transaction on Consumer Electronics, Vol. CE-24, No. 3, 1978, pp. 300-307. doi:10.1109/TCE.1978.267034
13. J. Gubelmann, P. A. Dal Fabro, M. Pastre and M. Kayal, "High-Efficiency Dynamic Supply CMOS Audio Power Amplifier for Low-Power Applications," Journal of Microelectronics, Vol. 40, No. 8, 2009, pp. 1175-1183. doi:10.1016/j.mejo.2009.03.003
14. M. Mijic, D. Masovic, D. Sumarac Pavlovic and M. Petrovic, "Statistical Properties of Music Signals," Audio Engineering Society 126th Convention, 2009.
15. T. Thiede, et al., "PEAQ the ITU Standard for Objective Measurement of Perceived Audio Quality," Journal of Audio Engineering Society, Vol. 48, No. 1, 2000, pp. 3- 29.
16. N. Srinivas and K. Deb, "Multiobjective Optimization Using Nondominated Sorting in Genetic Algorithms," Journal of Evolutionary Computation, Vol. 2, No. 3, 1994, pp. 221-248. doi:10.1162/evco.1994.2.3.221
17. T. El-Ghazali, "Metaheuristics: From Design to Implementation," Jonh Wiley and Sons Inc., Chichester, 2009.
18. R. M. Lewis and V. Torczon, "Pattern Search Algorithms for Bound Constrained Minimization," Journal SIAM on Optimization, Vol. 9, No. 4, 1999, pp. 1082-1099.

Chapter 3

PHOTONIC INTEGRATED SEMICONDUCTOR OPTICAL AMPLIFIER SWITCH CIRCUITS

R. Stabile and K.A. Williams

Eindhoven University of Technology The Netherlands

INTRODUCTION

The acceptance of pervasive digital media has placed society in the Exabyte era (10^{15} Bytes). However the data centres and switching technologies at the heart of the Internet have led to an industry with CO_2 emissions comparable to aviation (Congress 2007). Electronics now struggles with bandwidth and power. Electronic processor speeds had historically followed Gordon Moore's exponential law (Roadmap 2005), but have recently limited at a few thousand Megahertz. Chips now get too hot to operate efficiently at higher speed and thus performance gains are achieved by running increasing numbers of moderate speed circuits in parallel. A bottleneck is now emerging in the interconnection network. As interconnection is increasingly performed in the optical domain, it is increasingly attractive to introduce photonic switching technology. While there is still considerable debate with regard to the precise role for photonics (Huang et al., 2003; Grubb et al., 2006; Tucker, 2008; Miller 2010), new power-efficient, cost-effective and broadband approaches are actively pursued.

Supercomputers and data centers already deploy photonics to simplify and manage interconnection and are set to benefit from progress in parallel optical interconnects (Adamiecki et al., 2005; Buckman et al., 2004; Lemoff et al., 2004; Patel et al., 2003; Lemoff et al., 2005; Shares et al., 2006; Dangel et al., 2008). However, it is much more efficient to route the data over reconfigurable wiring, than to overprovision the optical wiring. Wavelength domain routing has been seen by many as the means to add such reconfigurability. Fast tuneable lasers (Gripp et al., 2003) and tuneable wavelength converters (Nicholes et al., 2010) have made significant progress, although bandwidth and connectivity remain restrictive so far. All-optical techniques have been considered to make the required step-change in processing speeds. Nonlinearities accessible with high optical powers and high electrical currents in semiconductor optical amplifiers (SOAs) create mixing products which can copy broadband information photonically (Stubkjaer, 2000; Ellis et al., 1995; Spiekman et al., 2000). When used with a suitable filter, these effects can be exploited to create photonic switches and even logic. However, the required combination of high power lasers, high current SOAs and tight tolerance filters is a very difficult one to integrate and scale. Hybrid electronic and photonic switching approaches (Chiaroni et al., 2010) are increasingly studied to perform broadband signal processing functions in the simplist and most power-efficient manner while managing deep memory and high computation functions electronically. This can still reduce network delay and remove power-consuming optical-electronic-optical conversions (Masetti et al., 2003; Chiaroni et al., 2004). The SOA gate has provided the underlying switch element for the many of these demonstrators, leading to a new class of bufferless photonic switch which assumes (Shacham et al., 2005; Lin et al., 2005; Glick et al., 2005) or implements (Hemenway et al., 2004) buffering at the edge of the photonic network. Such approaches become more acceptable in short-reach computer networking where each connection already offers considerable buffering (McAuley, 2003). Formidable challenges still remain in terms of bandwidth, cost, connectivity, and energy footprint, but photonic integration is now striving to deliver in many of these areas (Grubb et al, 2006; Maxwell, 2006; Nagarajan & Smit, 2007). This chapter addresses the engineering of SOA gates for high-connectivity integrated photonic switching circuits. Section 2 reviews the characteristics of the SOA gates themselves, considering signal integrity, bandwidth and energy efficiency. Section 3 gives a quantitative insight into the performance of SOA gates in meshed networks, addressing noise, distortion and crosstalk. Section 4 reviews the scalability of single stage integrated switches before considering recent progress in monolithic multi-stage interconnection networks in Section 5. Section 6 provides an outlook.

SOA GATES

SOA gates exhibit a multi-Terahertz bandwidth which may be switched from a high-gain state to a high-loss state within a nanosecond using low-voltage electronics. The electronic structure is that of a diode, typically with a low sub-Volt turn on voltage and series resistance of a few Ohms. Photonic switching circuits using SOAs have therefore been relatively straightforward to implement in the laboratory. The required electrical power for the SOA gate is largely independent of the optical signal, thus breaking the link between rising energy consumption and rising line-rate which plagues electronics. SOA gates and the underlying III-V technologies also bring the ability to integrate broadband controllable gain elements with the broadest range of photonic components. A wide range of optical switch concepts based on SOAs have already been proposed to facilitate nanosecond timescale path reconfiguration (Renaud et al., 1996; Williams, 2007) performing favourably with the even broader range of high speed photonic techniques (Williams et al., 2005). Now we review the state of the art for the SOA gate technology itself, highlighting system level metrics in terms of signal integrity, bandwidth and power efficiency.

Signal Integrity

The broadband optical signal into an amplifying SOA gate potentially accrues noise and distortion in amplitude and phase. Noise degrades signal integrity for very low optical input powers, while distortion can limit very high input power operation. The useful intermediate operating range, commonly described as the input power dynamic range (Wolfson, 1999), is therefore maximised through the reduction of the noise figure and increase in the distortion threshold. The signal degradation is generally characterised in terms of the additional signal power penalty required to maintain received signal integrity. Figure 1 quantifies power penalty degradation in terms of noise at low optical input powers and distortion at high optical input powers for the case of a two input two output 2x2 SOA switch fabric (Williams, 2006).

Noise originates primarily from the amplified spontaneous emission inherent in the on-state SOA gate. The treatment for optical systems has been most comprehensively treated for fiber amplifier circuits (Desurvire, 1994). The interactions of signals, shot noise, amplified spontaneous emission noise and the respective beat terms can require careful filtering and bandwidth management to ensure optimum performance. The alignment of optical signals with respect to the gain spectrum also impacts performance through the degree of population inversion. Noise may be managed through the minimisation of loss and the reduced requirement for high current amplifiers (Lord & Stallard, 1989). State of the art noise figures for fiber-coupled SOAs are of the order 6-8dB (Borghesani et al., 2003), depending on whether the structure is optimised for low-power input signals (preamplifiers) or power booster amplifiers (post-amplifiers). These values are higher than for fiber amplifiers, due to the losses in fiber to chip coupling and imperfect population inversion. The design focus has therefore been on reducing losses (Morito et al., 2005).

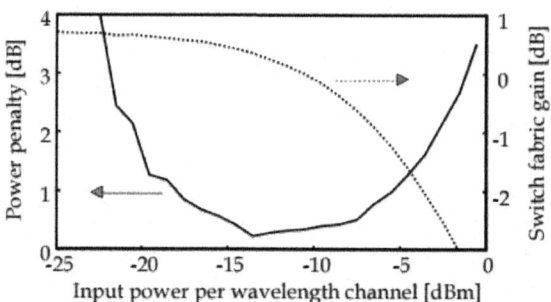

Figure 1: Simulated input power dynamic range for a 2x2 SOA switch fabric (Williams, 2006).

Distortion in the saturation regime results from the charge carrier depletion from the incoming data signal. When optical data signals are amplitude-modulated (on-off keyed), the signal can deplete charge carriers and therefore reduce gain on the timescale of the spontaneous lifetime. This leads to the time dependent patterning and therefore nonlinear distortions on the optical output signal waveform. This can be alleviated by changing the data format: Proposals range from wavelength keying (Ho et al., 1996; Kim & Chandrasekhar, 2000), wavelength domain power averaging (Mikkelsen et al., 2000; Shao et al., 1994), and wavelength coding (Roberts et al., 2005) for on-off keyed modulation. Increasingly popular constant power envelope formats (Wei et al., 2004; Cho et al., 2004; Ciaramella et al., 2008, Winzer, 2009) are also more resilient. Distortion is less evident for very low data rates where bit periods exceed the nanosecond time-scale spontaneous lifetime, and also for very high data rates where the longest sequence of bits are shorter than the spontaneous lifetime. Indeed, the optical transfer function can be considered as a notch filter and this mode of operation has already been exploited for noise suppression (Sato & Toba, 2001).

Pseudo random bit sequences are routinely used to assess data transmission. The longer 2^{31} patterns have been particularly important for point to point telecommunications links to stress-test all elements for the broadest bandwidth. The longest sequence of ones in a 2^{31} pattern remains at the same level for over 3ns for a 10Gbit/s sequence, and is thus sensitive to patterning (Burmeister & Bowers, 2006). However line rates of 100Gbit/s and above would lead to maximum length sequences shorter than the spontaneous lifetime. For higher line rates still, sophisticated optical multiplexing schemes are devised, and the concept of the pattern length becomes less meaningful: Wavelength multiplexing measurements commonly decorrelate replicas of the same signals (Lin et al., 2007), while optically multiplexed signals use calibrated interleavers available only for the shortest 2^7 pattern sequences (Albores et al., 2009). Packet switched test-beds impose more fundamental constraints: a 2^{31} sequence contains over two billion bits, far exceeding any likely data packet length. Codes for receiver power balancing and packet checking also limit the effective pattern lengths, and therefore shorter sequences are commonly used. Techniques to increase the distortion threshold are readily understood through a manipulation of the steady state charge carrier rate equation. Equation 1 approximates the rate of change of charge carriers (left) in terms of the injected current, stimulated amplification, and spontaneous emission (right). The steady state condition is defined when the derivative tends to zero ($dN/dt \to 0$).

$$dN/dt = I/eV - \Gamma dg/dn(N-N_0)P - N/\tau_s \to 0 \quad (1)$$

The terms in Equation 1 correspond to the injected current I into active volume V. N represents the charge carrier density, Γ is the optical overlap integral describing the proportion of amplified light which overlaps with the active layer. dg/dn is the differential gain and N_0 is the transparency carrier density. τ_s is the charge carrier lifetime. By defining a gain term $G = dg/dn$ $(N-N_0)$ it is possible to substitute out the unknown carrier density variable N in Equation 1 and derive an expression for gain saturation by rearranging equation (1):

$$G(1 + \Gamma \tau_s \, dg/dn \, P) = g(\tau_s I/eV - N_0) \quad (2)$$

In the linear limit, the photon density P tends to zero, and the right hand side variables may be approximated by one linear gain term $G_{linear} = g(\tau_s I/eV - N0)$. A general expression for gain G may thus be defined in terms of a linear gain G_{linear}, photon density P and a photon density saturation term such that $G = G_{linear}/(1+P/P_{saturation})$. Saturation is now simply defined in terms of optical overlap integral Γ, carrier lifetime τ_s and differential gain dg/dn (Equation 3) and it turns out that each of these parameters can be exploited to reduce distortion.

$$P_{saturation} = (\Gamma \tau_s \, dg/dn)^{-1} \quad (3)$$

The optical overlap integral is defined by the waveguide design which has been chosen to confine the carriers and the optical mode. While bulk active regions offer the highest confinement, quantum wells (in reducing numbers) allow for an increase in distortion threshold with output saturation powers of order +15dBm and higher being reported (Borghesani et al., 2003; Morito et al., 2003). Quantum dot epitaxies allow even further reductions in optical overlap for the highest reported saturation powers (Akiyama et al., 2005). Tapered waveguide techniques additionally offer improved optical power handling (Donnelly et al., 1996; Dorgeuille et al., 1996). Optimising optical overlap does however have implications for current consumption, electro-optic efficiency and signal extinction in the off-state.

The carrier lifetime can be speeded up using an additional optical pump (Yoshino & Inoue, 1996; Pleumeekers et al., 2002; Yu & Jeppesen, 2001; Dupertuis et al., 2000). A natural evolution of this, gain clamping (Tiemeijer & Groeneveld, 1995; Bachman et al., 1996; Soulage et al., 1994), has also been extensively studied as a means to increase the distortion threshold. Here the amplification occurs within a lasing cavity and so an out-of-band oscillation defines the carrier density N at the threshold gain condition through fast stimulated emission. Gain clamping can increase the distortion threshold by several decibels (Wolfson, 1999; Williams et al., 2002) and can even be extended to allow variable gain (Davies et al., 2002).

The differential gain term in equation 2 describes how the change in complex dielectric constant amplifies the optical signal. This parameter may be engineered through epitaxial design. The associated differential refractive index modulation, commonly approximated by a line-width broadening coefficient, can also be exploited to suppress distortion. Fast chirped components may be precision filtered from slower chirped components in the output signal to enhance the effective bandwidth (Inoue, 1997; Manning et al., 2007). While the approach does remove energy from the optical signal, it also enables some of the most impressive line rates in all-optical switching (Liu et al., 2007).

Bandwidth

SOA gates may be characterised by a number of time-constants and bandwidths. The Gigahertz speed at which the circuit may be electronically reconfigured is determined primarily by the spontaneous recombination lifetime and any speed-up technique employed (section 2.1). While this time constant has an impact on the durations of packets and guardbands in a packet-type network, this does not directly impact the signalling speed, where the multi-Terahertz optical gain bandwidth of the SOA becomes important. These limits are now discussed in the context of state of the art.

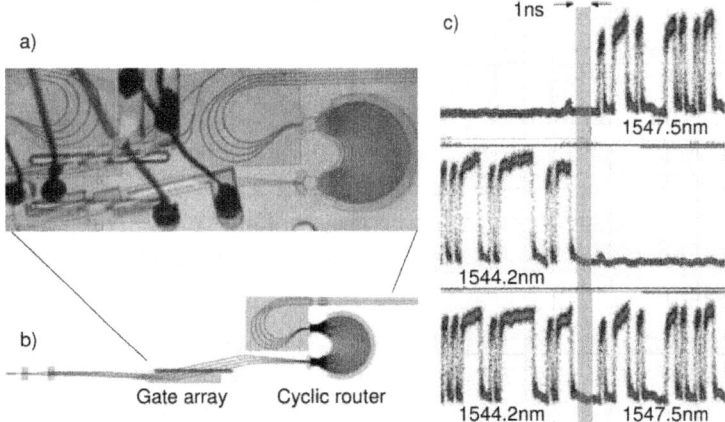

Figure 2: Dynamic routing with nanosecond switching windows for a SOA cyclic router (Rohit et al., 2010): a) The microscope photograph for the SOA gate array and arrayed waveguide cyclic router b) The waveguide arrangement fot the single input, multiple output circuit c) Time traces showing the selecting and routing of wavelength channels.

The electronic switching time from high gain to high loss is limited primarily by the spontaneous recombination lifetime with reports routinely in the nanosecond range (Dorgeuille et al., 1998; Kikichi et al., 2003; Albores-Mejia et al., 2010; Rohit et al., 2010; Burmeister & Bowers, 2006), enabling comparable nanosecond duration dark guard bands between data packets. Figure 2 shows how such fast switching speeds can be exploited in the routing of data in a SOA-gated router. Schemes for label based routing have been reported using comparable approaches (Lee et al., 2005; Shacham et al., 2005).

Real time current control has been considered as a means to ensure optimum operating characteristics of the individual SOA gates. Techniques range from the monitoring of the narrow-band tone (Ellis et al., 1988) and broadband data (Wonfor et al., 2001) on the SOA electrodes themselves through to customised monitor diodes (Tiemeijer et al., 1997) and integrated power monitoring (Newkirk et al., 1992; Lee et al., 2005). Hierarchical approaches have also been proposed to enable the management of photonic parameters independently of the digital switch state (White et al., 2007). The possibility to react to thermal transients within the circuit, and even enable self calibration is increasingly important as circuit complexity evolves. This abstraction of the physical layer becomes increasingly important as network level functions such as self-configuration are considered (Lin et al., 2005).

Signalling line-rates of up to 40Gbit/s have been demonstrated using SOAs in a transmission environment (Brisson et al., 2002), and also for integrated switch elements (Burmeister & Bowers, 2006). To extend beyond 40Gb/s requires optical multiplexing. Here SOAs have been demonstrated for in-line amplification for multiwavelength transmission (Reid et al., 1998; Jennen et al., 1998; Sun et al., 1999). The early experiments operated the SOAs within the saturation regime, but later demonstrations in the linear regime with reduced crosstalk enable hundreds of Gbit/s WDM transmission (Spiekman et al., 2000).

Optically transparent networking becomes feasible once the circuit elements become polarisation insensitive. Polarisation properties are engineered through the design of the waveguide dimensions and the radiative transitions in the active media. The latter are tailored using epitaxially defined strain. A broad range of reports have demonstrated polarisation independent operation for both bulk (Emery et al., 1997; Dreyer et al., 2002; Morito et al., 2000; Kakitsuka et al., 2000; Morito et al., 2003; Morito et al., 2005) and quantum wells SOAs (Godefroy et al., 1995; Kelly et al., 1997; Ougazzadeu, 1995; Tiemeijer et al., 1996).

Energy

The energy efficiency for an interconnection network is commonly quantified in terms of energy requirement per bit and includes the full end-to-end digital power usage. This concise metric allows for a cross-comparison with electronic switching fabrics, and assists with the road-mapping for CMOS technology. Figure 3 shows schematic arrangement for two example photonic interconnection networks with electronic and photonic switching. Photonic links remove transmission losses from the comparison, allowing a focus on the switch technologies themselves. At the time of writing, state of the art vertical cavity laser array transceivers with multimode fibers enabled energy efficiencies of a few picoJoules per bits, and distributed feedback lasers on silicon are being developed for reduced power consumption single mode fiber transceivers. Transceiver technologies dominate the interconnect power budget and a prime motivator for optical switch research has now become the replacement of large numbers of power consuming transceivers with a smaller, data agnostic switch circuit, to remove power draining OEO conversions and excess packaging.

Photonic integration reduces optical losses by minimising the number of on-off-chip connections. This additionally improves noise performance and reduce operating gain for the SOA gates. This is important as it is the current used for amplification, non-radiative and spontaneous recombination which ultimately determines energy consumption. If the nonradiative currents become too high, and Joule heating in the resistive p-layers of the SOA gates becomes significant, this can lead to a spiralling reduction in available gain, and the need for significant heat extraction. Spot-size conversion (Morito et al., 2003) is increasingly implemented to remove the losses between the SOA chip and the off-chip network elements, such as the fiber patch-cords.

Cooler-free operation is now mandatory for data communications transceivers, but remains unthinkable in many high performance telecommunications links. Integrated circuits exploiting semiconductor optical amplifiers are however well suited to uncooled operation due to the broad spectral bandwidth. Initial reports have been promising. Uncooled operation for a quantum dot SOA has been demonstrated for a wide temperature range up to 70 °C (Aw et al., 2008), providing 19dB of optical gain at high temperatures with negligible 0.1dB system penalty at 10Gb/s. Aluminium containing quaternaries, used for the highest performance uncooled 10Gb/s data communications lasers, have also been used for SOAs. These epitaxial designs allow for enhanced electronic confinement and therefore excellent electronic injection efficiency at high temperature. SOAs have also been operated at 45°C such that the packaged SOA module may operate with sub-Watt operating power over the temperature range 0-75°C (Tanaka et al., 2010).

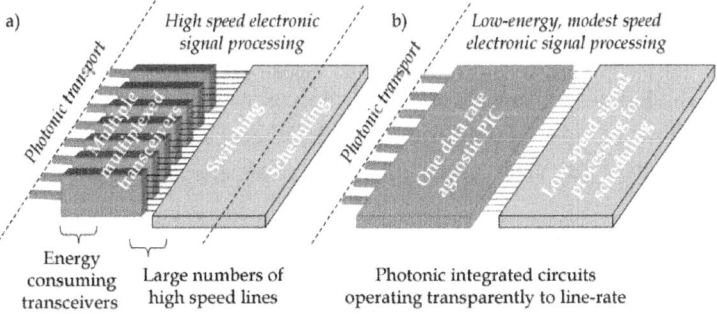

Figure 3: Schematic diagrams highlighting the motivation for hybrid photonic switch matrices with electronic switch (left) and photonic switch (right).

NETWORKS

High-connectivity, multi-port electronic switches exploit multi-stage interconnection networks (Dally & Townes, 2004; Kabacinski, 2005) and photonic networks are also set to benefit from such approaches. Figure 4 shows an example of a switch network proposed to allow the scaling of a SOA broadcast select architecture with four outputs per stage using the hybrid Clos/broadcast-and-select architecture (White et al., 2009).

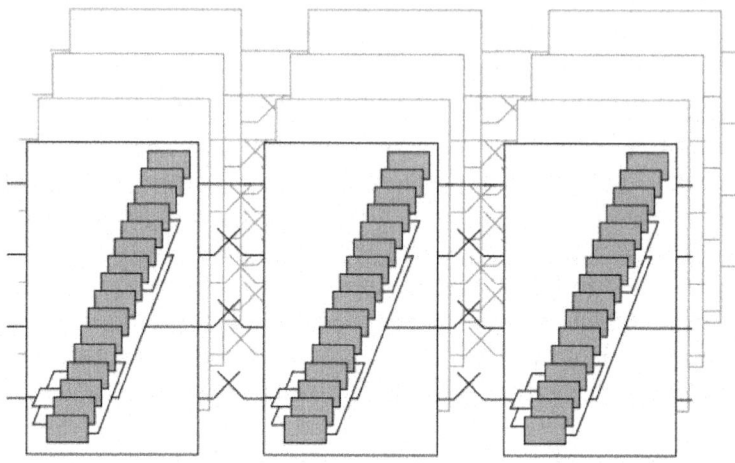

Figure 4: An example multi-stage switch architecture showing parallel scaling and serial interconnection of SOA gates.

A 4x4 broadcast and select switch using SOA gates is placed within each of the twelve switch cells. These are interconnected to each other in three stages to create the larger 16x16 network. Both serial and parallel interconnection of SOA gates is required for the multistage interconnection networks. The interactions between SOA elements in such an architecture is now considered, firstly in terms of signal evolution through the cascaded network, and secondly in terms of crosstalk from incompletely extinguished signals from interferer paths.

Cascaded Networks

The concatenation of multiple SOAs in amplified transmission and switching networks can lead to aggregated noise and distortion. The build-up of noise between stages can be minimised through reduction in gain and loss (Lord & Stallard, 1989). Reflections at the inputs and outputs of the SOA gates were particularly problematic in the early literature (Mukai et al., 1982; Grosskopf et al., 1988; Lord & Stallard, 1989), but can now be minimised through integration (Barbarin et al., 2005) and facet treatments (Buus et al., 1991). The residual distortion of signals (Section 2.1) can additionally build up with increasing numbers of SOA gates, leading to a reduction in the input power dynamic range, and ultimately the power penalty itself.

The largest cascaded networks of SOAs have been studied using recirculating loops, where a signal is switched into and out of a loop with an amplifier and a loss element. The signal circulates for predetermined numbers of iterations – often this is varied as part of the study – and is then assessed for signal degradation. Up to forty cascades have been feasible while maintaining an eye pattern opening – good discrimination between logical levels – for 10Gb/s data sequences (Onishchukov et al., 1998). Studies have also considered transmission over individual fiber spools and field installed fiber spans. Figure 5 summarises many of the leading reports into signal degradation with increasing number of SOAs. Data points are included for a pioneering research teams including those at Philips (Kuindersma et al., 1996; Smets et al., 1997; Jennen et al., 1998) and Bell Labs (Olsson, 1989; Ryu et al., 1989). The evidence suggests that power penalty can be modest for reasonably low levels of cascaded amplifiers, with a steady degradation in penalty as cascade numbers approach ten or more SOAs even when circuits are operated with high levels of gain. It is worth noting that much of this data predates the innovative low distortion amplifier designs developed over the last decade. Operating parameters can nonetheless become increasingly stringent with important implications for control systems (Section 2.2).

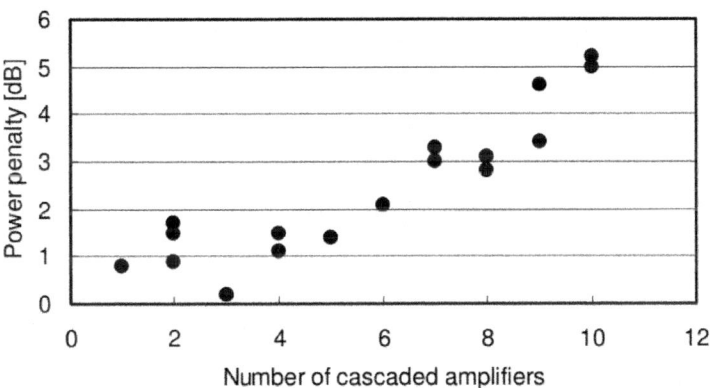

Figure 5: Power penalty in transmission experiments for cascaded semiconductor optical amplifiers.

Crosstalk

The aggregation of stray signals from disparate locations in a switch network leads to crosstalk. Contributions may be separated into coherent leakage, incoherent leakage, and cross gain modulation within co-propagating wavelength multiplexes.

Coherent crosstalk was identified as a particularly troublesome source of signal degradation for large-mesh, optically-transparent, telecommunications networks. Channels unintentionally combined with either remnants of themselves or other identical wavelengths lead to interferometric beating (Legg et al., 1994). Coherent crosstalk with long timescale fluctuations compromises threshold setting in receivers. The resulting beat noise incurs large power penalties and bit error floors (Gillner et al., 1999). If the path length differences are minimised to less than one bit period and the wavelengths are stable, as might be anticipated in a monolithic multistage network, phase difference becomes invariant and less problematic (Dods et al., 1997). Coherent crosstalk can incur an overhead of order 10dB on the crosstalk requirement (Goldstein et al., 1994; Goldstein & Eskildsen, 1995; Eskildsen & Hansen, 1997) and this has led some to suggest a –40dB extinction ratio requirement for telecommunication networks using an optical switch technology (Larsen and Gustavsson, 1997). Figure 6 summarises representative quantifying the role of crosstalk on signal degradation. Coherent crosstalk is identified with open symbols, while the closed symbols represent incoherent crosstalk measurements and calculations. The calculations performed by Buckman are also included for the cases of Gaussian and numerically determined distributions for incoherent crosstalk characteristics. It is evident from figure 6 that the level of crosstalk which may be accomodated is significantly higher for incoherent forms of crosstalk (Goldstein et al., 1994; Buckman et al., 1997; Yang & Yao, 1996; Jeong & Goodman, 1996; Albores-Mejia et al., 2009).

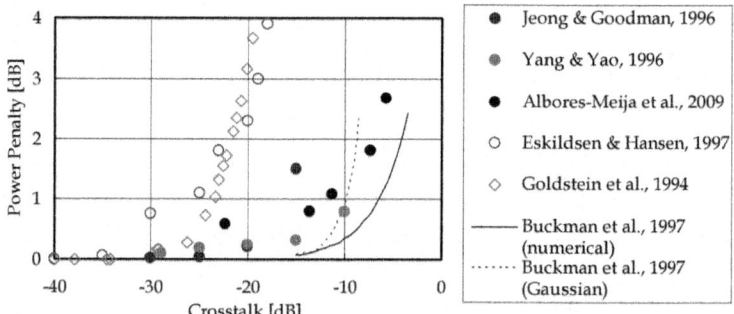

Figure 6: Crosstalk incurred penalty in SOA networks.

Switch extinction ratio is related to crosstalk at the circuit level. A worst case approximation for crosstalk build up in a given path is simply the sum of signal leakage contributions in each switch in the path (Saxtoft & Chidgey, 1993). Cumulated crosstalk ratio may be described as the product of the number of stages between an input and output N_{stages}, the number of interferer inputs at each stage with radix N_{radix}, and the extinction ratio of the switch element $X_{extinction}$:

$$\Sigma X_{crosstalk} = N_{stages} \cdot (N_{radix} - 1) \cdot X_{extinction} \qquad (4)$$

While the approach can be a useful guide for low channel counts, this can lead to overestimated power penalty at high channel counts (Buckman et al., 1997) due to statistical averaging (see for example Section 2.1). Nonetheless extinction ratios achieved for SOA gates are commonly reported in the 40dB range (Larsen & Gustavsson, 1997; Varazza et al., 2004, Tanaka et al., 2009; Albores-Mejia et al., 2010; Stabile et al., 2010). Inter-wavelength crosstalk has been studied across architectures. Many early switch architectures assumed one wavelength per switch element in multiwavelength fabrics, and this called for a multi-domain description of spatially- and spectrally-originating crosstalk (Gillner et al., 1999; Zhou et al., 1994; Zhou et al., 1996). Recent requirements for massive data capacities have led recent work to focus on multi-wavelength routing where interwavelength crosstalk can occur through cross gain modulation (Oberg & Olsson, 1988; Inoue, 1989; Summerfield & Tucker, 1999).

MULTI-PORT SWITCHES

Creating multi-port switches from SOA gates requires additional interconnecting passive circuit elements. As the techniques and technologies for creating integrated power splitters, low-loss wiring, low-radius bends, corner mirrors and waveguide crossings have evolved, the levels of integration have allowed connectivity to increase from two to four and eight output ports.

Two Port Switch Elements

The broadest range of switching and routing concepts have been demonstrated for the simplest two input two output multiport switches. The SOA gate based switches can be classified as interoferometric or as broadcast and select. The former should allow near complete coupling of optical power into the desired path, enabling the removal of unnecessary and undesirable energy loss. The latter allows a broader range of network functionality, including broadcast and multicast.

Interferometric schemes include the exploitation and frustration of multimode interference in matrices of concatenated 1x2 MMI switches (Fish et al., 1998), vertical directional couplers (Varazza et al., 2004) and gated arrayed waveguide grating based switches (Soganci et al., 2010). The first two approaches lend themselves well to cross-grid architectures and have been demonstrated at 4x4 connectivity. The incorporation of SOA gates with an interferometer also offers enhanced extinction ratio. The switched arrayed waveguide grating approach is also scalable, although only as a 1xN architecture.

Broadcast and select architectures have been more widely studied as they are intrinsically suited to conventional laser based processing methods and epitaxies. The SOA gates are able to overcome losses associated with the splitter network, allowing zero fiber-to-fiber insertion loss at modest currents. Selective area epitaxy has allowed the separate optimisation of active and passive circuit components required for insertion-loss-free operation (Sasaki et al., 1998; Hamamoto & Komatsu, 1995). The splitting and combining functions have been implemented using Y-couplers (e.g. Lindgren et al., 1990), multimode interference couplers (e.g. Albores-Mejia et al., 2009) or arrangements of total internal reflecting mirrors (e.g. Himeno et al., 1988; Gini et al., 1992; Burton et al., 1993; Sherlock et al., 1994; Williams et al., 2005). Chip footprints of below 1mm^2 have been acheived in this manner. Figure 7 shows the example of the mirrors created in an all active switch design interconnecting eight SOA gates in a cross-grid array. The input and output guides include a linearly tapered mode expander, which terminates at one of four splitters. The splitters comprised 45° totally internal reflecting mirror which partially intersect the guided mode. Part of the light is routed into the perpendicular guide and the remaining part is routed to the through path.

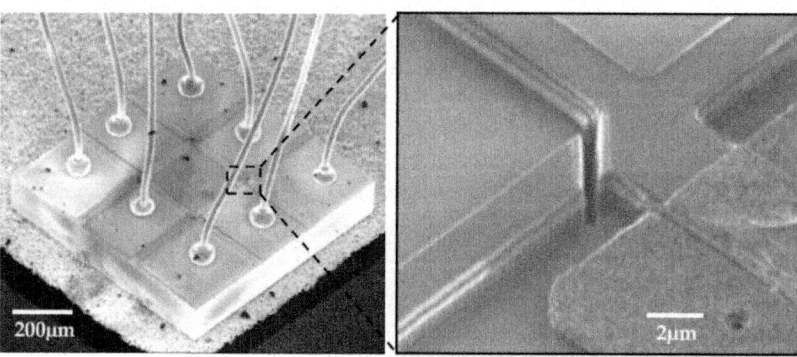

Figure 7: Two port integrated switch circuit (left) within a footprint of under 1mm^2 using (right) ultracompact total internal reflecting mirrors (Williams at al., 2005).

Microbends offer a route to even further size reductions, while addressing a tolerance to fabrication variability (Stabile & Williams, 2010). Whispering gallery mode operation is predicted to give order of magnitude relaxation in required tolerances with respect to single mode microbends. Polarization conversion can also be maintained below 1% with appropriately designed structures.

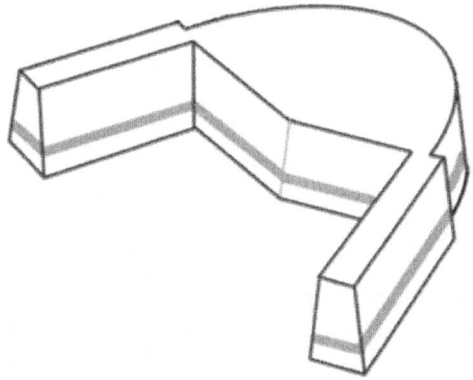

Figure 8: Schematic diagram for a fabrication tolerant whispering gallery mode bend for high density switch circuits (Stabile & Williams, 2010).

Quantum dot epitaxies have also been considered to exploit anticipated advantages for broadband amplification, low distortion and low noise (Akiyama et al., 2005). The first monolithic 2x2 switch demonstration has been performed for the 1300nm spectral window (Liu et al., 2007) showing negligible power penalty of <0.1dB for 10Gb/s data routing. The first demonstrations in the 1550nm window followed, showing excellent power penalties of order 0.2dB for 10Gbit/s data routing (Albores-Mejia et al., 2009). Multiple monolithically integrated 2x2 circuits have also been demonstrated with 0.4-0.6dB penalty showing only a weak signal degradation as quantum dot circuit elements are incorporated in larger switch fabrics (Albores-Mejia et al., 2008).

Four Port Switch Elements

Single stage four port switches have been implemented for a number of broadcast and select configurations (Gustavsson et al., 1992; Bachmann et al., 1996; Larsen & Gustavsson, 1997; van Berlo et al., 1995; Sasaki et al., 1998). Electrode counts of between sixteen and twentyfour result, depending on whether additional on-chip amplification is required to overcome circuit losses. This can add considerable complexity to circuit layout and is a potential limit to single stage scaling. The first transmission experiments were reported for 50 km distances at 2.488 Gbit/s, with less than 1 dB power penalty (Gustavsson et al., 1992) with an input power dynamic range of over 10dB. Wavelength division multiplexed transmission was also demonstrated with four 622 Mb/s wavelength channels spaced equally from 1548-1560nm (Almstrom et al.,

1996). Field trials at 2.5 Gbit/s were performed with three switch circuits in a 160 km fiber-optic link. The majority of studies have been restricted to modest data capacities between one input port and one output port (Gustavsson et al., 1992; Gustavsson et al., 1993; Djordjevic et al., 2004).

Multi-port dynamic routing has recently been demonstrated for a 4x4 switch using a roundrobin scheduler and nanosecond-speed control electronics (Stabile et al., 2010). Figure 8 shows the monolithic photonic circuit on the left, and the output signals on the right. The SOA gates are sequentially biased to enable the routing of the inputs to the outputs. The right hand figures show the time traces recorded for each of the outputs, showing data packets from each available input. Rotating priority (round-robin) path arbitration allows the simplest control algorithm with only one input clock signal, abstracting the photonic complexity from the logic control plane.

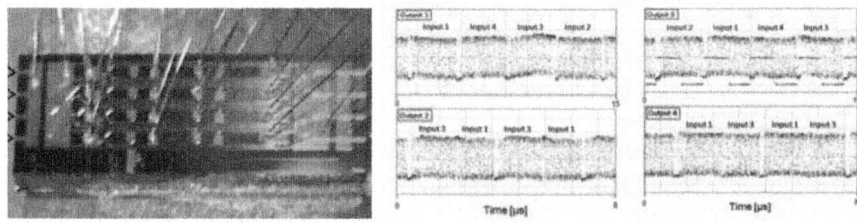

Figure 8: Four port integrated switch circuit within 4mm² showing dynamic multi-path routing (Stabile et al., 2010).

Multi-path routing has also been assessed for wavelength multiplexed inputs to three ports in a discretely populated switching fabric. Field programmable gate arrays enabled the synchronisation of switching and diagnostics. A power penalty in the range of 0.3–0.6 dB was observed due to multi-path crosstalk and a further power penalty in the range of 0.4–1.2 dB was incurred through dynamic routing (Lin et al., 2007). Connection scaling studies have allowed insight into the available power margins for SOA switch fabrics operating at high line wavelength division multiplexed line-rates. The potential for single-stage 8×8 switches at a data capacity of 10×10 Gbit/s is predicted with a 1.6dB power margin, identifying a potential route to Tbit/s switch performance in a singlestage low-complexity switch fabric (Lin et al., 2006).

Eight Port Switch Elements

Scaling to even higher levels of connectivity have been constrained by existing waveguide crossing and waveguide bend techniques, and this is most clearly evidenced by the dearth of single stage 8x8 switches. Researchers realising high connectivity single stage switches have therefore focussed efforts on 1x8 monolithic connectivity. Array integration has been explored as the first step towards large scale monolithic integration (Dorgeuille et al., 1998; Suzuki et al., 2001; Sahri et al., 2001; Kikuchi et al., 2003; Tanaka et al., 2010). The packaged array of 32 gain clamped SOA gates (Sahri et al., 2001) has enabled the most extensive system level assessments in telecommunications test-beds (Dittmann et al., 2003). Implementation of arrays of eight gates have also led to the early demonstrations of 8×8 optical switching matrices based on SOA gate arrays with 1.28Tbit/s (8×16×10Gb/s) aggregate throughput (Dorgeuille et al., 2000). These approaches rely on fiber splitter networks.

Quantum dot all-active epitaxial designs (Wang et al., 2009) have been implemented using multi-electrode amplifiers to create the separate SOA gates. The input channel is split to the eight output gates by means of three stages of on-chip 1x2 MMI couplers. The use of low splitting ratios is expected to allow more reproducible optical output power balancing. The excellent measured power penalties allow the cascading of two stages which should enable 1x64 functionality.

Active-passive regrown wafers (Tanaka et al., 2009) have also been used to create compact monolithic 1x8 switches. The thin tensile-strained MQW active layers used for the SOA gates allow for an optimisation of output saturation power, noise, and polarization insensitivity. A compact circuit footprint is facilitated by using a high density chip to fiber coupling and through the use of a field flattened splitter to create a uniform split ratio 1x8 in a highly compact 250 μm structure. This approach exhibits an on-state gain of 14.3 dB which is largely wavelength and temperature insensitive. A path to path gain deviation of order 3.0 dB is also achieved. Extinction ratios of order –70dB were reported with an extensive input power dynamic range of 20.5 dB for 10-Gbit/s signals. The high levels of gain overcome the additional off-chip splitter losses which are incurred when combining eight such circuits to construct an 8x8 switching fabric (Kinoshita, 2009).

MULTI-STAGE INTERCONNECTION NETWORKS

A broad range of multi-stage networks have been studied for photonic networks (Beneš, 1962; Wu & Feng, 1980; Spanke & Beneš, 1987; Hluchyj & Karol, 1991; Shacham & Bergman, 2007). The constraints imposed in SOA gate based networks lead to a preference for smaller numbers of stages (Williams et al., 2008; White et al., 2009). Simulations are presented to provide insight into the scalability of multi-stage photonic networks. Then examples of multistage networks are given for 2x2 and 4x4 building blocks, highlighting the state of the art for connectivity, the numbers of integrated stages and line-rate. Numerical simulations for the physical layer have been performed using travelling wave amplifier modelling which inherently accounts for noise and distortion and allows for wavelength multiplexed system simulation (Williams et al., 2008). Connectivity limits for Tbit/s photonic switch fabrics are studied by scaling the number of splitters in a three stage switch fabric: An intermediate loss between each SOA gate accounts for the radix of the switch element. A 3.5dB loss describes each 1x2 splitter or coupler element in the circuit. Figure 9 summarises the dimensioning simulations by presenting input power dynamic range as a function of the number of splitters per stage. Power penalty contours are given for 1dB and 2dB power penalties to show tolerated inter-stage losses and therefore connectivity.

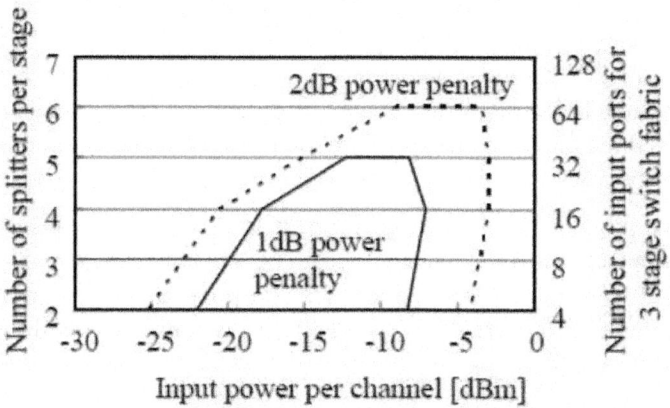

Figure 9: Simulated power penalty in increasing connectivity SOA gate switching networks Optical data rate at 10λx10Gbit/s using on-off keyed data format (Williams et al., 2008).

Input power dynamic range for 10λx10Gb/s wavelength multiplexed data is seen to reduce both with the number of switch stages and the optical loss between each switch stage. The dynamic range specified for a 1dB power penalty over three stages is observed to exceed 10dB for the four splitter architectures, which is equivalent to a three stage 16×16 switch. For the case of six splitters, a 5dB dynamic range for 2dB power penalty is indicative of viable performance for a 64×64 interconnect based on 8×8 switch stages. Large testbeds exploiting multiple stages of discrete SOA gates have supported these findings. Wavelength multiplexed routing in a 12×12 switch exploiting three stages of concatenated 1×2 SOAswitches enables Terabit class interconnection (Liboiron-Ladouceur et al., 2006). Two stages of SOA gates are implemented in a 64×64 wavelength routed architecture proposed for supercomputers (Luijten & Grzybowski, 2009).

Connectivity for integrated photonic circuits has recently been increased to record levels through the use of the Clos-Broadcast/Select architecture highlighted in Figure 4. Three stages of four 4x4 switch building blocks were integrated within the same circuit (Wang et al., 2009) to demonstrate the first 16×16 port count optical switches using an all-active AlGaInAs quantum well epitaxy. Paths in the circuit have enabled 10Gbit/s routing with 2dB circuit gain and a power penalty of 2.5dB. The electrical power consumption of the allactive chip is estimated to be 12W for a fully operational circuit, which corresponds to a modest power density of $0.3 W/mm^2$. The power consumption could be approximately halved by replacing the current active shuffle networks with their passive equivalents.

Capacity has also recently been increased to record 320Gb/s line-rates per path for a multistage photonic interconnection network (Albores-Mejia et al., 2010). This represents both the leading edge in the number of monolithically integrated switching stages and the highest reported line rates through a switching fabric. Bit error rate studies show only modest levels of signal degradation. The circuit is presented in Figure 10. The N-stage planar architecture includes up to four serially interconnected crossbar switch elements in one path, and is representative of a broader class of 2x2 based multistage interconnection networks. The step change in line rate is believed to be attributable to the use of the active-passive epitaxial regrowth, which allows the separate optimisation of gates and routing circuits.

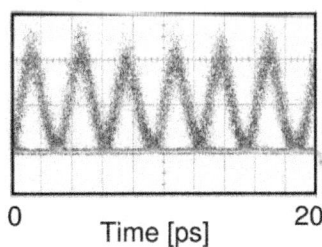

Figure 10: Photograph of a four port multistage interconnection network, and right, the eye diagrams after four stages of integrated crossbars for 320Gb/s (Albores-Mejia et al., 2010).

CONCLUSION

Integrated photonics is poised to become a key technology where the highest signalling speeds are required. The numbers of integrated optoelectronic components which can be integrated on a chip can rise significantly, and with this, the sophistication of circuit functions can be expected to grow. The critical parameters required for high capacity, high connectivity switching circuits have now been demonstrated, and the challenge is to devise architectures that are able to simultaneously match performance with energy efficiency and integration. A symbiotic relationship between massive bandwidth photonic circuits and intelligent electronic control circuits could well evolve to create a generation of ultrahigh speed signal processors.

REFERENCES

1. Adamiecki, A., M. Duelk and J.H. Sinsky, "25 Gbit/s electrical duobinary transmission over FR-4 backplanes", Electronics Letters, 41, 14, 826-827, (2005)
2. Akiyama, T., M. Ekawa, M. Sugawara, K. Kawaguchi, H. Sudo, A. Kuramata, H. Ebe and Y. Arakawa, "An ultrawide-band semiconductor optical amplifier having an extremely high penalty-free output power of 23 dBm achieved with quantum dots", Photonic Technology Letters, 17, 8, 1614-1616, (2005)
3. Albores-Mejia, A., K.A. Williams, T. de Vries, E. Smalbrigge, Y.S. Oei, M.K. Smit, S. Anantathanasarn, R. Notzel, "Scalable quantum dot optical Switch matrix in the 1.55 um wavelength range", Proceedings Photonics in Switching, Paper D-06-4 (2008)
4. Albores-Mejia, A., K.A. Williams, T. de Vries, E. Smalbrugge, Y.S. Oei, M.K. Smit and R. Notzel, "Integrated 2 x 2 quantum dot optical crossbar switch in 1.55 mm wavelength range", Electronics Letters, 45, 6, 313-314, (2009)
5. Albores-Mejia, A., K.A. Williams, F. Gomez-Agis, S. Zhang, H.J.S. Dorren, X.J.M. Leijtens, T de Vries, Y.S. Oei, M.J.R. Heck, L.M. Augustin, R. Notzel, D.J. Robbins, M.K. Smit, "160Gb/s serial line rates in a monolithic optoelectronic multistage interconnection network", proceedings 17th Annual IEEE Symposium on High-Performance Interconnects, New York, (2009)
6. Albores-Mejia, A., K.A. Williams, T. de Vries, E. Smalbrugge, Y.S. Oei, M.K. Smit, and R. Notzel, "Low power penalty monolithically-cascaded 1550nm-wavelength quantum-dot crossbar switches", proceedings Optical Fiber Conference, (2009)
7. Albores-Mejia, A.F. Gomez-Agis, H.J.S. Dorren, X.J.M. Leijtens, M.K. Smit, D.J. Robbins and K.A. Williams, "320Gb/s data routing in a monolithic multistage semiconductor optical amplifier switching circuits", proceedings European Conference on Optical Communications,

Invited paper, We.7.E.1, (2010)

8. Albores-Mejia, A., F. Gomez-Agis, S. Zhang, H.J.S. Dorren, X.J.M. Leijtens, T. de Vries, Y.S. Oei, M.J.R. Heck, R. Notzel, D.J. Robbins, M.K. Smit and K.A. Williams, "Monolithic multistage optoelectronic switch circuit routing 160Gb/s line-rate data," Journal of Lightwave Technology, to appear, (2010)

9. Almstrom, E., C.P. Larsen, L. Gillner, W.H. van Berlo, M. Gustavsson and E. Berglind, "Experimental and analytical evaluation of packaged 4x4 InGaAsP/InP semiconductor optical amplifier gate switch matrices for optical networks", Journal of Lightwave Technology, 14, 6, 996-1004, (1996)

10. Aw E.T., H. Wang, M.G. Thompson, A. Wonfor, R.V. Penty, I.H. White, A.R. Kovsh, "Uncooled 2x2 quantum dot semiconductor optical amplifier based switch", Proceedings Conference on Lasers and Electro-Optics, (2008)

11. Bachmann, M., P. Doussiere, J.Y. Emery, R. N'Go, F. Pommereau, L. Goldstein, G. Soulage and A. Jourdan, "Polarisation-insensitive clamped-gain SOA with integrated spotsize converter and DBR gratings for WDM applications at 1.55um wavelength", Electronics Letters, 32, 22, 2076-2078, (1996)

12. Barbarin, Y., E.A.J.M. Bente, C. Marguet, E.J.S. Leclere, J.J.M. Binsma and M.K. Smit, "Measurement of reflectivity of butt-joint active-passive interfaces in integrated extended cavity lasers", Photonics Technology Letters, 17, 11, 2265-2267, (2005)

13. Beneš, V. "On rearrangeable three stage connecting networks," Bell Systems Technology Journal XLI, 1481–1492 (1962).

14. Borghesani, A., N. Fensom, A. Scott, G. Crow, L. Johnston, J. King, L. Rivers, S. Cole, S. Perrin, S. Scrase, G. Bonfrate, A. Ellis, I. Lealman, G. Crouzel, L.H.K. Chun, A. Lupu, E. Mahe, P. Maigne, "High saturation output (>16.5dBm) and low noise figure (<6dB) semiconductor optical amplifier for C band operation", proceedings Optical Fiber Communications Conference, paper 534-536, (2003)

15. Brisson, C., S. Chandrasekhar, G. Raybon and K.F. Drew, "Experimental investigation of SOAs for linear amplification in 40Gb/s transmission systems", proceedings Optical Fiber Communications Conference, WV6, (2002)

16. Buckman L.A., L.P. Chen and K.Y. Lau, "Crosstalk penalty in all-optical distributed switching networks", Photonics Technology Letters, 9, 2, 250-252, (1997)

17. Buckman-Windover, L.A., J.N. Simon, S.A. Rosenau, K.S. Giboney, G.M. Flower, L.W. Mirkarimi, A. Grot, B. Law, C.-K. Lin, A. Tandon, R.W. Gruhlke, H. Xia, G. Rankin, M.R.T. Tan, and D.W. Dolfi, "Parallel-optical interconnects >100 Gb/s", Journal of Lightwave Technology, 22, 9, 2055, (2004)
18. Burmeister, E.F. and J.E. Bowers, "Integrated gate matrix switch for optical packet buffering", Photonics Technology Letters, 18, 1, 103-105, (2006)
19. Burton J.D., P.J. Fiddyment, M.J. Robertson and P. Sully, "Monolithic InGaAsP-InP laser amplifier gate switch matrix", Journal of Quantum Electronics, 29, 6, 2023-2027, (1993)
20. Buus, J., M.C. Farries and D.J. Robbins, "Reflectivity of coated and tilted semiconductor facets", Journal of Quantum Electronics, 27, 6, 1837, (1991)
21. Chiaroni, D., F. Masetti, D. Verchère, A. Jourdan, G. Luyts, B. Pawels, "A comparison of electronic and optical packet/burst switching fabrics", proceedings Optical Fiber Communication Conference, Los Angeles, California, WF1, (2004)
22. Chiaroni, D., R. Urata, J. Gripp, J.E. Simsarian, G. Austin, S. Etienne, T. Segawa, Y. Pointurier, C. Simonneau,, Y. Suzaki, T. Nakahara, M. Thottan, A. Adamiecki, D. Neilson, J.C. Antona, S. Bigo, R. Takahashi, V. Radoaca, "Demonstration of the interconnection of two optical packet rings with a hybrid optoelectronic packet router", proceedings European Conference on Optical Communications, PD3.5, (2010)
23. Cho P., Y. Achiam, G. Levy-Yurista, M. Margalit, Y. Gross and J. Khurgin, "Investigation of SOA nonlinearities on the amplification of DWDM channels with spectral efficiency up to 2.5b/s/Hz", Photonics Technology Letters, 16, 3, 918-920 (2004)
24. Ciaramella, E., A. D'Ericco and V. Donzella, "Using semiconductor optical amplifiers with constant envelope WDM signals", J. Quantum Electronics, 44, 5, 403-409, (2008)
25. Congress report on server and data center energy efficiency: Public law 109-431 U.S. Environmental Protection Agency ENERGY STAR Program, August 2, (2007)
26. Cunningham, D. and W.G. Lane, Gigabit ethernet networking, Macmillan (1999)
27. Dally W.J. and B. Townes, Principles and practices of interconnection networks, Morgan Kaufmann (2004)

28. Dangel, R., C. Berger, R. Beyeler, L. Dellmann, M. Gmür, R. Hamelin, F. Horst, T. Lamprecht, T. Morf, S. Oggioni, M. Spreafico, and B.J. Offrein, " Polymerwaveguide-based board-level optical interconnect technology for datacom applications", Transactions on Advanced Packaging, 31, 4, 759-757, (2008)
29. Davies, A.R., K.A. Williams, R. V. Penty, I. H. White, M. Glick and D. McAuley, "Variablegain operation of a gain-clamped sampled grating diode laser amplifier", proceedings on Conference on Lasers and Electro Optics, (2002)
30. Desurvire, E., Erbium doped fiber amplifiers, Wiley & Sons, (1994)
31. Dittmann, L., C. Develder, D. Chiaroni, F. Neri, F. Callegati, W. Koerber, A. Stavdas, M. Renaud, A. Rafel, J. Solé-Pareta, W. Cerroni, N. Leligou, L. Dembeck, B. Mortensen, M. Pickavet, N. Le Sauze, M. Mahony, B. Berde, and G. Eilenberger, "The European IST project DAVID: A viable approach toward optical packet switching", Journal on Selected Areas in Communications, 21, 7, 1026-1040 (2003)
32. Djordjevic, I. B., R. Varazza, M. Hill, and S. Yu, "Packet switching performance at 10 Gbit/s across a 4x4 optical crosspoint switch matrix", Photonics Technology Letters, 16, 1, 102–104 (2004).
33. Dods, S., J.P.R. Lacey and R.S. Tucker, "Homodyne crosstalk in WDM ring and bus networks", IEEE Photonics Technology Letters, 9, 9, 1285-1287, (1997)
34. Donnelly, J.P., J.N. Walpole, G.E. Betts, S.H. Groves, J.D. Woodhouse, F.J. O'Donnell, L.J. Missaggia, R.J. Bailey, A. Napoleone, "High-power 1.3-μm InGaAsP-InP amplifiers with tapered gain regions", Photonics Technology Letters, 8, 11, 1450-1452, (1996)
35. Dorgeuille, F., B. Mersali, M. Feuillade, S. Sainson, S. Slempkes, M. Foucher, "Monolithic InGaAsP-InP tapered laser amplifier gate 2 x 2 switch matrix with gain", Photonics Technology Letters, 8, 9, 1178-1180, (1996)
36. Dorgeuille, F., Lavigne, B., Emery, J.Y., Di Maggio, M., Le Bris, J., Chiaroni, D., Renaud, M., Baucknecht, R., Schneibel, H.P., Graf, C., Melchior, H., " Fast optical amplifier gate array for WDM routing and switching applications",proceedings Optical Fiber Communication Conference, (1998)
37. Dorgeuille, F., L. Noirie, J.P. Faure, A. Ambrisy, S. Rabaron, F. Boubal, M. Schilling and C. Artigue, "1.28 Tb/s throughput 8 x 8 optical switch based on-arrays of gainclamped semiconductor optical amplifier gates", proceedings Optical Fiber Communication Conference, 221-223, paper

PD18-1/221, (2000)

38. Dreyer K., C.H. Joyner, J.L. Pleumeekers, T.P. Hessler, P.E. Selbmann, B. Deveaud, B. Dagens and J.Y. Emery, "High gain mode adapted semiconductor optical amplifier with 12.4dBm saturation power at 1550nm", Journal of Lightwave Technology, 20, 4, 718-721, (2002)

39. Dupertuis, M. A., Pleumeekers, J. L., Hessler, T. P., Selbmann, P. E., Deveaud, B., Dagens, B., and Emery, J. Y., "Extremely fast high-gain and low-current SOA by optical speed-up at transparency", Photonic Technology Letters, 12, 11, 1453–1455, (2000)

40. Ellis, A. D., D. Malyon and W.A. Stallard, "A novel all electrical scheme for laser amplifier gain control", proceedings European Conference on Optical Communications, 487- 490, (1988)

41. Ellis, A.D., D.M. Patrick, D. Flannery, R.J. Manning, D.A.O. Davies, and D.M. Spirit, "Ultrahigh-speed OTDM networks using semiconductor amplifier-based processing nodes", Journal of Lightwave Technology, 13, 5, 761-770, (1995)

42. Emery J.Y., T. Ducellier, M. Bachmann, P. Doussiere, F. Pommereau, R. Ngo, F. Gaborit, L. Goldstein, G. Laube and J. Barrau, "High performance 1.55um polarisation insensitive semiconductor optical amplifier based on a low tensile strained bulk InGaAsP", Electronics Letters, 33, 12, 1083-1084, (1997)

43. Eskildsen, L. and P.B. Hansen, "Interferometric noise in lightwave systems with optical preamplifiers", Photonics Technology Letters, 9, 11, 1538-1540, (1997)

44. Fish, G.A. ,B. Mason, L.A. Coldren and S.P. DenBaars, "Compact 4x4 InGaAsP-InP optical crossconnect with a scalable architecture", Photonics Technology Letters, 10, 9, 1256-1258, (1998)

45. Gillner, L., C.P. Larsen and M. Gustavsson, "Scalability of optical multiwavelength switching networks: Crosstalk analysis", Journal of Lightwave Technology, 17, 1, 58-67, (1999)

46. Gini, E.I, G. Guekos, and H. Melchior, "Low loss corner mirrors with 45° deflection angle for integrated optics", Electronics Letters, 28, 5, 499-501 (1992)

47. Glick, M., M. Dales, D. McAuley, T. Lin, K.A. Williams, R.V. Penty, I.H. White, "SWIFT: A testbed with optically switched data paths for computing applications", proceedings 7th International Conference on Transparent Optical Networks, (2005)

48. Godefroy A., A le Corre, F. Clerot, S. Salaun, S. Loualiche, J.C. Simon,

L. Henry, C. Vaudry, J.C. Keromnes, G. Joulie and P. Lamouler, "1.55um polarisation insensitive optical amplifier with stain balanced superlattice active layer", Photonic Technology Letters, 7, 5, 473-475, (1995)

49. Goldstein, E.L., L.E. Eskildsen, and A.F. Elrefaie, "Performance implications of component crosstalk in transparent lightwave networks", Photonics Technology Letters, 6, 5, 657-660, (1994)

50. Goldstein, E.L. and L. Eskildsen, "Scaling limitations in transparent optical networks due to low-level crosstalk", Photonics Technology Letters, 7, 1, 93-94 (1995)

51. Gripp, J., M. Duelk, J. E. Simsarian, A. Bhardwaj, P. Bernasconi, O. Laznicka and M. Zirngibl, "Optical switch fabrics for ultra-high-capacity IP routers", Journal of Lightwave Technology, 21, 11, 2839-2850, (2003)

52. Grosskopf, G., R. Ludwig, and H.G. Weber, "Cascaded inline semiconductor laser amplifiers in a coherent optical fiber transmission system", Electronics Letters, 24, 9, 551-552, (1988)

53. Grubb, S. G., D. F. Welch, D. Perkins, C. Liou, and S. Melle, "OEO versus all-optical networks", proceedings Lasers and Electro-optics Society Annual Meeting, Montreal, invited paper TuH1 (2006)

54. Gustavsson, M. B. Lagerstrom, L. Thylen, M. Janson, L. Lundgren, A.C. Morner, M. Rask, and B. Stoltz, "Monolithically integrated 4 x 4 InGaAsP/lnP laser amplifier gate switch arrays", Electronics Letters, 28, 24, 2223-2225, (1992)

55. Gustavsson, M., M. Janson and L. Lundgren, "Digital transmission experiment with monolithic 4 x 4 InGaAsP/InP laser amplifier gate switch array", Electronics Letters, 29, 12, 1083-1085, (1993)

56. Hamamoto K., K. Komatsu, "Insertion-loss-free 2x2 InGaAsP/lnP optical switch fabricated using bandgap energy controlled selective MOVPE", Electronics Letters, 31, 20, 1779-1781, (1995)

57. Hemenway, R., R.R. Grzybowski, C. Minkenberg and R. Luijten, "Optical packet switched interconnect for supercomputer applications", OSA Journal of Optical Networking, 3, 12, 900-913, (2004)

58. Himeno, A., H. Terui, M. Kobayashi, "Loss measurement and analysis of high-silica reflection bending optical waveguides", Journal of Lightwave Technology, 6, 1, 41- 46, (1988)

59. Ho, K.P., S.K. Liaw and C. Lin, "Reduction of semiconductor laser amplifier induced distortion and crosstalk for WDM systems using light injection", Electronics Letters, 32, 24, 2210-2211, (1996)

60. Hluchyj M.G. and M.J. Karol, "ShuffleNet: An application of generalised

perfect shuffles to multihop lightwave networks", Journal Lightwave Technology, 9, 10, 1386-1397, (1991)

61. Huang, D., T. Sze, A. Landin, R. Lytel, and H. Davidson, "Optical interconnects: out of the box forever?", Journal of Selected Topics in Quantum Electronics, 9, 2, 614-623 (2003)

62. Inoue, K., "Crosstalk and its power penalty in multichannel transmission due to gain saturation in a semiconductor laser amplifier", Journal of Lightwave Technology, 7, 7, 1118-1124, (1989)

63. Inoue, K., "Optical filtering technique to suppress waveform distortion induced in a gainsaturated semiconductor optical amplifier", Electronics Letters, 33, 10, 885–886, (1997)

64. Jennen, J. G. L., R. C. J. Smets, H. de Waardt, G. N. van den Hoven, and A. J. Boot, "4x10Gbit/s NRZ transmission in the 1310 nm window over 80 km of standard single mode fiber using semiconductor optical amplifiers", Proceedings European Conference on Optical Communications, 235 - 236, (1998)

65. Jeong G. and J.W. Goodman, "Analysis of linear crosstalk in photonic crossbar switches based on on/off gates", Journal of Lightwave Technology, 14, 3, 359-365, (1996)

66. Kabacinski, W., Nonblocking electronic and photonic switching fabrics, Springer, (2005)

67. Kakitsuka T., Y. Shibata, M. Itoh, Y. Kadota and Y. Yoshikuni, "Influence of buried structure on polarisation sensitivity in strained bulk semiconductor optical amplifiers", Journal of Quantum Electronics, 38, 1, 85-92, (2002)

68. Kelly A.E., I.F. Lealman, L.J. Rivers, S.D. Perrin and M. Silver, "Low noise (7.2dB) and high gain (29dB) semiconductor optical amplifier with a single layer AR coating", Electronic Letters, 33, 6, 536-538, (1997)

69. Kikuchi, N., Y. Shibata, Y. Tohmori, "Monolithically integrated 64-channel WDM channel selector", NTT review , 1, 7, 43-49 (2003)

70. Kinoshita, S., "Monolithically-integrated SOA gate switch and its application to high-speed switching systems", proceedings Asia Communications and Photonics Conference, ThD2, (2009)

71. Kim, H.K., Chandrasekhar, S., "Reduction of cross-gain modulation in the semiconductor optical amplifier by using wavelength modulated signal", Photonics Technology Letters", 12, 10, 1412-1414, (2000)

72. Kuindersma, P.I., G.P.J.M. Cuijpers, J.G.L. Jennen, J.J.E. Reid, L.F. Tiemeijer, H. de Waardt, A.J. Boot, "10Gbit/s RZ transmission at 1309nm

over 420km using a chain of multiple quantum well semiconductor optical amplifier modules at 38km intervals", proceedings European Conference on Optical Communications, TuD2.1, (1996)

73. Larsen, C.P. and M. Gustavsson, "Linear crosstalk in 4x4 semiconductor optical amplifier gate switch matrix", Journal of Lightwave Technology, 15, 10, 1865-1870, (1997)

74. Lee S.C., R. Varazza, S. Yu, "Optical label processing and 10-Gb/s variable length optical packet switching using a 4 x 4 optical crosspoint switch", Photonics Technology Letters, 17, 5, 1085-1087 (2005)

75. Lee, S.C., J. Begg, R. Varrazza and S. Yu, "Automatic per-packet dynamic power equalization in a 4 × 4 active coupler-based optical crosspoint packet switch matrix", Photonics Technology Letters, 17, 12, 331- 332 (2005)

76. Legg P.D., D. Hunter, I. Andonovic and P. Barnsley, "Inter-channel crosstalk phenomena in optical time division multiplexed switching networks", Photonics Technology Letters, 6, 5, 661-663, (1994)

77. Lemoff, B.E., M.E. Ali, G. Panotopoulos, G.M. Flower, B. Madhavan, A.F.J. Levi, and D.W. Dolfi, "MAUI: Enabling fiber-to-the-processor with parallel multiwavelength optical interconnects", Journal of Lightwave Technology, 22, 9, 2043-2044, (2004)

78. Lemoff, B.E., M.E. Ali, G. Panotopoulos, E. de Groot, G.M. Flower, G.H. Rankin, A.J. Schmit, K.D. Djordjev, M.R.T. Tan, A. Tandon, W. Gong, R.P. Tella, B. Law, L.-K. Chia, and D.W. Dolfi, "Demonstration of a compact low-power 250-Gb/s parallel-WDM optical interconnect", Photonics Technology Letters, 17, 1, 220-222, (2005)

79. Liboiron-Ladouceur, O., B. A. Small, and K. Bergman, "The data vortex optical packet switched interconnection network", Journal of Lightwave Technology, 24, 1, 262– 270, (2006)

80. Lin, T., K. A. Williams, R. V. Penty, I. H. White, M. Glick and D. McAuley, "Self-configuring intelligent control for short reach 100Gb/s optical packet routing", proceedings Optical Fiber Communications Conference, Paper OWK5, (2005)

81. Lin, T., K. A. Williams, R. V. Penty, I. H. White, M. Glick, and D. McAuley, "Performance and scalability of a single-stage SOA switch for 10x10Gb/s wavelength striped packet routing", Photonics Technology Letters, 18, 5, 691-693, (2006)

82. Lin, T., K. A. Williams, R. V. Penty, I. H. White, M. Glick, "Capacity scaling in a multi-host wavelength-striped SOA-based switch fabric", Journal of Lightwave Technology, 25, 3, 655-663 (2007)

83. Lindgren, S., M.G. Oberg, J. Andre, S. Nilsson, B. Broberg, B. Holmberg, and L. Backbom, "Loss-compensated optical Y-branch switch in InGaAsP-InP", Journal of Lightwave Technology, 8, 10, 1591-1595 (1990)
84. Liu, S., Hu, X., Thompson, M.G., Sellin, R.L., Williams, K.A., Penty, R.V., White, I.H., and Kovsh, A.R.:, "Cascaded performance of quantum dot semiconductor optical amplifier in a recirculating loop", Proceedings European Conference on Optical Communications, paper Th4.5.6, (2006)
85. Liu, Y., E. Tangdiongga, Z. Li, H. de Waardt, A.M.J. Koonen, G.D. Khoe, X.W. Shu, I. Bennion, H.J.S. Dorren, "Error-free 320-Gb/s all-optical wavelength conversion using a single semiconductor optical amplifier", Journal of Lightwave Technology, 25, 1, 103-108, (2007)
86. Lord, A. and W.A. Stallard, "A laser amplifier model for system optimisation", Optical and Quantum Electronics, 21, 6, 463-470, (1989)
87. Luijten, R.P. and R. Grzybowski, "The OSMOSIS optical packet switch for supercomputers", proceedings Optical Fiber Communications Conference, OTuF3, (2009)
88. Manning R., R. Giller, X. Yang, R.P. Webb, D. Cotter, "Faster switching with semiconductor optical amplifiers", Proceedings Photonics in Switching, 145-146, (2007)
89. Masetti, F., D. Chiaroni, R. Dragnea, R. Robotham, D. Zriny, "High-speed high-capacity packet-switching fabric: a key system for required flexibility and capacity", OSA Journal of Optical Networking, 2, 7, 255-265, (2003)
90. Maxwell, G., "Low-cost hybrid photonic integrated circuits using passive alignment techniques", proceedings Lasers and Electro-optics Society Annual Meeting, invited paper MJ2, (2006)
91. Mikkelsen, B. and Raybon, G., "Reduction of crosstalk in semiconductor optical amplifiers by amplifying dispersed WDM signals (7x20Gb/s)", Optical Fiber Communications Conference, ThJ5-I, (2000)
92. Miller D.A.B., "Optical interconnects to electronic chips", Applied Optics, 49, 25, F59-F70, (2010)
93. McAuley, D., "Optical Local Area Network", in Computer Systems: Theory, Technology and Applications, A. Herbert and K. Sparck-Jones, Eds. Springer-Verlag, (2003)
94. Morito K., M. Ekawa, T. Watanabe and Y. Kotaki, "High saturation output power (+17dBm) 1550nm polarisation insensitive semiconductor optical amplifier", proceedings European Conference in Optical Communication,

paper 1.3.2, (2000)

95. Morito, K., Ekawa, M., Watanabe, T., and Kotaki, Y., "High output-power polarizationinsensitive semiconductor optical amplifier", Journal of Lightwave Technology, 21, 1, 176-181, (2003)

96. Morito, K., S. Tanaka, S. Tomabechi and A. Kuramata, "A broadband MQW semiconductor optical amplifier with high saturation output power and low noise figure", Photonics Technology Letters, 17, 5, 974-976, (2005)

97. Mukai, T., Y. Yamamoto, T. Kimura, "S/N and error rate performance in AlGaAs semiconductor laser preamplifier and linear repeater systems", Transactions Microwave Theory and Techniques, 30, 10, 1548-1556, (1982)

98. Nagarajan R., and M.K. Smit, "Photonic integration", IEEE Laser and Electro-Optic Society Newsletter, February (2007)

99. Newkirk, M.A., U. Koren, B.I. Miller, M.D. Chien, M.G. Young, T.L. Koch, G. Raybon, C.A. Burrus, B. Tell, K.F. Brown-Goebeler, "Three-section semiconductor optical amplifier for monitoring of optical gain", Photonics Technology Letters, 4, 11, 1258- 1260, (1992)

100. Nicholes, S.C., M.L. Masanovic, E. Lively, L.A. Coldren and D.J. Blumenthal, "An 8x8 monolithic tuneable optical router (MOTOR) packet forwarding chip", Journal of Lightwave Technology, 28, 4, 641-650, (2010)

101. Oberg M. G. and N. A. Olsson, "Crosstalk between intensity-modulated wavelengthdivision multiplexed signals in a semiconductor-laser amplifier," Journal of Quantum Electronics, 24, 1, 52-59, (1988)

102. Olsson, N.A. "Lightwave systems with optical amplifiers", Journal of Lightwave Technology, 2, 7, 1071-1081, (1989)

103. Onishchukov, G., V. Lokhnygin, A. Shipulin and P. Reidel, "10Gbit/s transmission over 1500km with semiconductor optical amplifiers", Electronics Letters, 34, 16, 1597- 1598, (1998)

104. Ougazzadeu A., "Atmospheric pressure MOVPE growth of high performance polarisation insensitive strain compensated MQW InGaAsP/InGaAs optical amplifier", Electronics Letters, 31, 15, 1242-1244, (1995)

105. Patel, R.R., S.W. Bond, M.D. Pocha, M.C. Larson, H.E. Garrett, R.F. Drayton, H.E. Petersen, D.M. Krol, R.J. Deri, and M.E. Lowry, "Multiwavelength parallel optical interconnects for massively parallel processing", Journal of Selected Topics in Quantum Electronics, 9, 2, 657-666, (2003)

106. Pleumeekers, J. L., Kauer, M., Dreyer, K., Burrus, C., Dentai, A. G., Shunk, S., Leuthold, J., and Joyner, C. H., "Acceleration of gain recovery in semiconductor optical amplifiers by optical injection near transparency", Photonic Technology Letters, 14, 1, 12–14, (2002)

107. Reid, J. J. E., L. Cucala, M. Settembre, R. C. J. Smets, M. Ferreira, and H. F. Haunstein, "An international field trial at 1.3 μm using an 800 km cascade of semiconductor optical amplifiers", Proceedings European Optical Communications Conference, 567 - 568, (1998)

108. Renaud, M., M. Bachmann, M. Erman, "Semiconductor optical space switches," Journal of Selected Topics in Quantum Electronics, 2, 2, 277-288 (1996)

109. Roadmap: International technology roadmap for semiconductors, http://public.itrs.net/, (2005)

110. Roberts, G.F., Williams, K.A, Penty, R.V., White, I. H, Glick, M., McAuley, D. Kang, D. J. and Blamire, M., "Multi-wavelength data encoding for improved input power dynamic range in semiconductor optical amplifier switches", proceedings European Conference on Optical Communications, (2005)

111. Rohit, A., K. A. Williams, X. J. M. Leijtens, T. de Vries, Y. S. Oei, M. J. R. Heck, L. M. Augustin, R. Notzel, D. J. Robbins, and M. K. Smit, "Monolithic multi-band nanosecond programmable wavelength router", Photonics Journal, 2, 1, 29-35 (2010)

112. Ryu, S., Taga H. Yamamoto S., Mochizuki K. Wakabayashi H., "546km, 140Mbit/s FSK coherent transmission experiment through 10 cascaded semiconductor laser amplifiers, Electronics Letters, 25, 25, 1682-1684, (1989)

113. Sahri, N., D. Prieto , S. Silvestre , D. Keller , F. Pommerau , M. Renaud , O. Rofidal , A. Dupas , F. Dorgeuille and D. Chiaroni, "A highly integrated 32-SOA gates optoelectronic module suitable for IP multi-terabit optical packet routers", Proceedings Optical Fiber Commununications Conference, PD32-1 (2001)

114. Sato K., H. Toba, "Reduction of mode partition noise by using semiconductor optical amplifiers", Journal of Selected Topics in Quantum Electronics, 7, 2, 328-333, (2001)

115. Saxtoft C. and P. Chidgey, "Error rate degradation due to switch crosstalk in large modular switched optical networks", IEEE Photonics Technology Letters, 5, 7, 828-831, (1993)

116. Sasaki, J., H. Hatakeyama, T. Tamanuki, S. Kitamura, M. Yamaguchi, N. Kitamura, T. Shimoda, M. Kitamura, T. Kato, M. Itoh, "Hybrid integrated

4x4 optical matrix switch using self-aligned semiconductor optical amplifier gate arrays and silica planar lightwave circuit", Electronics Letters, 34, 10, 986-987, (1998)

117. Shacham, A., B.A. Small, O. Liboirin-Ladouceur and K. Bergman, "A fully implemented 12x12 data vortex optical packet switching interconnection network", Journal of Lightwave Technology, 23, 10, 3066-3075, (2005)

118. Shacham, A. and K. Bergman, "Building ultralow-latency interconnection networks using photonic integration", IEEE Micro, 6-20, (2007)

119. Shao S.K. and M.S. Kao, "WDM coding for high-capacity lightwave systems", Journal of Lightwave Technology, 12, 1, 137-148, (1994)

120. Shares, L., J.A. Kash, F.E. Doany, C.L. Schow, C. Schuster, D.M. Kuchta, P.K. Pepeljugoski, J.M. Trewhella, C.W. Baks, R.A. John, L. Shan, Y.H. Kwark, R.A. Budd, P. Chiniwalla, F.R. Libsch, J. Rosner, C.K. Tsang, C.S. Patel, J.D. Schaub, R. Dangel, F. Horst, B.J. Offrein, D. Kucharski, D. Guckenberger, S. Hegde, H. Nyikal, C.-K. Lin, A. Tandon, G.R. Trott, M. Nystrom, D.P. Bour, M.R.T. Tan, and D.W. Dolfi, "Terabus: Terabit/second-class card-level optical interconnect technologies", Journal of Selected Topics in Quantum Electronics, 12, 5, 1032-1044, (2006)

121. Sherlock G., J.D. Burton, P.J. Fiddyment, P.C. Sully, A.E. Kelly, M.J. Robertson, "Integrated 2x2 optical switch with gain", Electronics Letters, 30, 2, 137-138, (1994)

122. Smets, R.C.J., J.G.L. Jennen, H. de Waart, B. Teichmann, C. Dorschky, R. Seltz, J.J.E. Reid, L.F. Tiemeijer, P.I. Kuindersma, A.J. Boot, "114km repeaterless, 10Gb/s transmission at 1310nm using an RZ data format", proceedings Optical Fiber Conference, ThH2 (1997)

123. Soganci, M., T. Tanemura, K. Takeda, M. Zaitsu, M. Takenaka and Y. Nakano, "Monolithic InP 100-port photonic switch", proceedings European Conference on Optical Communications, post-deadline paper, PD1.5, (2010)

124. Soulage, G., Doussiere, P., Jourdan, A., and Sotom, M., "Clamped gain travelling wave semiconductor optical amplifier as a large dynamic range optical gate", Proceedings European Conference on Optical Communications, Florence, Italy, 451–454, (1994)

125. Spanke, R. A. and V. E. Beneš, "N-stage planar optical permutation network," Applied Optics, 26, 7, 1226–1229 (1987)

126. Spiekman, L. H., J. M. Wiesenfeld, A. H. Gnauck, L. D. Garrett, G. N. van den Hoven, T. van Dongen, M. J. H. Sander-Jochem, and J. J. M. Binsma, "Transmission of 8 DWDM channels at 20 Gb/s over 160 km

of standard fiber using a cascade of semiconductor optical amplifiers", Photonics Technology Letters, 12, 6, 717-719, (2000)

127. Spiekman, L.H., A.H. Gnauck, J.M. Wiesenfeld and L.D. Garrett, "DWDM transmission of thirty two 10Gbit/s channels through 160km link using semiconductor optical amplifiers", Electronics Letters, 36, 12, 1046-1047, (2000)

128. Stabile R. and K.A. Williams, "Low polarisation conversion in whispering gallery mode micro-bends", proceedings European Conference on Integrated Optics, (2010)

129. Stabile, R., H. Wang, A. Wonfor, K. Wang, R.V. Penty, I.H. White and K. A. Williams, "Multipath routing in a fully scheduled integrated optical switch fabric", proceedings European Conference on Optical Communications, paper We.8.A.6, (2010)

130. Stubkjaer, K.E., "Semiconductor optical amplifier-based all-optical gates for high-speed optical processing", Journal of Selected Topics in Quantum Electronics, 6, 6, 1428- 1435, (2000)

131. Summerfield, M. A. and R.S. Tucker, , "Frequency-domain model of multiwave mixing in bulk semiconductor optical amplifiers", Journal of Selected Topics in Quantum Electronics, 5, 3, 839-850, (1999)

132. Sun, Y., A.K. Srivastava, S. Banerjee, J.W. Sulhoff, R. Pan, K. Kantor, R.M. Jopson and A.R. Chraplyvy, "Error free transmission of 32x2.5Gb/s DWDM channels over 125km using cascaded in-line semiconductor optical amplifiers", Electronics Letters, 35, 21, 1863-1865, (1999)

133. Suzuki, Y., K. Magari, Y. Kondo, Y. Kawaguchi, Y. Kadota, K. Yoshino, "High-gain array of semiconductor optical amplifier integrated with bent spot-size converter (BEND SS-SOA)", Journal of Lightwave Technology 19, 11, 1745-1750, (2001)

134. Tanaka S., S. Tomabechi, A, Uetake, M. Ekawa, K. Morito, "Highly uniform eight channel SOA-gate array with high saturation output power and low noise figure", Photonics Technology Letters, 19, 16, 1275-1277, (2007)

135. Tanaka, S., S.H. Jeong, S.Yamazaki, A. Uetake, S. Tomabechi, M. Ekawa, and K. Morito, "Monolithically integrated 8:1 SOA gate switch with large extinction ratio and wide input power dynamic range", Journal of Quantum Electronics, 45, 9, 1155-1162, (2009)

136. Tanaka S., N. Hatori, A, Uetake, S. Okumura, M. Ekawa, G. Nakagawa and K. Morito, "Compact, very-low-electric-power-consumption (0.84W) 1.3um optical amplifier module using AlGaInAs MQW-SOA", Proceedings European Conference on Optical Communications, Th10D3

(2010)

137. Tiemeijer, L.F.; Groeneveld, C.M.; "Packaged high gain unidirectional 1300 nm MQW laser amplifiers", proceedings Electronic Components and Technology Conference, 751, (1995)

138. Tiemeijer L.F., P.J.A. Thijs, T. van Dongen, J.J.M. Binsma and E.J. Jansen, "Polarization resolved, complete characterisation of 1310nm fiber pigtailed multiple-quantumwell optical amplifiers", Photonics Technology Letters, 14, 6, 1524-1533, (1996)

139. Tiemeijer, L.F., S. Walczyk, A.J.M. Verboven, G.N. van den Hoven, P.J.A. Thijs, T. van Dongen, J.J.M. Binsma, E.J. Jansen, "High-gain 1310 nm semiconductor optical amplifier modules with a built-in amplified signal monitor for optical gain control", Photonics Technology Letters, 9, 3, 309-311, (1997)

140. Tucker, R.S., "Optical packet switching: A reality check", OSA Journal of Optical Switching and Networking, 5, 2-9, (2008)

141. van Berlo, W., M. Janson, L. Lungren, A.C. Morner, J. Terlecki, M. Gustavsson, P. Granestrand, P. Svensson, "Polarization-insensitive 4x4 InGaAsP-InP laser amplifier gate switch matrix", Photonics Technology Letters, 7, 11, 1291-1293, (1995)

142. Varazza R., I.B. Djordjevic and S. Yu, "Active vertical-coupler-based optical crosspoint switch matrix for optical packet-switching applications", Journal of Lightwave Technology, 22, 9, 2034-2042, (2004)

143. Wang H., K.A. Williams, A. Wonfor, T. de Vries, E. Smallbrugge, Y.S. Oei, M.K. Smit, R. Notzel, S. Liu, R.V. Penty, I.H. White, "Low penalty cascaded operation of a monolithically integrated quantum dot 1x8 port optical switch", Proceedings European Conference on Optical Communications, (2009)

144. Wang H., Aw E.T., K.A. Williams, A. Wonfor, R.V. Penty, I.H. White, "Lossless multistage SOA switch fabric using high capacity 4x4 switch circuits", Proceedings Optical Fiber Conference (2009)

145. Wang, H., A. Wonfor, K.A Williams, RV. Penty, I.H. White, "Demonstration of a lossless monolithic 16x16 QW SOA switch", proceedings European Conference in Integrated Optics, (2010)

146. Wei X., Y. Su, X. Liu, J. Leuthold and S. Chandrasekhar, "10Gb/s RZ-DPSK transmitter using a saturated SOA as a power booster and limiting amplifier", Photonics Technology Letters, 16, 6, 1582-1584, (2004)

147. White, I.H., K.A. Williams, R.V. Penty, T. Lin, A. Wonfor, E.T. Aw, M.

Glick, M. Dales and D. McAuley, "Control architecture for high capacity multistage photonic switch circuits", OSA Journal of Optical Networking, 6, 2, 180-188 (2007)

148. White, I.H., E.T. Aw, K.A. Williams, H. Wang, A. Wonfor and R.V. Penty, "Scalable optical switches for computing applications", OSA Journal of Optical Networking, Invited paper, 8, 2, 215–224 (2009)

149. Williams, K.A., R.V. Penty, I.H. White and D. McAuley, "Advantages of gain clamping in semiconductor amplifier crosspoint switches", proceedings Optical Fiber Communication Conference (2002)

150. Williams, K.A., G.F. Roberts, T. Lin, R.V. Penty, I.H. White, M. Glick and D. McAuley, "Monolithic 2x2 optical switch for wavelength multiplexed interconnects", Journal of Selected Topics in Quantum Electronics, Special issue on integrated optics and optoelectronics, 11, 78-85 (2005)

151. Williams, K.A., "High capacity switched optical interconnects for low-latency packet routing", IEEE Photonics Society Benelux Symposium, (2006)

152. Williams, K.A., "Integrated semiconductor optical amplifier based switch fabrics for highcapacity interconnects", OSA Journal of Optical Networking, Invited paper, 6, 2, 189-199 (2007)

153. Williams, K.A., E.T. Aw, H. Wang, R.V. Penty, I.H. White, "Physical layer modelling of semiconductor optical amplifier based Terabit/second switch fabrics", Numerical Simulation of Optical Devices, Post-deadline paper ThPD5, (2008)

154. Winzer, P.J., "Modulation and multiplexing in optical communication systems", IEEE Photonics Sociey Newsletter, 23, 1, 4-10, (2009)

155. Wolfson, D., "Detailed theoretical investigation and comparison of the cascadability of conventional and gain-clamped SOA gates in multiwavelength optical networks", Photonics Technology Letters 11, 11, 1494–1496 (1999)

156. Wonfor, A., S. Yu, R.V. Penty and I.H. White, "Constant output power control in an optical crosspoint switch allowing enhanced noise performance operation", in European Conference on Optical Communications, 136-137 (2001)

157. Wu C. and T. Feng, "On a class of multistage interconnection networks", 29, 8, 694-702, (1980)

158. Yang S. and Y.G. Yao, "Impact of crosstalk induced beat noise on the size of laser amplifier based optical space switch structures", Photonics Technology Letters, 8, 7, 894-896, (1996)

159. Yoshino, M. and Inoue, K. "Improvement of saturation output power in a semiconductor laser amplifier through pumping light injection", Photonics Technology Letters, 8, 1, 58–59, (1996)
160. Yu, J. and Jeppesen, P. "Improvement of cascaded semiconductor optical amplifier gates by using holding light injection", Journal of Lightwave Technology, 19, 5, 614–623, (2001)
161. Zhou J., M. J. O'Mahony, and S.D. Walker, "Analysis of optical crosstalk effects in multiwavelength switched networks", Photonics Technology Letters, 6, 2, 302-304, (1994)
162. Zhou J., R. Cadeddu, E. Casaccia and M.J. O'Mahony, "Crosstalk in multiwavelength optical crossconnect networks", Journal of Lightwave Technology, 14, 6, 1423-1435, (1996)

Chapter 4

A DIGITAL AUTO-ZEROING CIRCUIT TO REDUCE OFFSET IN SUB-THRESHOLD SENSE AMPLIFIERS

Peter Beshay[1], Joseph F. Ryan[2] and Benton H. Calhoun[1]

[1] The Charles L. Brown Department of Electrical and Computer Engineering, University of Virginia, Charlottesville, VA 22904, USA

[2] Intel Corporation, Hillsboro, OR 97124, USA

ABSTRACT

Device variability in modern processes has become a major concern in SRAM design leading to degradation of both performance and yield. Variation induced offset in the sense amplifiers requires a larger bitline differential, which slows down SRAM access times and causes increased power consumption. The effect aggravated in the sub-threshold region. In this paper, we propose a circuit that reduces the sense amp offset using an auto-zeroing scheme with automatic temperature, voltage, and aging tracking. The circuit enables flexible tuning of the offset voltage. Measurements taken from a 45 nm test chip show the circuit is able to limit the offset to 20 mV. A 16kB SRAM is designed using the auto-zeroing circuit for the sense amps. The reduction in the total read energy and delay is reported for various configurations of the memory.

INTRODUCTION

Variation induced offset in the sense amplifiers requires a larger bit-line differential, which slows down SRAM access times and causes increased power consumption. In the sub-threshold (sub-V_T) region of operation, in particular, the effect is more dominant because the threshold voltage (V_T) variation has an exponential effect on the drive current. It is shown in [1] that the offset gets worse in sub-threshold relative to strong inversion as technology scales. Furthermore, increasing the sizes of devices in the sense amplifier does not yield the reduction of input referred offset according to $1/(WL)^{0.5}$ that is achieved for strong inversion operation. In addition it causes increased cell instability and a severely degraded read-current [1,2]. Several attempts have been made before to tackle the problem of offset voltage in sense amplifiers (SAs) including redundancy [3], transistor upsizing [4], digitally controlled compensation [5] and dynamic compensation [6]. Our approach to eliminating offset is a digital auto-zeroing (DAZ) scheme inspired by analog amplifier offset correction [7]. The main advantages of the approach are the near-zero offset after cancellation, offset tuning, and the automatic temperature, voltage, and aging tracking achievable using a repeated offset calibration phase, which makes the design useful in the sub-threshold region due to the high offset voltage sensitivity to supply voltage and temperature variations. In Verma and Chandrakasan [3], SA redundancy is used. It requires the SRAM bit-line from each column be connected to N different SAs. One SA will be selected whose offset is bound by the high and low logic levels of bit-line. This scheme statistically reduces the reliance on SAs with high offset, but it adds area and test time calibration to select the SA to use from each set. In Pileggi [4], transistor upsizing is used. A linear response surface model was developed to relate the SA offset voltage to the threshold voltages variations. The model is used to determine the statistical optimized transistors sizing. The optimized design resulted in a 25% decrease in the standard deviation of the offset voltage at a cost of 10% increase in active area. In Ryan and Calhoun [1], a methodology is proposed for sizing sub-VT SAs to minimize offset using SAs for ultra-low power operation. In Bhargava [5] a digitally controlled compensation is used. The scheme is applied to a latch-style and Strong-ARM SA topologies. Measured results from a 45nm test-chip show that the standard deviation of the offset is reduced by 5X. In Sachdev [6], dynamic compensation is used. A group of transistors are selectively coupled to high and low voltage levels via multi-phase timing. This results in a voltage level on nodes of interest that is a function of transistor mismatch. The voltage levels act to compensate for the transistor mismatch. This scheme is similar to the auto-zeroing scheme presented in this work. However the presented scheme uses a compensation capacitor, charge pump and feed-back circuit. Hence, the calibration phase is not necessarily needed prior to every sensing cycle. This improves the SA power consumption as will be illustrated in Section 5.

Section 2 describes the implementation of the auto-zeroing offset compensation scheme. Section 3 illustrates the voltage, temperature and aging tracking. Section 4 illustrates the offset tuning. Section 5 provides details of the power consumption and methods proposed to minimize it. Section 6 illustrates the offset sensitivity. Section 7 illustrates the offset compensation across technologies. Section 8 shows the improvements in read energy and delay gained by utilizing the DAZ SA in a 16kB SRAM. Section 9 provides a comparison to other offset compensation schemes. Section 10 provides measurements of the DAZ SA offset voltage from a 45 nm bulk CMOS test chip.

MISMATCH COMPENSATION USING AUTO- ZEROING CIRCUITRY

Our auto-zeroing scheme uses a split-phase clock and charge pump feedback circuit. Figure 1a shows a conventional latch-based sense amp with PMOS inputs (e.g., to support near-V_{ss} sensing on a low swing bus). Figure 1b shows the auto-zeroing circuit attached to the sense amp. The same scheme can apply to a SA with NMOS inputs in an SRAM. The charge pump circuit is shown in Figure 2. ENI and ENO are the input voltage differential and offset tuning phases respectively. ENR1 and ENR2 are reset phases. During ENR1, a zero differential input is applied to the sense amp. The ENO phase then occurs, and the SA resolves based on its intrinsic offset.

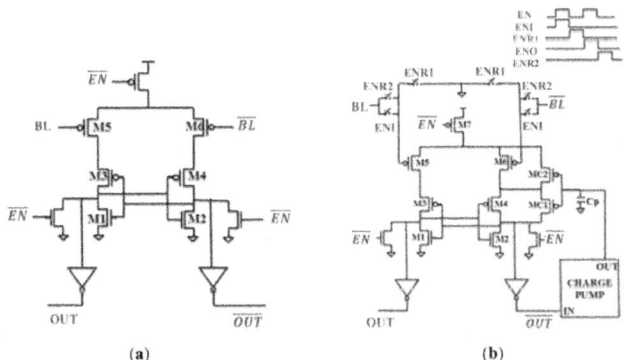

Figure 1: (a) Latch-based sense amp for near-V_{ss} inputs; (b) Auto-zeroing circuit attached to the sense amp [7].

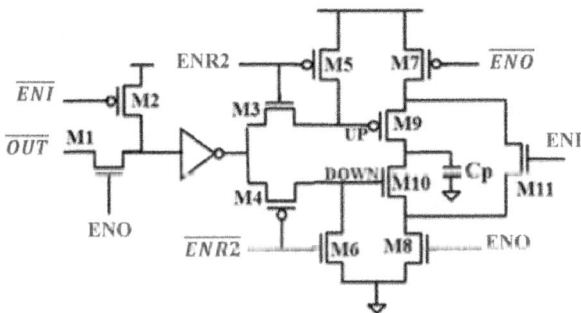

Figure 2: Charge pump circuit for adjusting the voltage on Cp. [7].

The sense amp output is fed to the charge pump circuit that charges the capacitor, Cp, up or down. During ENR2, the differential input is applied to the sense amp. ENI then occurs, and the SA resolves based on the differential input. Note that phases ENR1 and ENO can be omitted or included based on how often re-calibration is needed. Transistors MC1 and MC2 control the drive strength of the right side of the sense amp to compensate for the offset.

The charge pump controls the drive current in both transistors to equalize the strength of the SA right and left sides to reduce the offset. The offset is compensated with minimal capacitive loading at the output and is independent of input DC bias (V_{INDC}). A supply voltage and clock frequency of 0.5 V and 1MHz are used in the simulations. The output voltage of the sense amp and the voltage on Cp are illustrated in Figure 3 for an input differential of −10 mV. The initial voltage on Cp is zero. This causes an intrinsic positive offset voltage that set the SA output voltage to 1. Simulations indicate that the voltage on Cp required for a zero offset is 142 mV. For a 10 mV offset, the voltage on Cp can vary within ±12 mV.

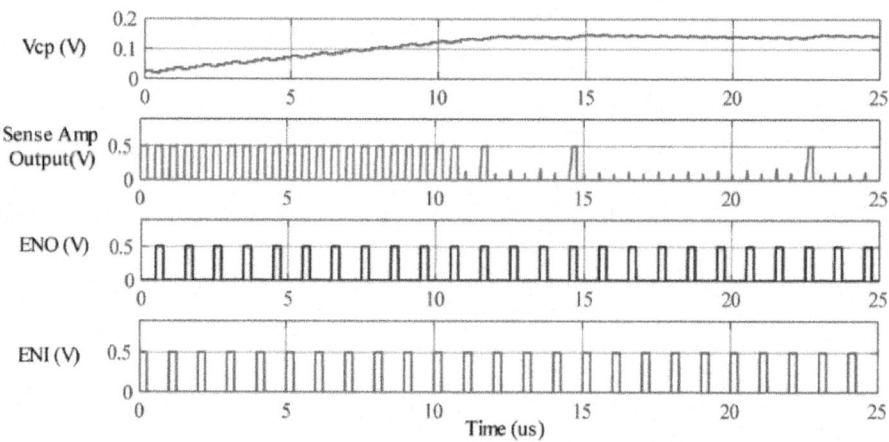

Figure 3: Simulated output voltage of the sense amp and Cp voltage for a −10 mV differential input voltage at 0.5 V and 1MHz in 45 nm CMOS.

This imposes a minimum and maximum limit on Cp voltage to 130 mV and 154 mV in order to maintain an offset less than 10 mV. The deviation of Cp voltage from to the value corresponding to zero offset (142 mV in this case) is plotted inFigure 4 for desired final offset voltages of 5 mV, 10 mV, 15 mV, and 20 mV. Low offset voltages are usually realized using a higher value of Cp. In Figure 3, the offset compensation is completed when the voltage on Cp settles to its final value within the 130 mV to 154 mV range. The sense amp then resolves its output correctly to 0 during the input phase. A zero differential voltage is applied to the SA input during the offset phase. This sets the SA output to "1" when Cp voltage drops below 142mV and "0" otherwise. In this design, rate at which Cp charges up is higher than its charge down rate. This helps to minimize the power consumption as will be discussed in Section 5.

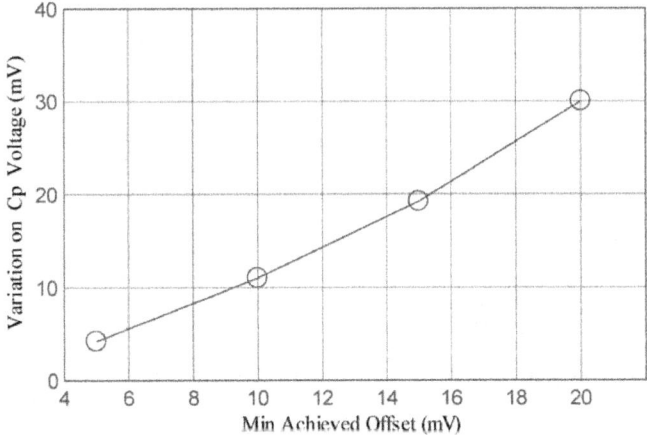

Figure 4: Variation on Cp voltage *vs.* Minimum Achieved Offset.

VOLTAGE, TEMPERATURE, AND AGING TRACKING

To demonstrate temperature, voltage, and aging tracking, the offset voltage that remains after compensation is calculated for various voltages and temperatures as shown in Figure 5a. Simulations in a commercial 45 nm process show that the circuit maintains a constant offset across temperature. The accuracy of voltage tracking depends on the supply voltage. Higher supply voltage causes more charge to be pumped to Cp during each offset calibration cycle, and this larger change in charge leads to a coarser resolution, as Figure 5a illustrates.

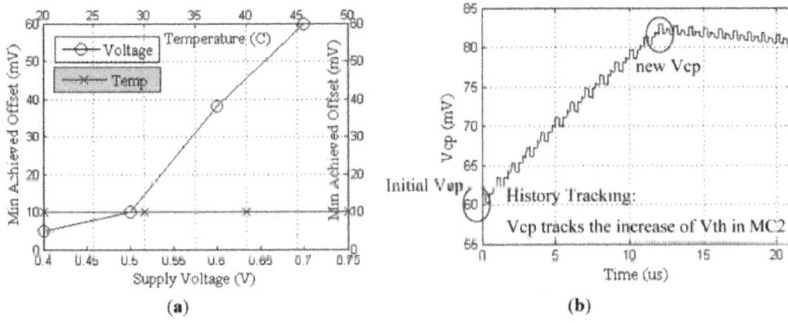

Figure 5: (a) Voltage and temperature tracking; (b) Aging tracking [7].

The auto-zeroing scheme also has the ability to compensate for any changes in device characteristics after circuit deployment. One common cause for such changes is effective threshold voltage shifting due to Bias Temperature Instability (BTI), hot carrier injection, or other aging effects. To demonstrate how this circuit can compensate for such changes, Figure 5b shows the capacitor voltage after an abrupt increase in the threshold voltage of MC2, to emulate an aging effect. The charge pump boosts the voltage on Cp to decrease the drive strength of MC2 in response and rapidly restores the compensated offset voltage.

OFFSET TUNING

We define settling time as the difference between the time when the zero differential-input is first applied and the time when the voltage of the output capacitor settles as shown in Figure 6. Changing the size of the output capacitor (Cp) affects the amount of charge added during the offset compensation phase (ENO) and so controls both the offset and the settling time. Figure 7 demonstrates the trade-off between accuracy (min achieved offset) and settling time using different values of output capacitors.

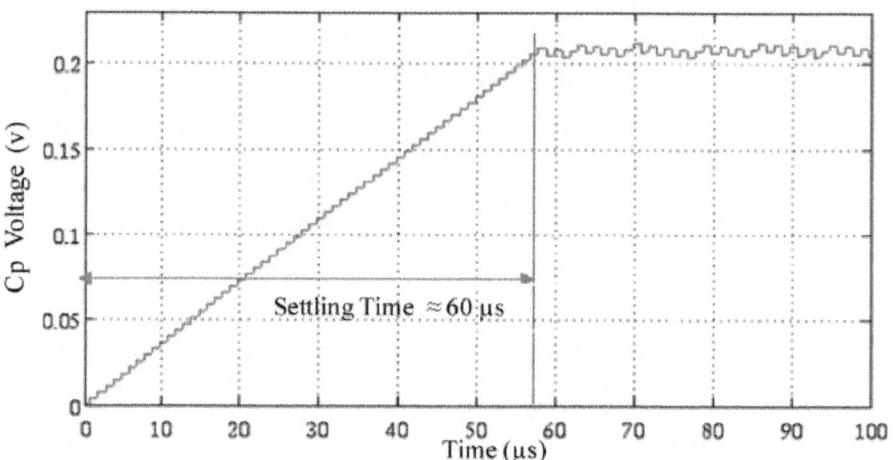

Figure 6: Settling time simulation [7].

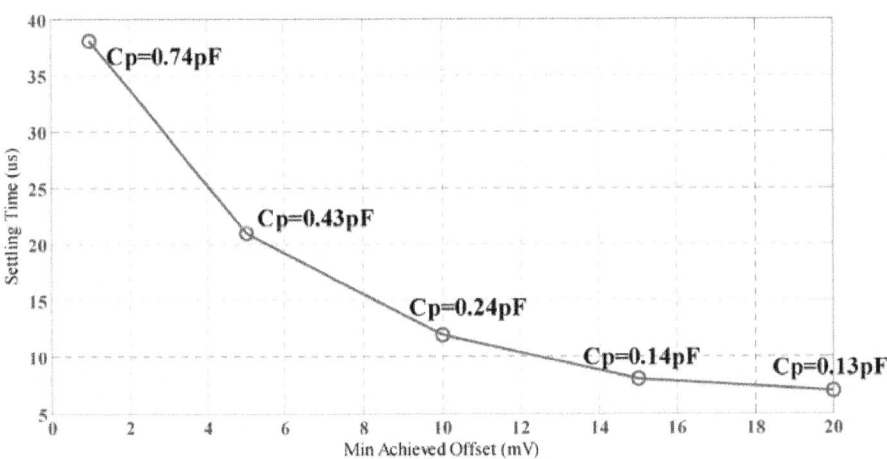

Figure 7: Min achieved offset *vs.* settling time for different values of output capacitor. [7].

POWER CONSUMPTION

The main contribution to the power consumed by the DAZ SA comes from the continuous calibration. Decreasing the number of cycles of calibration phase (ENI and ENO) relative to the input phase decreases the switching power of the feedback circuit and the power consumed in charging and discharging (Cp) but is limited by the leakage at the output capacitor (Cp). The overhead area of the scheme includes the area of the timing circuit, the charge pump circuit, and the output capacitance (Cp). For an offset voltage of 1mV, a 0.74pF output capacitance is needed. In this case, the area of Cp can dominate the total area overhead. In Figure 8, the offset calibration phase occurs once every 15 clock cycles. The maximum calibration period or the minimum number of offset calibration cycles needed is limited by the leakage on Cp. Simulation results indicated a maximum calibration period of 200 μs. This high period makes the difference in power consumption between the DAZ SA and the Latch SA insignificant. The total power consumption of the DAZ SA and the Latch SA is 2.02 nW and 2 nW respectively. The minimum number of offset calibrations is independent of the required offset or the value of Cp, but it depends on the charge pump current. In this design the charging current is 0.5 μA. High current allows fewer number of calibration cycles. Increasing the charge pump current however increases the dynamic power consumption.

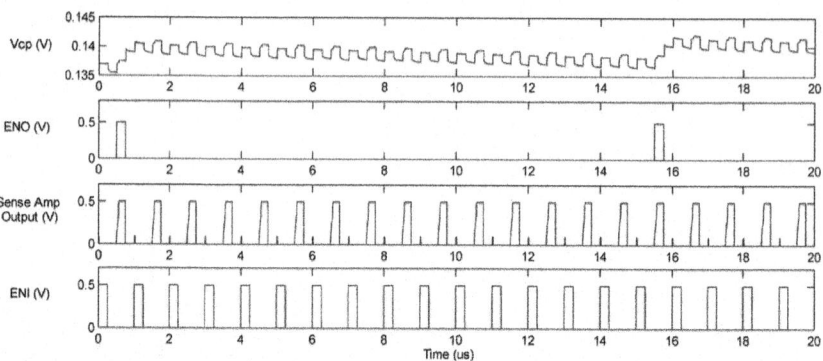

Figure 8: Offset compensation clock period for a 10 mV offset voltage.

Shorting the output virtual nodes of the charge pump through M11 can decrease the leakage by reducing V_{DS} of the switches and improve settling time as shown in Figure 9. The switching power can also be decreased by strengthening M9 in the charge pump circuit relative to M10 to avoid the continuous toggling of the sense amp output during offset compensation phase (ENO) after settling as shown in Figure 10. Strengthening M9 has the downside effect of increasing the settling time when Cp is moving to lower voltages; the time Cp takes to discharge will increase. However, the compensation usually starts with zero-initial voltage on the capacitor Cp that makes the settling time mostly dependent on the charging rate.

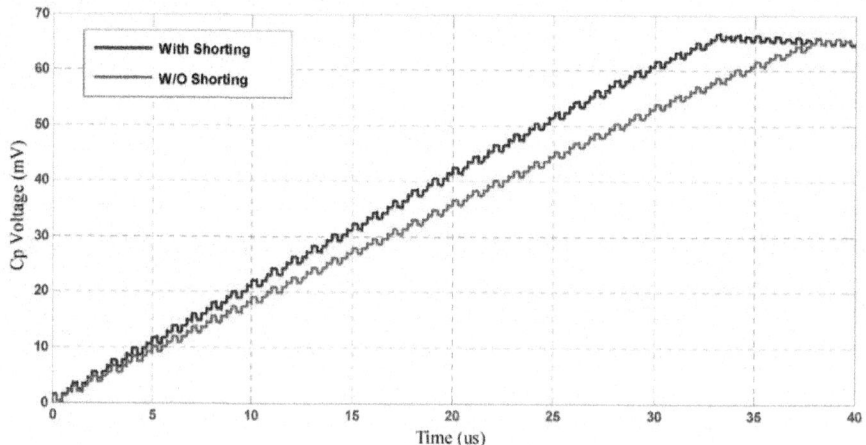

Figure 9: Voltage on Cp with and without shorting the virtual supply nodes of the charge pump [7].

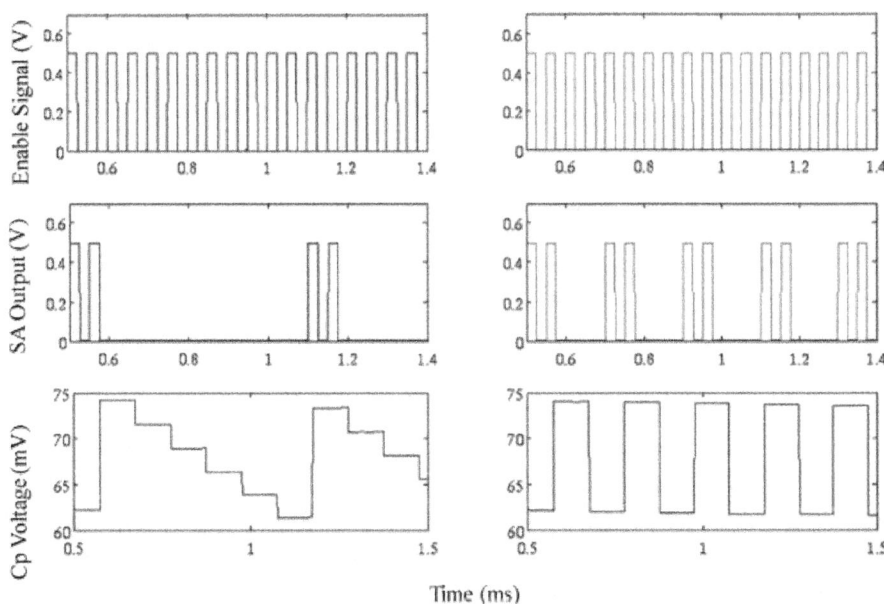

Figure 10: Strengthening the pull up transistor of the charge pump (left column of sims) reduces the rate at which the SA output switches high relative to equal strength devices (right column). This reduces power consumption [7].

OFFSET SENSITIVITY

The sensitivity of the offset compensation depends on the split phases, charge pump circuit, and the output capacitance. The accuracy of the split phases has the dominant influence on the resolution. A small overlap between ENO and ENR2 phases can dramatically degrade the accuracy by connecting M1 and M2 to the supply rails during charging. That leads to a significant increase in the charge pump rate degrading the accuracy as shown in Figure 11b, where the min achieved offset is plotted against the error in split phase timing, measured as the percentage of time overlap between ENO and ENR2. The scheme is also sensitive to variations in the M9 and M10 transistors in the charge pump circuit. They are responsible for charging/discharging Cp, and so the one with more drive strength determines the final offset value. Figure 11a shows the sensitivity of the offset voltage to the output capacitance Cp.

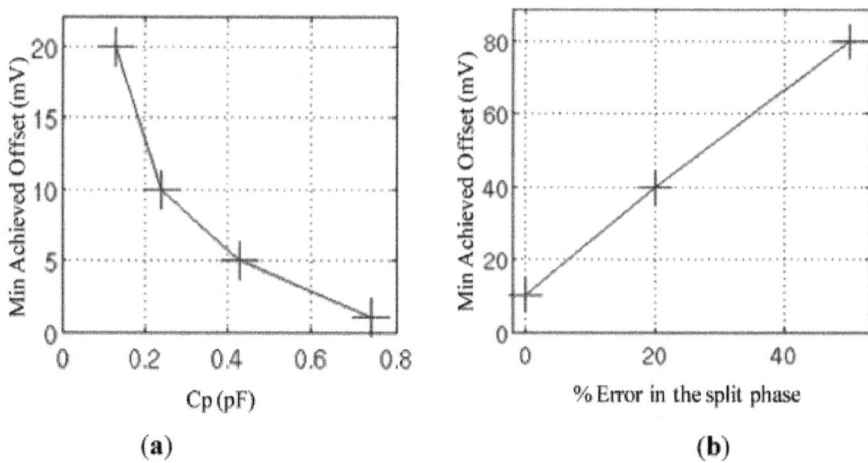

Figure 11: (a) Offset voltage sensitivity to output capacitance; (b) sensitivity to split phases [7].

The offset voltage is also sensitive to the frequency of the split phase. The increase in the split phase frequency increases the enable signal switching and degrades the compensated offset voltage.

OFFSET COMPENSATION ACROSS TECHNOLOGY NODES

The intrinsic offset voltage of the sense amplifier relies on the technology of the design. The effect of process variability on the final offset of the auto-zeroing circuit is marginal as explained in Section 1 due to the repeated compensation. To illustrate the variation of the SAs offset across technologies, 1000 Monte-Carlo simulations were performed to evaluate the offset voltage of a latch-based SA using 45, 65, 90 and 130 nm commercial technology models. Figure 12 illustrates the 3σ value of the SA offset voltage across technologies. The results indicate that the largest offset voltage belongs to 45 nm technology, followed by 32 nm, 65 nm, 90 nm and 130 nm respectively. Although the offset behavior is not monotonically increasing with technology scaling, the plot indicates a trend of increased offset in emerging technologies, *i.e.*, 32, 45 nm. Since offset voltage is increasing with newer technologies, compensation becomes more essential as technology scales. The next section demonstrates the benefits gained from using the DAZ SA in a 16kB SRAM. The total energy and delay of the SRAM is calculated and compared to the uncompensated SA case.

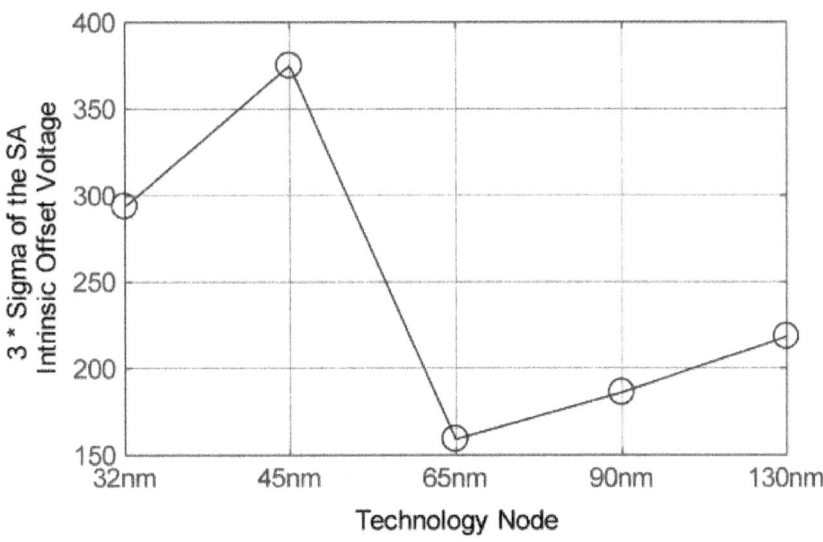

Figure 12: 3σ of the intrinsic voltage of the SA across technology nodes.

16 KB SRAM DESIGN

In this section, we investigate the effect of utilizing the DAZ SA in a 16kB SRAM memory. The power consumption of the DAZ SA is higher than the uncompensated one due to the clock generator, charge pump and the buffer stages needed for the non-overlapping clock. The sense amplifier delay is also higher due to the high capacitive loading. Reducing the sense amp offset reduces the necessary bit-line swing, which decreases both the precharge and bitcell energy and delay during the read operation. The reduction in the read energy and delay depends on the number of banks, rows and words per row of the memory. The energy and delay of the 16 kB SRAM is calculated using a 20 mV DAZ SA for all possible configurations and plotted in Figure 13. Each point is annotated with (B, R, W) where B is the number of banks, R is the number of rows, and W is the number of words per row. The results indicate a significant improvement in both the energy and delay for cases with large numbers of rows and small improvement or degradation for cases with small number of rows. The design point of 1 bank, 512 rows and 2 words per row has the biggest improvement of 10% in energy and 24% in delay. The design point of 4 banks, 32 rows and 8 words per row has the biggest degradation of 6% in energy and 5% in delay. The DAZ SA created 3 new optimal design points (1, 256, 4), (1, 128, 8) and (8, 128, 1). The improvement in energy and delay for (1, 256, 4) is 12% and 13% respectively. For (1, 128, 8) it improves the energy by 13% and degrades the delay by 5%. The energy of (1, 128, 8) with DAZ SA is the lowest. This could not be achieved using a Latch SA. The system level parameters of the SRAM that satisfy the energy/delay requirements changed (*i.e.*, design point (16, 64, 1) is the minimum energy/delay point with uncompensated SA. Using DAZ SA, the min delay/energy design changed to (1, 128, 8).

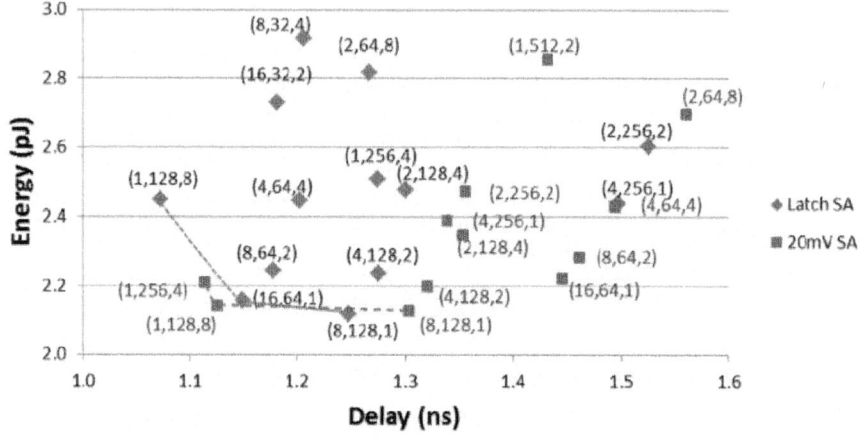

Figure 13: Design space of 16kB SRAM Memory with uncompensated and 20 mV digital auto-zeroing (DAZ) SA.

COMPARISON TO OTHER OFFSET COMPENSATION SCHEMES

The main advantages of the scheme are the continuous calibration that makes it specifically useful for sub-threshold operation and the flexibility to tune the offset voltage. The latter provides different design options that can be utilized in the SRAM design process. Approaches like redundancy [3], transistor upsizing [4], and digitally controlled compensation [5] do not support continuous calibration and hence would not be tolerant to voltage and temperature variation. The approach in [6] provides continuous calibration. The power consumed in this approach is only for compensation clock phase generation. There is no charge pump or additional circuitry. However this approach requires the calibration phase to essentially occur before every sensing cycle. As explained in Section

5, the DAZ SA can perform compensation every N number of cycles. High charge pump current can be used to increase N at a cost of higher dynamic and leakage power of the charge pump circuit. The power consumption of [6] is compared to that of DAZ SA with the offset calibration phase occurring every cycle (cycle period = 1 μs) and every 200μs with a controllable offset phase. The controllable offset phase logic is employed to force calibration every cycle at the beginning. The logic then enables calibration every N cycles once the voltage on Cp settled to its final value. The results are shown in Table 1. The settling time of both schemes is compared. The DAZ SA with controllable offset phase consumes the lowest power with a 12 μs settling time.

Table 1: Power consumption of dynamic offset compensation and auto-zeroing circuit

Offset Compensation Scheme	Power Consumption	Settling time
DAZ SA	6 nW	12 μs
Dynamic Compensation [6]	4 nW	0.5 μs
DAZ SA with controllable offset phase	2.5 nW	12 μs

45 NM TEST CHIP MEASUREMENTS

A test chip fabricated in 45 nm technology is used to verify the scheme. The chip contains one regular SA array for benchmarking and another array that uses SAs with the auto-zeroing circuitry, with Cp equal to 32fF. The chip micrograph is shown in Figure 14. Figure 15 shows the layout of a single DAZ sense amplifier. The layout of the output capacitor is shown on the left side of Figure 15, consuming an area of 2.97 μm × 3.9 μm. The sense amplifier and the charge pump layout are shown on the right side of Figure 15, consuming an area of 4.39 μm × 5.29 μm. The supply voltage is set to 0.6V during measurements to mitigate the effect of noise on the measured results. The control signals are supplied to the auto-zeroing circuit at 1 MHz. Figure 16 shows the measured offset distribution of both banks. The positive terminal of the SAs is connected to 0.45 V. The negative terminal of the SAs is swept from 0.3 V to 0.6 V in increments of 5 mV. The SAs are enabled during each increment, and measurements of the SAs outputs are recorded. This information is then used to construct the SAs offset distribution in Figure 16. The measured mean (μ) and standard deviation (σ) of the uncompensated SA banks is −31 mV and 45 mV respectively. The auto-zeroing circuitry reduced the value of μ to −13mV and lowered σ to 9.3 mV. This indicates an 80% improvement in σ. The scheme limits the absolute value of the maximum offset to 50 mV.

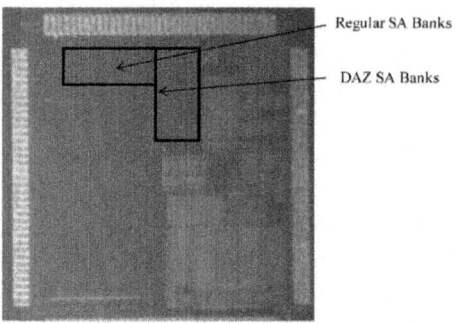

Figure 14: Chip Micrograph.

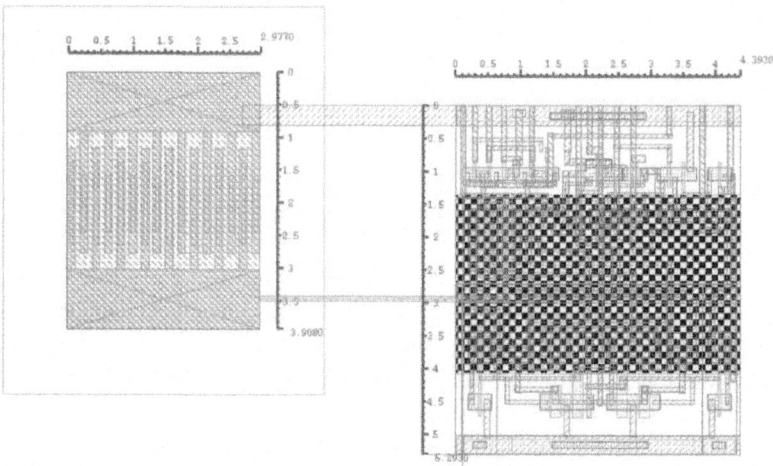

Figure 15: DAZ Sense Amplifier Layout.

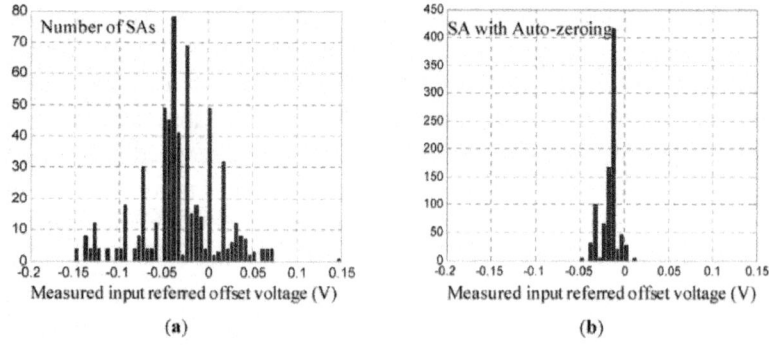

Figure 16: Measured offset voltage distribution (**a**) regular SA; (**b**) SA with auto-zeroing circuitry [7].

To verify the offset sensitivity to split phases, the offset of a sample DAZ SA is measured for different split phase frequencies. Figure 17 shows the offset voltage values for different split phase frequency.

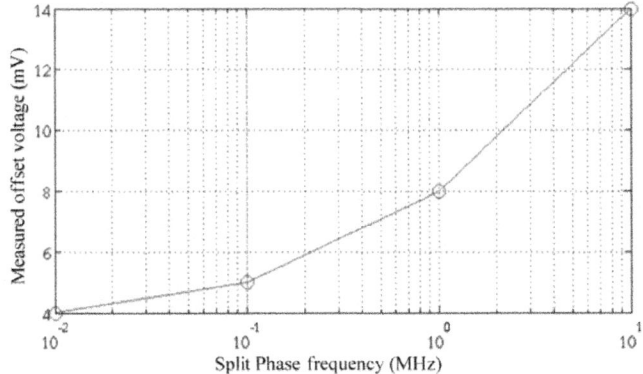

Figure 17: Measured offset voltage *vs.* split phase frequency [7].

CONCLUSIONS

We proposed a circuit that is capable of improving sense-amp offset to near zero, which is valuable for sub-threshold operation due to the heightened effect of mismatch. Simulations of the design (0.5 V, 1 MHz) show a compensated offset voltage of 1mV, settling time of 37 μs, and total power consumption of 12 nW. Measurements from a test chip fabricated in 45 nm technology showed the circuit's ability to improve σ of the offset voltage by 80% and limited the absolute maximum value of the offset voltage to 50 mV using a 1 MHz split phase frequency and 32fF output capacitance. Using the circuit in a 16 kB SRAM showed a reduction in the total energy and delay of 10% and 15% respectively. The trade-off between the sense amp compensated offset and power consumption is demonstrated. This makes the circuit able to provide the offset/power values that can generate the optimal SRAM design.

REFERENCES

1. Ryan, J.F.; Calhoun, B.H. Minimizing Offset for Latching Voltage-Mode Sense Amplifiers for Sub-Threshold Operation. In Proceedings of the 9th International Symposium on Quality Electronic Design, San Jose, California, USA, 17–19 March 2008; pp. 127–132.

2. Wang, A.; Chandrakasan, A.P.; Kosonocky, S.V. Optimal Supply and Threshold Scaling for Subthreshold CMOS Circuits. In Proceedings of

the IEEE Computer Society Annual Symposium on VLSI, Pittsburgh, PA, USA, 25–26 April 2002; pp. 5–9.
3. Verma, N.; Chandrakasan, A.P. A 256 kb 65 nm 8T subthreshold SRAM employing sense-amplifier redundancy. *IEEE J. Solid State Circ.* 2008, *43*, 141–149.
4. Pileggi, L.; Keskin, G.; Li, X.; Mai, K.; Proesel, J. Mismatch Analysis and Statistical Design at 65 nm and Below. In Proceedings of the IEEE Custom Integrated Circuits Conference, San Jose, California, USA, 21–24 September 2008; pp. 9–12.
5. Bhargava, M.; McCartney, M.P.; Hoefler, A.; Mai, K. Low-overhead, Digital Offset Compensated, SRAM Sense Amplifiers. In Proceedings of the IEEE Custom Integrated Circuits Conference, San Jose, California, USA, 13–16 September 2009; pp. 705–708.
6. Sachdev, M.; Sharifkhani, M.; Shah, J.S.; Rennie, D. Sense-amplification with Offset Cancellation for Static Random Access Memories. U.S. Patent Application 12/757,033, 8 April 2010.
7. Beshay, P.; Calhoun, B.H.; Ryan, J.F. Sub-threshold Sense Amplifier Compensation Using Auto-zeroing Circuitry. In Proceedings of the 2012 IEEE Subthreshold Microelectronics Conference, Waltham, Massachusetts, USA, 9–10 October 2012; pp. 1–3.

Chapter 5

EVOLVABLE METAHEURISTICS ON CIRCUIT DESIGN

Felipe Padilla[1,2], Aurora Torres[1], Julio Ponce[1], María Dolores Torres[1], Sylvie Ratté[2] and Eunice Ponce-de-León[1]
[1]Aguascalientes University, México
[2]École de Technologie Supérieure Canada

INTRODUCTION

Evolutionary computation algorithms are stochastic optimization methods; they are conveniently presented using the metaphor of natural evolution: a randomly initialized population of individuals evolves following a simulation of the Darwinian principle. New individuals are generated using genetic operations such as mutation and crossover. The probability of survival of the newly generated solutions depends on their fitness (Michalewicz et al., 1995). Evolutionary algorithms (EAs) have been successfully used to solve different types of optimization problems (Back, 1996). In the most general terms, evolution can be described as a two-step iterative process, consisting of random variation followed by selection.

The structure of any evolutionary computation algorithm is shown in the figure 1.

```
procedure evolutionary algorithm
t ←0
initialize P(t)
evaluate P(t)
while (not termination-condition) do
begin
        t←t + 1
        select P(t) from P(t  1)
        alter P(t)
        evaluate P(t)
end
```

Figure 1: Structure of any evolutionary algorithm

The term evolutionary computation is used to describe techniques such as genetic algorithms, evolution strategies, evolutionary programming and genetic programming. The different approaches are distinguished by the genetic structures under adaption and the genetic operators that generate new candidate solutions (Cordon et al., 2001).

Evolvable hardware (EHW) is an exquisite combination of evolutionary computation and electronic hardware. While the most common techniques of evolutionary computation are genetic algorithms and genetic programming, electronic hardware implies not only digital but analog circuits also. This field has earned importance since the early 1990's because of the advent of reconfigurable hardware. The ultimate objective of this field is to design and construct intelligent hardware, capable of online adaptation (Yao and Higuchi, 1999). The first classification of evolvable hardware can be found in (De Garis, 1993). In this work De Garis established there are extrinsic and intrinsic EHW. While Extrinsic EHW simulates evolution by software and downloads to hardware only the best configuration; intrinsic EHW simulates evolution directly in hardware. Nowadays the scope of this discipline has grown vastly. According to Zebulum (Zebulum, 1996), evolvable hardware can be classified by several criterion like hardware evaluation, evolvable computation approach, application area and evolvable platform. In regard to its application area EHW in divided in: Circuit design, robotics and control, pattern recognition, fault tolerance and very large scale integration (VLSI). We are interested in discuss about the first one.

Circuit design is the art of constructing a sized circuit from user specifications (Das and Vemuri, 2009). This task is divided according to the kind of circuits that are handled in digital and analog circuit design. Nowadays there are different algorithms that can be used to solve problems of optimization of circuits like: Genetic Programming, Genetic Algorithm, Estimation of the Distribution Algorithms, Ant Colony Optimizations, Others. The more amenable nature of digital circuits made researchers like Louis (Louis, 1993) and Koza (Koza, 1992) to focus first on the production of functional logic circuits. Afterwards, the goal was not only to obtain functional circuits, but optimum ones. The work of Louis (Louis, 1993) was pioneer on the use of genetic algorithms on the design of combinational circuits; Thompson et al (Thompson et al., 1996) were the first in coding logic gates and its connections. Other outstanding researches on digital design are Higuchi et al. (Higuchi et al., 1996) specially focused on intrinsic evolution based on neural networks; Hernández and Coello (Hernández and Coello, 2003) first worked with genetic algorithms and later with genetic programming and Information Theory. A very interesting case is the use of ACO on the optimization of combinatorial circuits (Mendoza, 2001).

The analog synthesis world also has numerous successful implementations of different metaheuristics like genetic algorithms (Lohn and Colombano, 1998), (Zebulum et al., 2000), (Goh and Li, 2001), (Das and Vemuri, 2007), (Khalifa et al., 2008), (Torres et al., 2010); genetic programming (Koza et al., 1997), (Hu et al., 2005)(Chang et al., 2006) and estimation of the distribution algorithms (Torres et al., 2009). Analog circuit synthesis is a process composed of two phases: the selection of a suitable topology and the sizing of all its components (Torres et al., 2010). While topology consists on the determination of the type of components and its connections; sizing refers to the selection of the components values. Further on this document, will be discuss some of the mentioned approaches. Others types of evolutionary algorithms are based in biological systems in which complex collective behaviour emerges from the local interaction of simple components. Some examples of these algorithms are Swarm Intelligence, Ant Colony, Bees Algorithm, etc. We will speak of an ant colony, this algorithm is based in the foraging behaviour of some species of ants. Ant colonies are capable of finding the shortest paths between their nest and food sources, through a substance denominated pheromone.

OPTIMIZATION ALGORITHM

Actual trends in VLSI technology are towards integration of mixed analog-digital circuits as a complete system-on-a-chip. Most of the knowledge intensive and challenging design effort spent in such systems design is due to the analog building blocks (Balkir et al., 2004). Analog design has been traditionally a difficult discipline of integrated circuits (IC) design. In circuit design optimization, a circuit and its performance specifications are given and the goal is to automatically determine the device sizes in order to meet the given performance specifications while minimizing a cost function, such as a weighted sum of the active area or power dissipation (Baghini et al., 2007). This is a difficult and critical step for several reasons: 1) most analog circuits require a custom optimized design; 2) the design problem is typically under constrained with many degrees of freedom; and 3) it is common that many (often conflicting) performance requirements must to be taken into account, and tradeoffs must be made that satisfy the designer (Rutenbar et al., 2007). Fuzzy techniques have been successfully applied in a variety of fields such as automatic control data classification, decision analysis, expert systems, computer vision, multi-criteria evaluation, genetic algorithms, ant colony systems, optimization, etc. Works showing the possibility of application of fuzzy logic in computer aided design (CAD) of electronic circuits started to appear in late 1980s and early 1990s. An argument for fuzzy logic application in CAD is derived from the nature of the algorithm used for solving design problems. The majority of algorithms for synthesis use heuristics that are based on human knowledge acquired through experience and understanding of problems. Another important source of knowledge is numerical data. Fuzzy logic systems are appropriate in such situations because they are able to deal simultaneously with both types of information: linguistic and numerical.

Also, fuzzy systems being universal appoximators can model any nonlinear functions of arbitrary complexity. This is very useful in modelling complex circuit functions of high accuracy at low cost, necessary in performance evaluation. Design optimization of an electronic circuit is a technique used to find the design parameter values (length and width of MOS transistors, bias current, capacitor values, etc.) in such a way that the final circuit performances (de gain, gain-bandwidth, slew rate, phase margin, etc.) meet as close as possible the design requirements. There is no general design procedure independent of the circuit; also, there is no formal representation to connect the circuit functions on its structure in a consistent manner. The major obstacle consists in the peculiarity of the analog signals: the continuous domain of the signals` amplitude and their continuous time dependency. Hereby the analog circuit design is known like an iterative, multi-phase task that necessitates a large spectrum of knowledge and abilities of designers.

GENETIC ALGORITHMS

Genetic algorithms originally were called "reproductive plans" by John Holland (Holland, 1975), and were the first emulators of the genetic evolution that produced practical results. In 1989, when Goldberg (Goldberg, 1989) published his book, mentioned more than 70 successful applications of this paradigm that continues winning popularity nowadays. According to Coello (Coello, 1996), a good definition of genetic algorithm was established by Koza in his book of 1992 (Koza, 1992), he says the following: "The genetic algorithm is a highly parallel mathematical algorithm that transforms a group (population) of individual mathematical objects (that usually have the form of chains of characters of fixed longitude), each one with an associate aptitude value, in new populations (for example the following generation) using modelling of operations under the Darwinian principle of the reproduction and survival of the "most capable", naturally, after the occurrence of the genetic operators (sexual recombination)".

Ponce de León (Ponce de León, 1997) summarizes the mechanism of operation of the simple genetic algorithm in the following way; "it is generated a population of n structures aleatorily (chains, chromosomes or individuals) and then, some operators act transforming the population. The transformation is carried out by means of the application of three operators; once this culminates, it is said that a generational cycle has finished". The three operators Ponce references are: selection, crossover and mutation. The genetic algorithm in the form like Holland illustrates it (Holland, 1975) has the following characteristic elements:

- Representation of binary chains.
- Proportional selection.
- Crossover like the main method to produce new individuals.

After the Holland's proposal, have been carried out different modifications; either by means of the use of different representation outlines, or until certain modifications to the selection operators, crossover, mutation and elitism. The diagram shown in the following figure presents the simplest version in the genetic algorithm, well-known as SGA (for the initials in English of "Simple Genetic Algorithm").

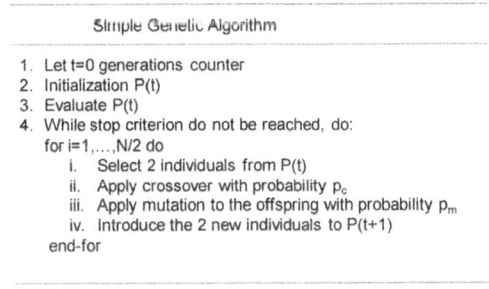

Figure 2: Pseudocode of SGA

Although the general mechanism of this algorithm is extremely simple, it can be demonstrated by means of Markov's chains that the evolutionary algorithms that use elitist selection mechanisms, will converge to a good global solution of certain functions whose domain can be an arbitrary space (Torres, 2010). Günter Rudolph in 1996, generalized the previous developments in theory of the convergence for binary search spaces and Euclidian ones to general search spaces (Rudolph, 1996).

Genetic Algorithms in Automated Analog Design

Due to the high level of complexity that implies the task of designing and also to the strong dependence that this task has with the knowledge and experience experts; the automatic design of analogical circuits is a challenge and a necessity. Some researchers of the area believe that the automation of the design should be preceded by a change in the process of current design, for example, governed by the execution of the restrictions (Jerke, 2009). The fact is that nowadays, it has not still been possible to automate this process in a complete way. One of the metaheuristics that have shown better benefits in the realization of this task are the genetic algorithm and the genetic programming; this space belongs to the genetic algorithm. Lohn and Colombado (Lohn and Colombado, 1998) used the genetic algorithm to design two analog filters, one of low complexity and one of medium complexity. The contribution of these researchers resides in that they demonstrated that it was feasible to use a very simple lineal representation. They proposed a code outline in which each element was represented by a fixed number of bytes called bytecodes in which they included an operation code that dictated the connection of each element and three bytes more they used to code its value. Koza on the other hand, continued making use of the genetic programming in the synthesis of computational circuits (Koza, 1997b) and controllers, filters and other kind of circuits (Koza, 1997).

According to Ricardo Zebulum and his collaborators (Zebulum et al., 1998), the Evolutionary Electronics is an area that seeks to find new techniques of automatic design based on Darwinian concepts. The authors of the mentioned work, made the comparison of three different methodologies in the design of electronic filters. Their work was put on approval with two cases of study: A low-pass filter discussed in (Koza, 1996) and a filter pass-band with band in passing between 2000 and 3000 Hz and the bands of rejection above 4000 and below 1000Hz. The methodologies on approval were the following: "Outline of representation of variable longitude in combination with an evolutionary algorithm that restricts the topology of the filter (parallel meshes of two elements each one). For the simulation, an own tool was used in C, based on Laplace´s analysis. "Outline of representation of fixed longitude in combination with an evolutionary algorithm that doesn't restrict the topology of the circuit. To analyse the circuits they used Smash and SPICE, obtaining the same results.

"Outline of representation of variable longitude in combination with an evolutionary algorithm that doesn't restrict the topology of the circuit. For the simulation of the circuits they used as much Smash as SPICE, obtaining the same results. In this work, Zebulum and his collaborators demonstrated that making use of an evolutionary algorithm based on the "Genetic Algorithm of Adaptation of Species (SAGA) of Harvey (Harvey, 1993), they could be obtained results comparable with those obtained using genetic programming, as for the answer in frequency of the obtained circuits using much smaller populations. This work concludes settling down that as for time, the first methodology was better, however this can explain to you for the rigidity of the used topology that allowed the use of a tool of quicker simulation. In spite of the success of this work, all the methodologies had inducer circuits whose values were so big as a result (2.2H for example) that are not very practical. On the other hand, investigators as Grimbleby and their collaborators (Grimbleby et al., 1995) they were working with mechanisms of numeric optimization in combination with genetic algorithms for the synthesis of analogical circuits using a chromosome of fixed length and a type of null component to fight with the variable size of the real circuits.

The XXI century has also been witness of numerous efforts made toward the automation of the synthesis of the analogical circuits, for example, in the year 2000, Zebulum et al. (Zebulum et al., 2000), established some advantages of variable length representation systems. Among other things, they argued that when using a fixed size, it is not only required expert knowledge of the problem, but the potential of the evolutionary algorithms is also limiting. That same year, they also proved an outline of representation of variable longitude that they understood passive elements, connected nodes and disconnected nodes. The authors emphasize the use of resistances and capacitors with programmable values in their architecture. These investigators intend to work the two phases of the evolution of an electric circuit (topology and adjustment of the parameters) in a sequential way, instead of making it simultaneously. In the year 2001, the investigating Goh and Li (Goh and Li, 2001) they began to outline some of the weaknesses that persisted in the process of design of analogical circuits that they were commented later by investigators as Khalifa and their collaborators (Khalifa et al., 2008), (Das, 2008) among others. The weaknesses that these investigators declare that they should be assisted, the reduction of the enormous computational effort that implies the evaluation of big generations of circuits that they don't always produce results and the reduction of the breach between the evolved circuits and those that finally are taken to the physical implementation, due to the restrictions of commercial physical devices. Other equally important aspects are related with the elaboration of tools that due to their complexity, they require expert personnel's manipulation or with a considerable level of knowledge (Krasnicki, 2001); as well as the execution in teams whose level of sophistication is outside of the reach of a great number of people.

Table 1: Relevant research on analog circuit synthesis using Genetic Algorithms (Torres, 2010)

Year	Author	Application
1993	Horrocks and Spittle	Active low-pass filter
1994	Horrocks and Khalifa	Low-pass filter
1995	Grimbleby	High-pass filter
1996	Horrocks and Khalifa	Low-pass filter
1998	Lohn and Colombano	Low-pass filter
1998	Zebulum et al.	Low-pass filter Band-pass filter
1999	Krasnicki et al.	OP-AMP
2000	Ando and Iba	Passive filters
2000	Zebulum et al.	Passive filters
2001	Goh and Li	Low-pass filter High-pass filter
2007	Das and Vemuri	Low-pass filter
2008	Khalifa et al.	Low-pass filter High-pass filter
2008	Das and Vemuri	OP-AMP
2010	Torres et al.	Low-pass filter

ESTIMATION OF DISTRIBUTION ALGORITHMS

Estimation of distribution algorithms (EDA's) constitute a relatively new field of the Evolutionary Computation (Larrañaga, 2002) that replaces genetic operators (crossover and mutation) for the estimation of the distribution of the selected individuals and the sampling from the distribution to obtain the new population. The objective of this paradigm is to avoid the use of arbitrary operators as crossover and mutation, to modeling explicitly the most promising solutions for sampling solutions from its distribution.

Pseudocode of the algorithm EDA

Step 1: Random generation of M individuals (initial Population)

Step 2: Repeat the steps 3-5 for the generation l=1, 2,... until an stop criterion is reached

Step 3: Select N <= M individuals from Dl-1 according to a selection method

Step 4: Estimate the distribution of probability pl(x) from the group of selected individuals

Step 5: Sample M individuals (new population) from pl(x)

EDAs can be classified according to two fundamental approaches. The first is the level of interdependences of variables, and the second is the type of involved variables. With regard to the level of interdependences EDAs are divided in 3, when the variables are independent, when there are bivaluated dependences and when there are multiple dependences. With regard to the type of involved variables, they can be discrete, continuous or mixed. The easiest version of an EDA is the "Univariate Marginal Distribution Algorithm" (UMDA) introduced by Mühlenbein (Mühlenbein and Paad, 1996). This algorithm works on the supposition of complete independence among variables. Pseudocode of this algorithm in presented in figure 3.

UMDA_AC

1. Begin
2. $D_0 \leftarrow$ Generate M individuals at random
3. Repeat for $l = 1, 2, ...$ until the stopping criteria met
    a) $D_{l-1}^{Se} \leftarrow$ Select $N \leq M$ individuals from D_{l-1} according to the selection method
    b) $p_l(x) = p(x | D_{l-1}^{Se}) = \prod_{i=1}^{n} p_i(x_i) = \prod_{i=1}^{n} \dfrac{\sum_{j=1}^{N} \delta_j (X_i = x_i | D_{l-1}^{Se})}{N} \leftarrow$
    Estimate the joint probability distribution
    c) $D_{l-1}^{Se} \leftarrow$ Sample M individuals from $p_l(x)$

Figure 3: Pseudocode for UMDA (Larrañaga, 2002).

Another very common approach for the estimation of the distribution supposing independence among the variables is the algorithm PBIL ("Population-based incremental learning") (Baluja, 1994) that contrary to UMDA, doesn't estimate a new model in each generation, but refines it.

The main problem of the distribution of the estimation algorithms, is to estimate the model; because as it gets more complicated, the dependences among the variables are captured in a better way, however, its estimation becomes more expensive (Larrañaga, 2002). Regarding models that consider bivariated dependences (dependences among pairs of variables), the most outstanding methods according its use in the literature are those that use chains like the "MIMIC" algorithm (Mutual Information Maximizing Input Clustering Algorithm) (De Bonet et al., 1996), those that use trees, as the case of the COMIT (Baluja and Davies, 1997) that uses the method of Chow and Liu [Chow 1968] based on the concept of mutual information and the BMDA (Pelikan, 1999), in which Pelikan and Mühlenbein propose a factoring of the distribution of joint probability. This algorithm is based on the construction of an acyclic directed graph of dependences that is not necessarily connected.

Finally, the most common n-varied models are those that allow estimating a model in a Bayesian-net form. This approach has originated a great variety of algorithms according to the learning method, according to the nature of the variables (discrete or continuous), according to the imposed restrictions, etc. (Larrañaga, 2002). The great success genetic algorithms (GAs) have shown on several synthesis problems, has motivated some researches to explore the EDA's world in analog circuit synthesis. Next table show some examples.

Table 2: Relevant works on analog circuit synthesis by means of Estimation of the Distribution Algorithms

Year	Author	Application	Used metaheuristic
2002	Mühlenbein et al.	Low-pass	UMDA
2007	Zinchenko et al.	Mixed circuit	UMDA
2009	Torres et al.	Filters	UMDA
2010	Torres et al.	Filters	MITEDA

From table 2 it can be seen UMDA is the most common approach implemented on the analog circuit synthesis, nevertheless, MITEDA represents an effort on exploring the behavior of more complex EDAS. This algorithm was developed inspired by the COMIT and it uses the concept of mutual information used by Baluja and Davies (Baluja, 1997) to build the tree of dependences. Later this tree is sampling in order to create new generations. This algorithm represents the first tool that considers bi-valuated dependencies used in the design of analogical circuits we know until this moment.

ANT COLONY OPTIMIZATION

The Ant Colony Optimization Algorithm is a meta-heuristic bio-inspired in the behavior of real ant colonies. The first algorithm which can be classified within this framework was presented in 1991 by Marco Dorigo. In his PHD thesis with Title: "Optimization, learning, and Natural Algorithms", modeling the way real ants solve problems using pheromones. Real ants are capable of finding the shortest path from a food source to their nest. The ants deposit a concentration of pheromone in theirs paths, and they follows with more probability the way with more concentration of pheromone that it was previously deposited by other ants, the essential trait of ACO algorithms is the combination of a priori information about the structure of a promising solution with a posteriori information about the structure of previously obtained good solutions. In the Ant Colony Algorithms a number of artificial ants (agents) build solutions for an optimization problem and exchange information on their quality via a scheme of global communication that is reminiscent of the one adopted by real ants. When exist paths without any amount of pheromone, the ants explore the neighbourhood area in a totally random way. In presence of an amount of pheromone, the ants follow a path with a probability based in the pheromone concentration. The ants deposit additional pheromone concentrations during his travels. Since the pheromone evaporates, the pheromone concentration in non-used paths tends to disappear slowly. To find the shortest path, a moving ants lay some pheromone on the ground, so an ant encountering a previously trail can detect it and decide with high probability to follow it. As a result, the collective behavior that emerges is a form of a positive feedback loop where the probability with which each ant choose the next path increases with the number of ants that previously chose the same path.

The Ant Colony System (ACS) models the behavior of ants, which are able to find the shortest path from their nest to a food source. Although individual ants move in a quasirandom form, performing relatively simple tasks, the entire colony of ants can collectively accomplish sophisticated movement patterns. Ants accomplish this by depositing a substance called a pheromone as they move. This chemical trail can be detected by other ants, which are probabilistically more likely to follow a path rich in pheromone. This trail information can be utilized to adapt to sudden unexpected changes to the terrain, such as when an obstruction blocks a previously used part of the path.

Application of Ant Colony to The Design of Combinatory Logic Circuits

To apply Ant Colony Algorithm to the design of logic circuits, in (Mendoza, 2001) is shown as the design of logic circuits with ACO. In the case of the logic circuits, the treatment of the problem does not seem to be so immediate.

Circuit Representation

The circuits are represent used a bidimensional matrix. Where each element of the matrix is a triplet of the type [Entrance 1, Entrance 2, Type of floodgate] (see figure 5). Was used five types of floodgates: AND, OR, NOT, XOR and WIRE, although this last one is not a floodgate, but rather it is a connection (a wire) that unites an element of certain column with another one of the previous column. Each element of the matrix receives its entrances solely of the exits of the previous column.

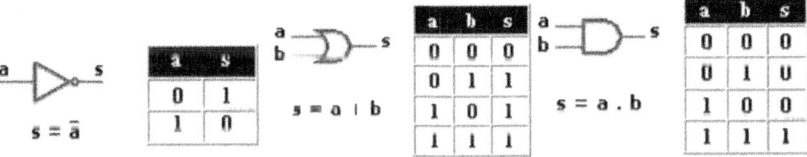

Figure: Basic floodgate Not, Or, And

The first column directly receives its entrances of the table really of the given circuit. The last column provides the exits of the circuit. The first N rows corresponds to the N exits of the circuit. This form to represent a circuit has been used successfully. In the following figure are shown the basic floodgate.

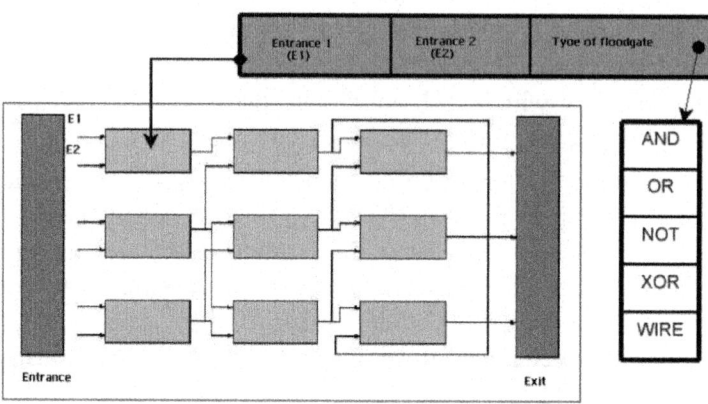

Figure 5: Matrix used to circuit representation.

Implementation

The route of an ant or agent will be a complete circuit. While each ant crosses a route, it constructs a circuit. In the TSP the ants find the route in terms of distance, do it here in terms of the number of floodgates. A state or city is a column, which is made up of several elements to which it is called substates to them, being these each one of the floodgates of a column and the number of combinations of possible entrances of each floodgate of this column. The first N substates (N is the number of exits in the circuit) is chosen with a selection factor P, and the others are chosen randomly. The distance between cities or states is measured as the increase or diminution from the successes to the exits of the circuit when changing from a level to another one. Unlike the problem of the TSP, in a same route (circuit), they do not have to visit all the states. The pheromones keep in a matrix called Trails. The length of this matrix corresponds to the number of exits of the circuit. Each element of Trails is a three-dimensional matrix as well. Next it is explained what they represent each one of the dimensions of the element. The first dimension of this matrix corresponds to the combination of possible entrances to the floodgate and goes from 0 to 6. The possible combinations of entrances, independent of the incoming number of the table really.

The second dimension corresponds to the number of floodgate, that is to say, goes of 0 to the number of floodgates except one (NumGates-1). The third dimension corresponds to the number of successes that take until the level (column) previous and really goes of 0 to the number of lines in the table, because the number of successes that can be had in any level is between 0 and the number of lines of the true table.

The Construction of a Solution (route)

As it was already mentioned before, a state is a column of the matrix, each element of the column is a floodgate with its respective entrances and their exit. Because of that, the election of a state is a process that becomes by parts (floodgate by floodgate), reason why we will call to each floodgate (element of the column) a substate. A state a combination of three elements (floodgate, IN1, IN2). In order to choose a substate of anyone of the first N rows, a value is assigned to him to each one of the possible combinations, call selection factor P, with which it will compete remaining in that position. The distance is a heuristic value and is given by the number of successes that the portion of the circuit constructed until the moment produces with respect to exit 1 of the True table. This is analogous to the distance in the TSP. Once it has assigned a factor of selection to all the combinations, is chosen what of them remains in the position in game. This is repeated with all the substates that belong to one of the rows that represent an exit of the circuit. The other substates, are chosen randomly. This is repeated until arriving at the last state from the circuit or column of the matrix. When all the ants finish their route, the pheromone signs are updated. This becomes in two steps: 1. First the amount is due to update pheromone in the ways, simulating the pheromone evaporation of the ways by the artificial ants to the passage of time. 2. The ways are due to update or to increase according to the routes constructed by each ant in the algorithm. This becomes of the following form: If the circuit result of the route is not valid (that it does not produce all the exits).

Multiobjective Optimization

A population based evolutionary multiobjective optimization approach (Coello, 2009) to design combinatorial circuit was proposed for first time by Coello and Hernández in 2000 (Coello and Hernández, 2000). This approach reduced the computational effort required by genetic algorithm to design circuit at gate level. The main motivation was the reduction of fitness function evaluations while keeping the capabilities of the GA to generate novel designs. The main ideas behind MGA algorithm are:

1. Circuit representation as a matrix (originally proposed by Louis in 1991 (Louis and Rawlins, 1991)) and an n-cardinality alphabet.
2. Incremental method to resized of matrix used to fit a circuit.

3. Fitness function in two stages. At the beginning only validity of the circuit outputs is taken into account, and at the ending the fitness function is modified such that any valid designs produced are rewarded for each WIRE gate that they include. (WIRE gate indicates a null operation, that is, the absence of gate)

4. Use a multi-objective optimization technique (Fonseca and Fleming, 1995) (Coello, 1999). In general, it redefines the single-objective optimization of as a multiobjective optimization problem in which we will have m+1 objectives, where f is the number of constraints. There is a new vector, $\bar{v} = (f, f_1, ..., f_n))$, where is the objective function $f_1, ..., f_n$ are the original constraints of the problem. An ideal solution : XX would thus have $f_i(X) = 0$ for $i = 1, ..., m$, and $f(X) \leq f(Y)$ for all feasible Y ; (assuming minimization). For combinatorial logic circuit design this technique consists on using a population based multiobjective optimization technique such as VEGA (Schaffer, 1984) to handle each of the outputs of the circuit as an objective. At each generation, the population is split in to m+1 sub-populations, $m=2^n$ (outputs), n: inputs of the circuit. The main mission of each sub-population is to match its corresponding output with the value indicated by the user in the truth table. After one of these objectives is satisfied, its corresponding sub-population is merged with the rest of the individuals in what becomes a joint effort to minimize the total amount of mismatches produced (between the encoded circuit and the truth table). Once a feasible individual is found, all individuals cooperate to minimize its number of gates (Coello and Hernández, 2002). The MGA algorithm outperformance the GA algorithm in quality of solution and decreased the evaluation amount of fitness function. This approach made a path in solving evolutionary design of combinational logic circuits.

Formulation of Multiobjective Optimization Problem

The multiobjective optimization problem can be formulated as follows (Coello and Hernández, 2000):

A General Multiobjective Optimization Problem (MOP): Find the vector $\bar{x}^* = [x_1^*, ..., x_n^*]^T$ which will satisfy the m inequality constraints:

$g_i(\bar{x}) \geq 0, i = 1, ..., m$ (1)

the p equality constraints

$$h_i(\vec{x}) = 0, i = 1, \dots, p \qquad (2)$$

and optimizes the vector function

$$\vec{f}(\vec{x}) = [f_1(\vec{x}), \dots, f_k(\vec{x})]^T \qquad (3)$$

Where $\vec{x} = [x_1, \dots, x_n]^T$ is the vector of decision variables.

That is, we wish to determine from among all $\vec{x} = [x_1, \dots, x_n]^T$, which satisfy the inequality and equality constraints above, the particular $\vec{x}^* = [x_1^*, \dots, x_n^*]^T$ which yields the optimum values of all the k objective functions of the problem. Let be Ω the set defined as all vectors $\vec{x} = [x_1, \dots, x_n]^T$, that do not violate the constraints.

Pareto Optimality Definition: We say that $\vec{x}^* = [x_1^*, \dots, x_n^*]^T \in \Omega, \Omega \subseteq \mathbb{R}^n, f_i: \mathbb{R}^n \to \mathbb{R}$, is *Pareto optimal* if for every $\vec{x} = [x_1, \dots, x_n]^T$, and $I = \{1, \dots, k\}$ either,

$$\bigwedge_{i \in I}(f_i(\vec{x}) = f_i(\vec{x}^*)) \qquad (4)$$

Or, there is at least one $i \in I$ such that

$$f_i(\vec{x}) > f_i(\vec{x}^*) \qquad (5)$$

$\vec{x}^* = [x_1^*, \dots, x_n^*]^T$ is Pareto optimal if there exists no feasible vector $\vec{x} = [x_1, \dots, x_n]^T$ which would decrease some criterion without causing a simultaneous increase in at least one other criterion.

Pareto Dominance Definition: A vector $\vec{u} = (u_1, \dots, u_n)$ is said to dominate $\vec{v} = (v_1, \dots, v_n)$ (denoted by $\vec{u} \preccurlyeq \vec{v}$) if and only if $\vec{u}$ is partially less than $\vec{v}$, i.e., $\forall i \in \{1, \dots, k\}, u_i \leq v_i \wedge \exists i \in \{1, \dots, k\}: u_i < v_i$.

Pareto Optimal Set Definition: For a given $\mathcal{MOP}, \vec{f}(\vec{x}) = [f_1(\vec{x}), \dots, f_k(\vec{x})]^T$, the Pareto optimal set $(\mathcal{P}^*)$ is defined as:

$$\mathcal{P}^* = \{\vec{x} \in \Omega\ /\ \nexists \vec{x}' = [x_1', \dots, x_n']^T \in \Omega (\vec{f}(\vec{x}') \preccurlyeq \vec{f}(\vec{x}))\} \qquad (6)$$

Pareto Front Definition: For a given $\mathcal{MOP}, \vec{f}(\vec{x}) = [f_1(\vec{x}), \dots, f_k(\vec{x})]^T$ and Pareto optimal set $\mathcal{P}^*$, the *Pareto front* $(\mathcal{PF}^*)$ is defined as:

$$\mathcal{PF}^* = \{\vec{u} = \vec{f}(\vec{x}) = [f_1(\vec{x}), \dots, f_k(\vec{x})]^T\ /\ \vec{x} \in \mathcal{P}^*\} \qquad (7)$$

APPLICATION

Due to the enormous success genetic algorithms has proved on the field of circuit design, this section has the purpose of show how this metaheuristic could be used for the synthesis of analog circuits. In order to implement a genetic algorithm for the artificial evolution of any kind of process, is indispensable to find a way to represent a solution of the given problem, to find the way to generate possible solutions, to be able to evaluate the quality of the solutions and to have a group of operators that let transform one solution into another. Figure 6 shows the general flow used to implement a genetic algorithm in the analog circuit design according to Azizi (Azizi, 2001).

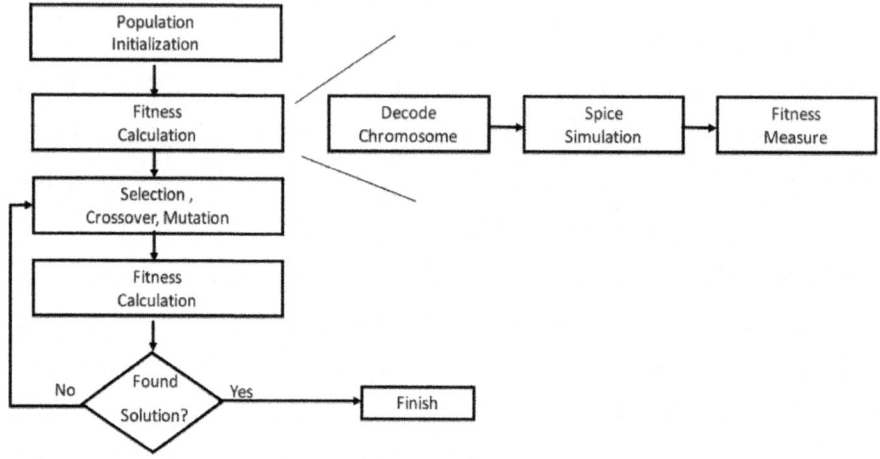

Figure 6: Genetic Algorithm Flow for Analog Circuit Synthesis.

Representation Mechanism

In order to initialize the first population, the programmer has to establish how each solution is going to be represented and how the population can be generated.

A genetic encoding for artificial evolution of analog networks must be capable of representing both; the topology and the sizing of the network (Mattiussi and Floreano, 2007). While topology refers to the way each element is going to be connected to each other; sizing refers to the type and dimension of each element on a net. Other important aspects of the representation mechanism are its ability to capture any kind of circuit and the chance to reduce the process and time inverted in translate the circuit into a netlist (net description list). The representation mechanism has also to be flexible enough to be used with a wide range of components values but sufficiently short to be computational handling. (Torres et al., 2009). Torres et al (2009), reported a representation mechanism for passive elements of two terminals. This mechanism uses a gene of six parts to represent an analog element as figure 7 shows. Each circuit is a linked list of several genes.

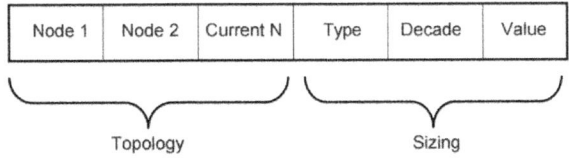

Figure 7: Gene description.

While node 1 and node 2 refers to the terminals of an electrical device; current N is a pointer that is going to be used to build the network. Type, decade and value are the parameters that completely characterize a specific element (Torres, 2009). These parameters use integer coding according to table 3.

Table 3: Sizing encoding system

Type	Decade	Value
C(0)	$10^{-6} - 10^{-9}$ (0-3)	E6 (0-5)
R(1)	$10^{+3} - 10^{+6}$ (0-3)	E12 (0-11)
L(2)	$10^{-3} - 10^{-6}$ (0-3)	E12 (0-11)

Next figure shows an element and its corresponding gene. We refers to initial node, that represents the beginning of an analog circuit.

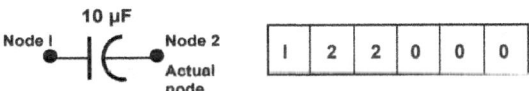

Figure 8: An element of circuit and its corresponding gene.

Generation Mechanism

Once, a representation mechanism has been selected, the generation routine need to be established. The generation mechanism proposed by Torres et al. (2009) is based on an operation code randomly generated. The operation code establishes the connection that has to be done in the construction process of an admitted topology. The process begins in "Initial node" and ends when certain termination criterion is reached. This criterion could be one of two possibilities: the connection is done with the "Final Node" or the circuit reaches a preset amount of elements. Next figure describes how the generation mechanism works (CNode refers to the current node, and INode corresponds to Initial Node) (Torres et al., 2009).

Generation mechanism
1. begin
2.   CNode <- INode
3.   while(Not meet termination criterion)
• Node1 = Cnode
• Generate OP-Code
• Execute_connection (Update Node2 and Cnode)
• Generate Type, Decade and Value
4.   end_while
5. end

Figure 9: Algorithm for the generation of each solution.

The circuit creation process performed by the former algorithm is very flexible. Once the operation code has been chosen and the connection has been done, type, decade and value of each element are generated. All operation codes used and their meaning are depicted in table 4.

Table 4: The operation code of the generation mechanism (Torres et al. 2010)

Op code	Instruction
0	Connect to grown
1	Connect to final node
2	Connect to x node
3	Connect to new node

Evaluation Function

Evolvable process depends on the ability to distinguish good and bad solutions, because it consists in continuously improve solutions from one generation to another. Therefore, a fitness function that describes how close a circuit is from the target is needed. Within the scope of analog circuit design, filters and amplifiers are the most frequently discussed. Fitness function used on the synthesis of low-pass filter will be presented below. Filters are circuits that block certain frequencies or bands of frequencies (Curtis, 2003). A low pass filter is the one that let pass low frequencies while blocks high frequencies. Next figure, illustrates the frequency response of an ideal and a real filter.

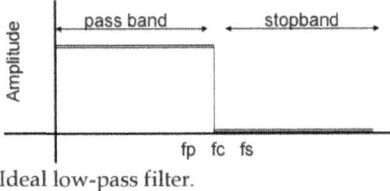
Ideal low-pass filter.

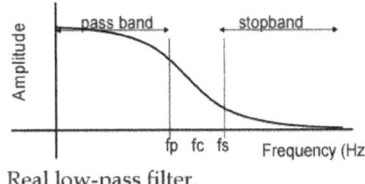
Real low-pass filter.

Figure 10: Frequency response of an ideal and a real low-pass filter.

The fitness function used by Torres et al, is based on the measurement of the distance between the ideal and the real (evolved) filter. This function is an adaptation of the one used by Koza (Koza et al., 1997) and Hilder and Tyrrell (Hilder and Tyrrell, 2007) among other researchers. This function is the sum of errors between the ideal frequency response and the actual candidate, along N sampling points. Equation 8 describes the fitness measure for filters.

$$F = \frac{1}{1+\xi} \quad (8)$$

Where :

$$\xi = \sum_{i=1}^{N} \lambda(\varepsilon_i) * \varepsilon_i \quad (9)$$

$$\varepsilon_i = \left| M(f_i)_{Target} - M(f_i)_{Actual} \right| \quad (10)$$

"ξ" represents the error over the N points of frequency. If the deviation from target magnitude is inacceptable according to the frequency band, then a penalty factor "λ" has to be assigned to the error function.

A sample error function "ξ" give us the absolute deviation between the actual output response and the target response over the "i" sampling point. M(fi)Target denotes target magnitude at a fi frequency, M(fi)Actual is the magnitude of the actual evolved circuit at a fi frequency and fi is the sampling frequency.

Transformation of a Solution

Finally, when representation, generation and evaluation of candidate solutions have been solved, the programmer needs to find a group of operators to transform one solution into another. Starting from two parents chosen by any selection routine, an offspring is produce through two possible operators: crossover and mutation. There are several selection algorithms; one of the more popular is the roulette-wheel. Roulette-wheel selection is an operator used for selecting potentially useful solutions for recombination. The fitness level of each solution is used to associate a probability of selection. If fi is the fitness of individual i in the population, its probability of being selected is $pi = \frac{fi}{\sum_{i=1}^{n} fj'}$, where n is the number of individuals in the population.

Crossover operation, introduces new solutions into the genetic algorithm starting from previous circuits; this operator is the responsible of changing some parts of a circuit by parts from another one. According to Dastidar et al., (Dastidar et al.,2005) and Das and Vemuri (Das and Vemuri, 2007), the use of some suitable connectivity rules, can reduce the unwanted search space not only for active, but for passive circuit synthesis. The crossover operator proposed by Torres et al., generates topological modifications because it alters the connection order of the offspring. This operator can be applied to one or two crossover points.

Next figure shows how this operator can be executed on the condense chromosome of two progenitors, using the representation mechanisms proposed by Torres et al. In the figure "T" refers to ground connection and "F" represents the final node of the analog circuit. This condense representation of each solution only has connection nodes and type of each element.

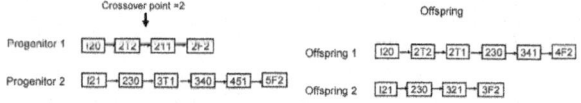

Figure 11: Crossover operator (Torres et al., 2010)

Mutation is an operator that traditionally introduces new solutions modifying only one chromosome. There are several ways to implement mutation, Goh and Li (Goh and Li, 2001) show a nice group on operators. The mutation exhibit in this section was proposed by Torres et al. This mutation operator is executed at gene level; and it works by altering a randomly chosen gene with another randomly generated. A mutated gene corresponds to a different type of element with different value, but connected to the same pair of nodes (Torres et al., 2010). Next figure shows an example of the use of this operator.

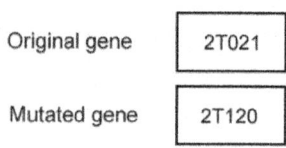

Figure 12: Mutation operator.

Using all elements discuss in this section, the interested reader can implement an effective genetic algorithm for the automated synthesis of a passive filter.

Conclusion and the Future Research

Nowadays exist applications in real life problems, where is possible used evolvable metaheuristics based on populations to the circuit design process, in this chapter was present some algorithms and applications, like Genetic Algorithms, Estimation of the distribution algorithms, Ant colony optimization. As shown there are multiple metaheuristics that can be used to circuit design trough different representations. We describe how is the representation with Genetic Algorithms. Since avoiding non valid topologies and non simulable networks, implies a very high reduction on time and computational resources in our problem; mainly three algorithms were compared at designing a low pass filter; a genetic algorithm (GA-AC), Ant Colony Systems (ACO-AC) and an estimation of distribution algorithm (UMDA-AC). Experimental results demonstrated that the group of mechanisms used in theses algorithms, worked better with GA-AC than with UMDA-AC and ACO-AC, according to the Pearson's Chisquared tests with respect to the generation of low rate of non spice-simulable circuits. Although UMDA-AC and ACO-AC performed faster the execution, and found a better individual on 200 generations' execution; statistically it cannot be said, the time difference is significant. With respect to the number of fitness evaluations, it can be said with statistical base, that UMDA-AC performs less evaluations than GA-AC per execution. In order to improve the performance of this algorithms, next step is the creation of a tool that blends the strengths of each metaheuristic. The work team is already working on the design of some new operators to be inserted on the EDA-AC and ACO-AC. GA-AC could be improved by enhancing the algorithm with some mechanisms of diversity control, like other kind of operators and another type of selection, in order to improve its exploration and delays its convergence. As future work is to continue working with various tools and algorithms that allow us to improve new circuit design. A new Artificial Intelligence that can be in charge of these systems, continues being distant into the horizon, in the same way that we still lack of methods to understand the original and peculiar things of each form to represent circuits.

REFERENCES

1. Back, T.: (1996). Evolutionary Algorithms in Theory and Practice. Oxford University Press, New York.
2. Baghini, M.; Kanphade, R.; Wakade, P.; Gawande, M.; Changani, M.;

Patil, M. (2007). GPbasedDesign and Optimization of a Floating Voltage Source for Low-Power andHighly Tunable OTA Applications, WSEAS Transactions on Circuits and Systems,Issue 10, Volume 6, October 2007, pp. 588-582.
3. Balkir, S.; Dundar, G.; Alpaydin, G. (2004). Evolution Based Synthesis of Analog Integrated Circuits and Systems, IEEE NASA/DoD Conference on Evolution Hardware,EH`04. Pp 26-29.
4. Baluja, S. (1994). Population-Based Incremental Learning: A Method for Integrating GeneticSearch Based Function Optimization and Competitive Learning. Technical ReportTR CMU-CS 94-163, Carnegie Mellon University.
5. Baluja, S. and Davies, S. (1997) Combining Multiple Optimization Runs with Optimal Dependency Trees. Technical Report TR CMU-CS-97-157, Carnegie Mellon University.
6. Chang, S.; Hou, H. and Su, Y. (2006). Automated Passive Filter Synthesis Using an Novel Tree Representation and Genetic Programming. IEEE ransactions on Evolutionary Computation, Vol. 10. No.1, February 2006. Pp. 93-100.
7. Coello, C. (1996). An Empirical Study of Evolutionary Techniques for Multiobjective Optimization in Engineering Design, PhD thesis, Department of Computer Science, Tulane University, New Orleans, Louisiana, USA..
8. Coello, C. A. A Comprehensive Survey of Evolutionary-Based Multiobjective Optimization Techniques. Knowledge and Information Systems. An International Journal, 1(3):269–308, August 1999.
9. Coello, C. A. and Hernández, A. (2002) Design of combinational logic circuits through an evolutionary multi-objective optimization approach. Artificial Intelligence for Engineering, Design, Analysis and Manufacture, 16(1): 39-53.
10. Coello, C. A. Hernández, A. and Buckles, B. P. (2000) Evolutionary Multiobjective Design of Combinational Logic Circuits, eh, pp.161-172, The Second NASA/DoD Workshop on Evolvable Hardware (EH'00).
11. Coello, C. A., Lamont and, G. B., and Van Veldhuizen, D. A. (2007) Evolutionary Algorithms for Solving Multi-Objective Problems, Second Edition, Springer, New York, ISBN 978-0-387-33254-3.
12. Cordon, O.; Herrera, F.; Homann, F. and Magdalena, L. (2001). Genetic Fuzzy Systems: Evolutionary Tuning and Learning of Fuzzy Knowledge Bases, World Scientic.

13. Das A. and Vemuri R. (2007). An Automated Passive Analog Circuit Synthesis Framework using Genetic Algorithms. IEEE Computer Society Annual Symposium on VLSI ISVL '07. pp. 145-152.
14. Das, A. (2008) Algorithms for Topology Synthesis of Analog Circuits. Doctoral thesis. University of Cincinnati. November.
15. Das, A. and Vemuri, R. (2009). A Graph Grammar Based Approach to Automated MultiObjective Analog Circuit Design. Design, Automation and Test in Europe Conference and Exhibition 2009. IEEE Conferences., pp. 700-705.
16. De Bonet, J.; Isbell, C. and Viola, C. (1996) MIMIC: Finding Optima by Estimating Probability Densities. Proceeding of Neural Information Processing Systems. Pp. 424-430.
17. De Garis, H. (1993). Evolvable Hardware: Genetic Programming of a Darwin Machine, in Artificial Neural Nets and Genetic Algorithms, Albretch, R.F., Reeves, C.R., and Steele, N.C., Eds., Springer-Verlag, New York.
18. Dorigo, M. (1991). Positive Feedback as a Search Strategy. Technical Report. No. 91-016. Politecnico Di Milano, Italy.
19. Fonseca, C. M. and Fleming, P. J. Genetic Algorithms for Multiobjective Optimization: Formulation, Discussion and Generalization. In S. Forrest, editor, Proceedings of the Fifth International Conference on Genetic Algorithms, pages 416–423, SanMateo, California, 1993. University of Illinois at Urbana-Champaign, Morgan Kauffman Publishers.
20. Goh, C. and . Li (2001). GA Automated Design and Synthesis of Analog Circuits with Practical Constraints. Proc. IEEE Int. Conf. Evol. Computation. pp. 170-177.
21. Goldberg, D. (1989) Genetic Algorithms in Search Optimization & Machine Learning. Addison-Wesley .
22. Grimbleby, J. (1995) Automatic Analogue Network Synthesis Using Genetic Algorithms, Proceedings of the First IEE/IEEE International Conference on Genetic Algorithms in Engineering Systems (GALESIAS-95), pp.53-58, UK.
23. Harvey, I. (1993). The Artificial Evolution of Adaptive Behaviour, PhD Thesis, University of Sussex, School of Cognitive and Computing Sciences, September, 1993.
24. Hernandez A. and Coello C. (2003). Evolutionary Synthesis of Logic Circuits UsingInformation Theory. Artificial Intelligence Review 20: 445–471, Kluwer Academic Publishers. Printed in the Netherlands.

25. Higuchi, T.; Iwata, M.; Kajitani, I.; Murakawa, M.; Yoshizawa, S. and Furuya, T. (1996). Hardware Evolution at Gate and Function Levels. In Proceedings of the International Conference on Biologically Inspired Autonomous Systems: Computation, Cognition and Action, Durham, North Carolina.
26. Holland, J. (1975). Adaptation in Natural and Artificial Systems. University of Michigan Press. Ann Arbor, MI, 1975; MIT Press, Cambridge, MA 1992.
27. Hu, J.; Zhong, X. and Goodman, E. (2005). Open-ended Robust Design of Analog Filters Using Genetic Programming Proceedings of the 2005 conference on Genetic and evolutionary computation, June 25-29, Washington, DC, USA, pp. 1619-1626.
28. Jerke, G. and Lienig, J. (2009) "Constraint-driven Design — The Next Step Towards Analog Design Automation". Proceedings of the 2009 international symposium on Physical design. San Diego, California, USA. Pp. 75-82. 2009.
29. Khalifa, Y.; Khan, B. and Taha, F. (2008). Multi-objective Optimization Tool for A Free Structure Analog Circuits Design Using Genetic Algorithms and Incorporating Parasitics. Hindawi Publishing Corporation. Journal of Artificial Evolution and Applications. Volume 2008, PP. 0-9.
30. Koza, J. (1992) Genetic Programming. On the Programming of Computers by Means of Natural Selection. MIT Press. Cambridge, Massachussetts, 1992.
31. Koza, J.; Bennett, F.; Andre, D. and Keane, M.. (1996) Toward Evolution of Electronic Animals Using Genetic Programming. Artificial Life V: Proceedings of the Fifth International Workshop on the Synthesis and Simulation of Living Systems. Cambridge, MA: The MIT Press.
32. Koza, J.; Bennethh, F.; Lohn, J.; Dunlap, F.; Keane M. and Andre D. (1997b) Automated synthesis of computacional circuits using genetic programming" in Proc. 1997 IEEE Conf. Evolutionary Computation. Piscataway, NJ: IEEE Press, pp. 447–452, 1997.
33. Koza, J.; Bennethh, F.; Andre, D. and Keane, M. (1997). Automated Synthesis of Analog Electrical Circuits by Means of Genetic Programming. IEEE Transactions on Evolutionary Computation, Vol 1, No.2, pp. 109-128.
34. Krasnicki, M.; Phelps, R.; Hellums, J.; McClung, M.; Rutenbar, R. and Carley, L. (2001). ASF: a practical simulation based methodology for the synthesis of custom analog circuits, In Proceedings of ICCAD 2001, pp. 350–357.

35. Larrañaga P. and Lozano, J. (2002) Estimation of Distribution Algorithms: A New Tool for Evolutionary Computation". Kluwer Academic Publishers.
36. Lohn, J. and Colombano, S. (1998). Automated Analog Circuit Synthesis using a Linear Representation. Proc. of the Second Int'l Conf on Evolvable Systems: From Biology to Hardware, Springer-Verlag, Berlin, pp. 125-133.
37. Louis, S. (1993). Genetic Algorithms as a Computational Tool for Design. PhD Thesis, Department of Computer Science, Indiana University.
38. Louis, S. J. and Rawlins, G. J. E. (1991) Using Genetic Algorithm to Design Structures. Technical Report 326. Computer Science Department, Indiana University, Bloomington, Indiana.
39. Mendoza, B. (2001). Uso de del Sistema de la Colonia de Hormigas para Optimizar Circuitos Lógicos Combinatorios. Tesis de Maestría en Inteligencia Artificial de la Universidad Veracruzana. México.
40. Michalewicz, Z.; Dasgupta, D.; Le Riche, R. and Schoenauer M. (1995). Evolutionary Algorithms for Constrained Engineering Problems.
41. Muhlenbein, H. and Paad G. (1996). From Recombination of Genes to the Estimation of Distributions I. Binary Parameters, in H.M.Voigt, et al., eds., Lecture Notes in Computer Science 1411: Parallel Problem Solvingfrom Nature - PPSN IV, pp. 178- 187.
42. Pelikan M. and Mühlenbein, H. (1999). The Bivariate Marginal Distribution Algorithm. Advances in Soft Computing-Engineering Design and Manufacturing. Pp. 521-535.
43. Ponce de León, E., (1997) Algoritmos Genéticos y su Aplicación a Problemas de Secuenciación". PhD. Tesis. Centro de Inteligencia Artificial. Instituto de Cibernética, Matemática y Física.
44. Rudolph, G. (1996) Convergence of Evolutionary Algorithms in General Search Spaces, In Proceedings of the Third IEEE Conference on Evolutionary Computation.
45. Rutenbar, R.; Gielen, G.; Roychowdhury, J. (2007). Hierarchical Modeling, Optmization, and Synthesis for System-level Anlog and RF Designs, Proc. of the IEEE, Vol. 95, Issue 3, March 2007, pp. 640-669.
46. Schaffer, J. D. (1984) Multiple Objective Optimization with Vector Evaluated Genetic Algorithms. PhD thesis, Vanderbilt University.
47. Shragowitz, E.; Lee,J.; Kang, Q. (1998). Application of Fuzzy Logic in Computer-Aided VLSI Design, IEEE Trans. on Fuzzy Systems, Vol. 6, No 1, February 1998, pp. 163-172.

48. Thompson, A.; Harvey, I. and Husbands, P. (1996). Unconstrained evolution and hard consequences. In E. Sanchez and M. Tomassini, editors, Toward EvolvableHardware: The Evolutionary Engineering Approach (Lecture Notes in Computer cience, Vol. 1062), pages 136--165, Heidelberg, Germany, Springer-Verlag.
49. Torres, A. (2010). Metaheurísticas Evolutivas en el Diseño de Circuitos Analógicos. Tesis Doctoral. Universidad Autónoma de Aguascalientes.
50. Torres, A.; Ponce de León, E.; Hernández, A.; Torres, M.D. and Díaz, E. (2010). A Robust Evolvable System for the Synthesis of Analog Circuits. Computación y Sistemas, Revista Iberoamericana de Computación. April-June, Vol. 13, No.4, pp. 295-312.
51. Torres, A.; Ponce de León, E.; Torres, M. D.; Díaz E. and Padilla, F. (2009). "Comparison of Two Evolvable Systems in the Automated Analog Circuit Synthesis". Artificial Intelligence, MICAI 2009. Eighth Mexican International Conference on, vol.1, pp 3- 8, 8-13.
52. Yao, X. and Higuchi, T. (1999) "Promises and Challenges of Evolvable Hardware", IEEE Transactions on Systems, Man and Cybernetics_Part C. Applications and Reviews,Vol 29, No.1.
53. Zebulum, R.; Pacheco M. and Vellasco M. (1996). Evolvable Systems in Hardware Design: Taxonomy, Survey and Applications, Proceedings of The First International Conference on Evolvable Systems: From Biology to Hardware (ICES'96), Lecture Notes in Computer Science 1259, pp. 344-358, Tsukuba, Japan, October 7-8.
54. Zebulum, R.; Pacheco M. and Vellasco, M. (1998). Comparison of different evolutionary methodologies applied to electronic filter design. In Proc. Of IEEE. Inttl. Conf. On Evolutionary Computation. May.
55. Zebulum, R.; Vellasco, M. and Pacheco, M. (2000) Variable length representation in evolutionary electronics. Evol. Comput., vol. 8, no. 1, pp. 93–120.

Chapter 6

DESIGN OF A SWITCHED CAPACITOR NEGATIVE FEEDBACK CIRCUIT FOR A VERY LOW LEVEL DC CURRENT AMPLIFIER

Hiroki Higa, Naoki Nakamura

Faculty of Engineering, University of the Ryukyus, Okinawa, Japan

ABSTRACT

To miniaturize a very low level dc current amplifier and to improve its output response speed, the switched capacitor negative feedback circuit (SCNF), instead of the conventionally used high-ohmage resistor, is presented in this paper. In our system, a switched capacitor filter (SCF) and an offset controller are also used to decrease vibrations and offset voltage at the output of the amplifier using SCNF. The theoretical output voltage of the very low level dc current amplifier using SCNF is obtained. The experimental results show that the unnecessary components of the amplifier's output are much decreased, and that the response speed of the amplifier with both the SCNF and SCF is faster than that using high-ohmage resistor.

INTRODUCTION

Response speeds of the measuring instruments are limited by those of very low level dc current amplifiers [1,2], when very small currents are measured by mass spectroscopes and radiation detectors. This implies that the amplifiers are required to observe rapid transient phenomena. The very low level dc current amplifier for measuring small currents generally consists of an amplifier having high input impedance and a high-ohmage negative feedback resistor. The amplifier with high-ohmage resistor has unavoidable effects of the stray capacitances across its terminals. This factor causes the amplifier to have a complicated frequency characteristic, which results in poor responses of the very low level dc current amplifier [1-3]. Some shielding techniques [4-6] have been reported for the purpose of decreasing these capacitive components. In spite of the fact that these methods have been employed, it was difficult to realize drastic improvements of the response speeds of the very low level dc current amplifier. Neither are the amplifiers with shielding methods appropriate for miniaturization. A positive feedback circuit [7] had also been used as another approach to decrease the stray capacitances. The amplifier however was unstable and began to oscillate in this case. The resultant high speed response of the amplifier has not been achieved so far.

To explore the optimum solution to the above problems, the switched capacitor negative feedback circuit (SCNF) has been developed. The switched capacitor (SC) circuit is equivalent to a resistor and is suitable for miniaturization. A theoretical output voltage of the very low level dc current amplifier using SCNF with a switched capacitor filter (SCF) was investigated in this paper. Furthermore, the output voltage of the amplifier was experimentally demonstrated.

CIRCUIT DESCRIPTION

Figure 1 summarizes a very low level dc current amplifier, including SCF and a small current source. C_g, R_g and K are the input capacitance, input resistance, and amplification factor of the amplifier having a high input resistance, respectively. In the experiment, we utilized a triangular wave voltage produced by the function generator V_g and the differentiating capacitor C_s (reactance attenuator) to obtain a square wave current I_s with a high output impedance as an input signal to the amplifier. C_o is the output capacitance to the ground of C_s.

The SCNF and SCF are shown in Figure 2. The SCNF is composed of several analog switches (S_1, S_2, $\cdots$, S_7) and capacitors. These switches are controlled by two nonoverlapping clock signals. S_1 is synchronous with S_2 and S_3, and S_4 is synchronous with S_5, S_6 and S_7, respectively. An output voltage of the very low level dc current amplifier has vibrations due to charge and discharge actions of the SC circuit.

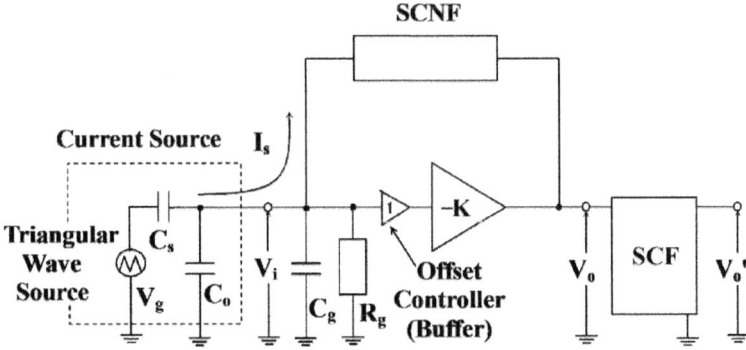

Figure 1: Circuit configuration of very low level dc current amplifier. SCNF and SCF stand for switched capacitor negative feedback circuit and switched capacitor filter, respectively.

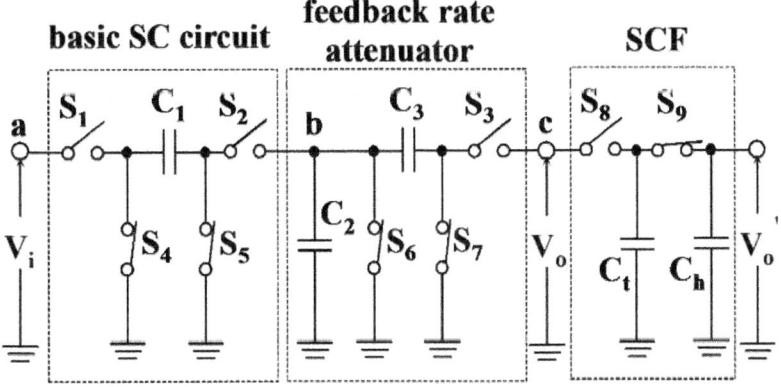

Figure 2: SCNF and SCF.

Thus, as shown in Figure 2, the SCF was connected to the output of the amplifier. The switches S_8 and S_9 are synchronous with S_1 and S_4, respectively.

CIRCUIT ANALYSIS

This section describes an equivalent resistance of SCNF and theoretical output voltage of the very low level dc current amplifier. In the following analysis, it is assumed that T_1 is the time when S_1, S_2, and S_3 are closed (S_4, S_5, S_6, and S_7 are opened), and T_2 is the time when S_1, S_2, and S_3 are opened (S_4, S_5, S_6, and S_7 are closed). Further, it is assumed that a small current I_s is dc current because the clock frequency of the SC circuit, f_s, is much higher than the frequency of I_s.

Equivalent Resistance of SCNF

From Figure 2, the voltage at node b, V_b, is represented by

$$V_b = \frac{C_3}{C_2+C_3}V_o \qquad (1)$$

and an electric charge q_1 at C_1 is

$$q_1 = C_1(V_b - V_i).$$

From Equation (1) and the relationship that $V_o = -KV_i$, the electric charge q_1 at C_1 can be rewritten as

$$q_1 = C_1\left(\frac{C_3}{C_2+C_3}V_o + \frac{V_o}{K}\right) \approx C_1 V_o \frac{C_2 + KC_3}{K(C_2+C_3)}, \qquad (2)$$

for $K \gg 1$. Because the electric charge q_1 at C_1 during T_2 is totally discharged, the quantity of the charge that is transported from node a to node b is equivalent to q_1. Thus, a current, I, flowing from node a into node b during one clock cycle T_s is

$$I = \frac{q_1}{T_s} = V_o \frac{C_1}{T_s} \cdot \frac{C_2 + KC_3}{K(C_2+C_3)}. \qquad (3)$$

Since the current to be measured in the very low level dc current amplifier I_s flows into the SC circuit, $I_s = I$. From the relationship that $V_o = R_{feq} I_s$ [1], the equivalent resistance of SCNF R_{feq} is given by

$$R_{feq} = \frac{T_s}{C_1} \cdot \frac{K(C_2+C_3)}{C_2 + KC_3}, \qquad (4)$$

while the equivalent resistance of the SC circuit [8] R_{sc} is represented by

$$R_{sc} = \frac{T_s}{C_1} = \frac{1}{C_1 f_s}, \qquad (5)$$

where f_s is the clock frequency. The attenuation factor of the attenuator X [9] becomes

$$X = \frac{C_2 + KC_3}{K(C_2+C_3)}. \qquad (6)$$

Hence, from Equations (4) to (6), R_{feq} can be obtained as

$$R_{feq} = \frac{1}{XC_1 f_s} = \frac{R_{sc}}{X}. \qquad (7)$$

From Equation (6), it is observed that X is dependent on the ratio of capacitances of C_2 and C_3.

Theoretical Output Voltage of the Amplifier

Output voltage of the very low level dc current amplifier using SCNF has vibrations due to charge and discharge actions of the SC circuit. In this section, theoretical output voltage of the very low level dc current amplifier is discussed.

The equivalent circuit of the SC circuit is shown in Figure 3(a). It is found from Equations (5) and (7) that the SCNF is equivalent to the capacitor of XC_1 and four switches S_1, S_3, S_4, and S_7. The equivalent circuit of the very low level dc current amplifier is shown in Figure 3(b). The box labeled "*" in Figure 3(b) stands for the SCNF and this figure shows that the equivalent circuit of the SCNF shown in Figure 3(a) is connected with the equivalent circuit of the amplifier at the terminals between nodes a and c.

Millman's theorem is applied to Figure 3(b), and the voltage V_i is represented by

$$V_i = \frac{I_s + j\omega XC_1(-KV_i)}{1/R_g + j\omega(C_o + C_g + XC_1)},$$

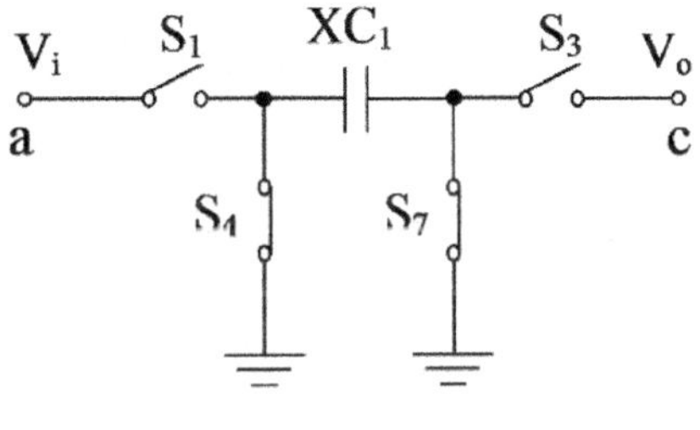

(a)

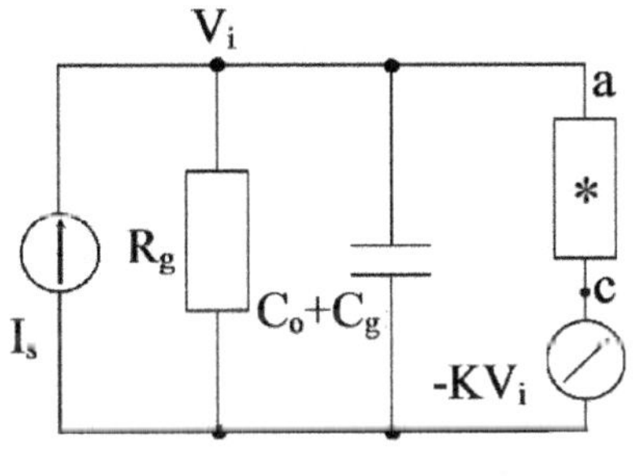

(b)

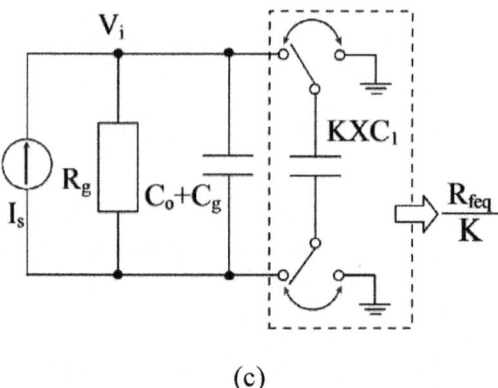

(c)

Figure 3: Equivalent circuits of (a) SCNF and (b) very low level dc current amplifier. Figure 3(c) shows simplified input equivalent circuit of Figure 3(b).

where ω is the angular frequency. The input admittance of the very low level dc current amplifier Y_{in} is

$$Y_{in} = \frac{I_s}{V_i} \approx \frac{1}{R_g} + j\omega\left(C_o + C_g + KXC_1\right), \qquad (8)$$

for $K \gg 1$. Using Equation (8), a simplified input equivalent circuit of the very low level dc current amplifier using SCNF can be drawn as shown in Figure 3(c).

An enlarged input voltage waveform of the amplifier at the positive final steady-state, V_i, is illustrated with the help of clock waveforms as shown in Figure 4. T_s is the clock cycle of the switches. T_1 and T_2 are $(1-\alpha)T_s$ and αT_s, respectively. Let the input voltage of the amplifier at $t = nT_s$ be $V_i(n)$. Subscript symbols "+" and "−" mean just after and just before the time event occurs, respectively. For example, $V_o(n+1)_-$ means the voltage just before $t = (n+1)T_s$. The amplitude of the input voltage for a cycle, V_m, is

$$V_m = \frac{1}{C_o + C_g + KXC_1} \int_{nT_s}^{(n+(1-\alpha))T_s} I_s\, dt$$

$$+ \frac{1}{C_o + C_g} \int_{(n+(1-\alpha))T_s}^{(n+1)T_s} I_s\, dt$$

$$= \frac{C_o + C_g + \alpha KXC_1}{\left(C_o + C_g\right)\left(C_o + C_g + KXC_1\right)} I_s T_s. \qquad (9)$$

Since electric charges of the SC circuit are conserved just before and after $t = nT_s$, the following equation is obtained

$$(C_o + C_g)V_i(n)_- = (C_o + C_g + KXC_1)V_i(n)_+.$$

The input voltage just before $t = nT_s$ is

$$V_i(n)_- = \left(1 + \frac{KXC_1}{C_o + C_g}\right)V_i(n)_+. \qquad (10)$$

From Figure 4 and Equation (10), the voltage V_m is

$$V_m = V_i(n)_- - V_i(n)_+ = \frac{KXC_1}{C_o + C_g}V_i(n)_+. \qquad (11)$$

From Equations (9) and (11), the input voltage just after $t = nT_s$ can be obtained as

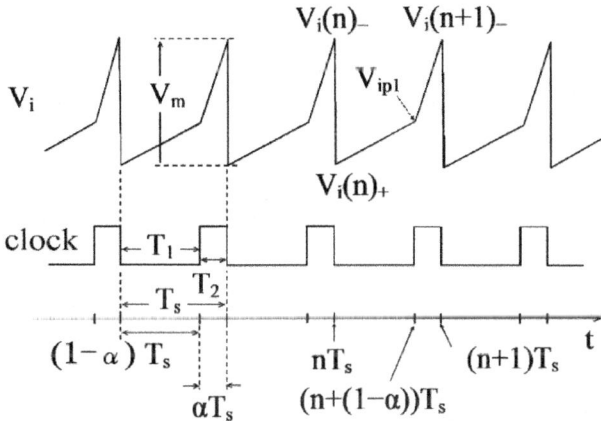

Figure 4: Relationship between enlarged input voltage and clock waveforms.

$$V_i(n)_+ = \frac{I_s T_s}{KXC_1} \cdot \frac{C_o + C_g + \alpha KXC_1}{C_o + C_g + KXC_1}. \qquad (12)$$

Thus the peak voltage T_{ip1} during T_1 is

$$V_{ip1} = V_i(n)_+ + \frac{1}{C_o + C_g + KXC_1} \int_{nT_s}^{(n+(1-\alpha))T_s} I_s dt$$

$$V_{ip1} = V_i(n)_+ + \frac{1}{C_o + C_g + KXC_1} \int_{nT_s}^{(n+(1-\alpha))T_s} I_s \, dt$$

$$= V_i(n)_+ + \frac{1-\alpha}{C_o + C_g + KXC_1} I_s T_s.$$

(13)

Substituting Equation (12) into Equation (13) gives the following equation:

$$V_{ip1} = \frac{I_s T_s}{KXC_1} = I_s \frac{R_{feq}}{K}.$$

(14)

Therefore, the peak output voltage of the amplifier during T_1, V_{op1}, can be written as

$$V_{op1} = -KV_{ip1} = -I_s R_{feq}.$$

(15)

It is seen from Equation (15) that the theoretical output voltage of the very low level dc current amplifier using SCNF can be obtained by sampling V_{op1}, and that the equivalent resistance, R_{feq}, is independent of duty ratio of the clock signal.

In this paper, the SCF was used to sample V_{op1} from the output voltage of the very low level dc current amplifier using SCNF for the following reasons. Using a sample-and-hold circuit generally requires a clock generator that completely differs from two non-overlapping clock signals utilized by the SCNF, and using a low-pass filter provides for not theoretical output voltage, but approximately half amplitude of output voltage of the amplifier at a final steady-state. On the other hand, using the SCF allows for sharing the two non-overlapping clock signals. Both the SCF and SCNF can be also manufacturable by the same process. Therefore, the SCF is useful from the viewpoint of miniaturization.

SIMULATION RESULTS

Using the electronic circuit simulator PSpice (Cadence Design System, Inc.), transient analyses of the very low level dc current amplifier using SCNF were carried out. K and C_g were set to 1300 and 17 pF, respectively. The equivalent resistance of the SC circuit R_{RC} was set to 1 MΩ by using C_1 of 10 pF and

f_s of 100 kHz. The attenuation factor X of 1/100 was also set by using both C_2 of 1000 pF and C_3 of 9.3 pF. Thus, the total equivalent resistance R_{feq} of the SCNF was 100 MΩ. The duty ratio of the clock cycle, α, was also set to 0.5. To evaluate response speeds of the very low level dc current amplifier, a square wave current I_s with a time period of 5 ms and an amplitude of 10 nA was input to the amplifier. From Equation (15), output voltage of 1 V should be obtained as the theoretical output voltage of the amplifier. A switch model [10] used in the computer simulation is shown in Figure 5. The symbols G, D, and S stand for gate, drain and source of a MOS-FET. Each analog switch is composed of a combination of an nMOS and pMOS, as shown in Figure 5(a). Assuming that parasitic capacitances between two terminals exist, as shown in Figures 5(b) and (c), the transient analyses by computer simulation were conducted.

First, it is assumed that nMOS has exactly the same parasitic capacitances as pMOS has. The parasitic capacitance values are shown in Table 1. These values were determined by trial and error. Figures 6(a) and (b) show the simulation results in the case of using the parasitic capacitances shown in Table 1. It can be seen that the output voltage of the amplifier has vibrations due to charge and discharge actions of the SCNF, and that the vibrations of output of the amplifier causes black areas in the output waveform (see Figures 6(a) and (b)). Thus, it is difficult to measure an input current from the output waveform of the amplifier. Average values of V_{op1} from 10 ms to 12.5 ms and from 12.5 ms to 15 ms in Figure 6(a) were obtained as +1.0 V and −1.0 V, respectively. The simulation result in the case of using the SCF at the output of the amplifier is shown in Figure 6(c). In this case, the peaks of the output voltage during T_1 were sampled by the SCF. The rise time of the output waveform of the SCF is 20.0 μs. As defined in general, the rise time is the time required for the output waveform to rise from 10% to 90% of its final steady-state value. It is clear from Figure 6(c) that using the SCF drastically reduces vibrations as well as unnecessary components, and that the input current I_s can be obtained by measuring the amplitude of its output voltage.

Secondly, on the assumption that nMOS and pMOS have totally different parasitic capacitances, as shown in Table 2, computer simulations of transient analyses of the very low-level dc current amplifier using SCNF were also performed. The parasitic capacitance values indicated

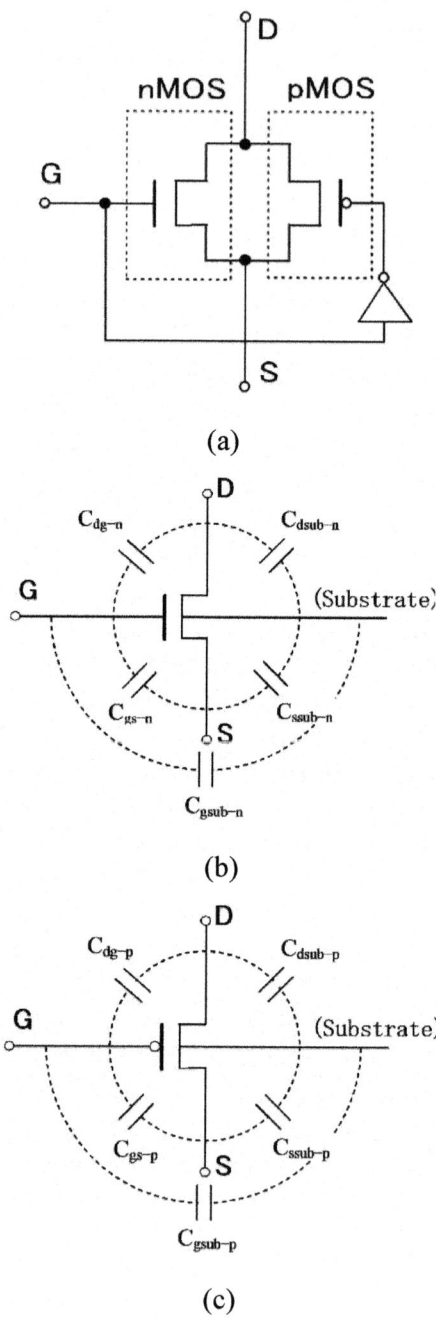

Figure 5: Switch model used in PSpice simulation. (a) Configuration of CMOS switch, (b) nMOS and (c) pMOS switch models with parasitic capacitances.

Table 1: Values of parasitic capacitances in the case of assumption that nMOS and pMOS have the same parasitive capacitive components

parasitic capacitance [pF]				
C_{dg-n}, C_{dg-p}	C_{gs-n}, C_{gs-p}	C_{dsub-n}, C_{dsub-p}	C_{ssub-n}, C_{ssub-p}	C_{gsub-n}, C_{gsub-p}
0.10	0.09	0.07	0.07	0.06

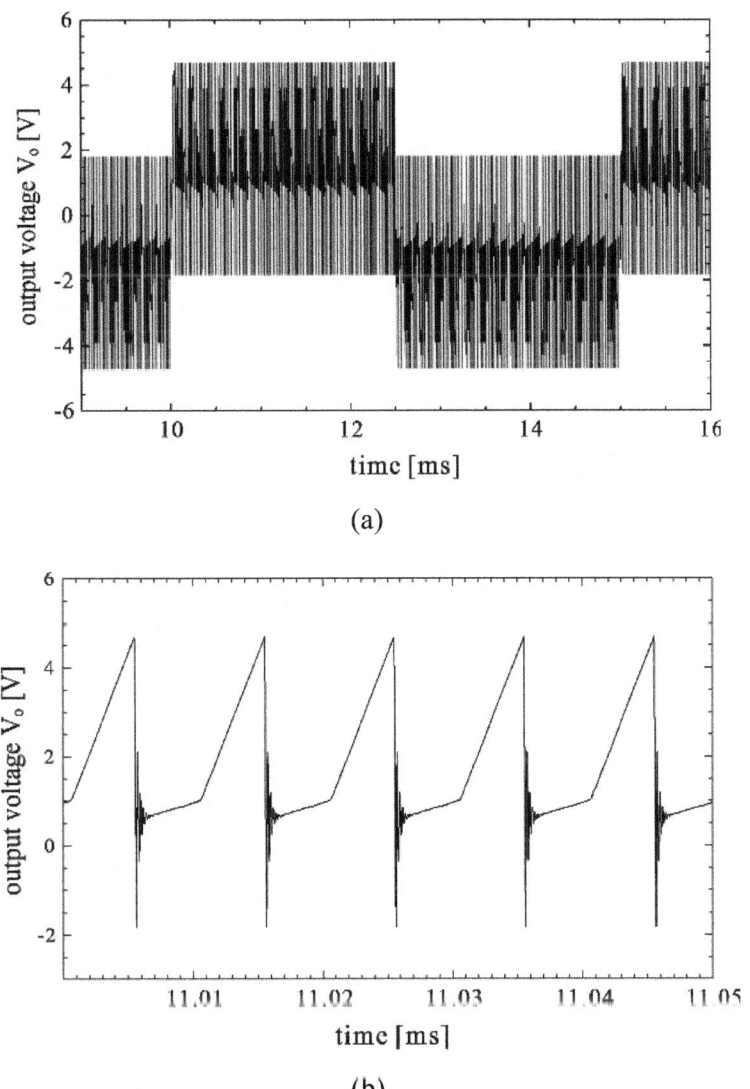

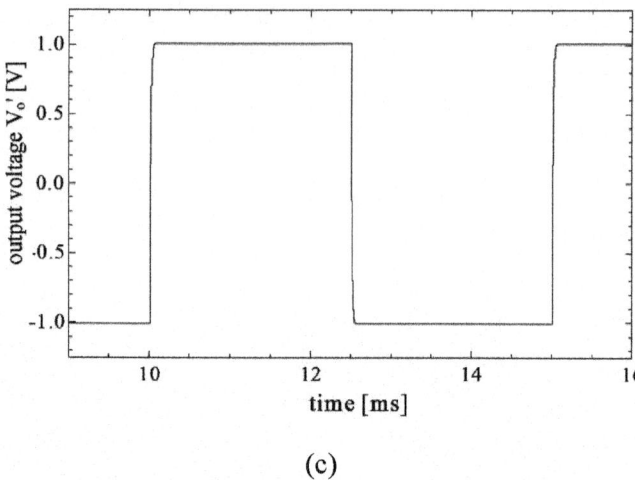

(c)

Figure 6: Simulation results in the case of using parasitic capacitances shown in Table 1. (a) Output waveform of very low level dc current amplifier using SCNF, (b) its enlarged waveform at a positive final steady-state, and (c) output waveform of SCF. The rise time of output waveform in Figure 6(c) is 20.0 μs.

Table 2: Values of parasitic capacitances in the case of assumption that nMOS and pMOS have totally different parasitic capacitive components

	parasitic capacitance [pF]				
	C_{dg-n}	C_{gs-n}	C_{dsub-n}	C_{ssub-n}	C_{gsub-n}
S_1	0.08	0.13	0.06	0.05	0.05
S_2	0.09	0.10	0.09	0.05	0.06
S_3	0.10	0.14	0.08	0.06	0.06
S_4	0.07	0.14	0.07	0.09	0.06
S_5	0.07	0.11	0.06	0.07	0.04
S_6	0.10	0.13	0.08	0.07	0.06
S_7	0.10	0.14	0.07	0.07	0.06
	C_{dg-p}	C_{gs-p}	C_{dsub-p}	C_{ssub-p}	C_{gsub-p}
S_1	0.15	0.09	0.07	0.07	0.06
S_2	0.11	0.07	0.06	0.06	0.08
S_3	0.12	0.09	0.06	0.07	0.06
S_4	0.15	0.06	0.06	0.07	0.06
S_5	0.14	0.12	0.10	0.08	0.09
S_6	0.14	0.09	0.04	0.05	0.06
S_7	0.14	0.08	0.10	0.09	0.08

in Table 2 were also determined by trial and error. Figures 7(a) and (b) show the simulation results in the case of using the parasitic capacitances shown in Table 2. It can be found from Figure 7(a) that the output voltage of the amplifier has also vibrations, as shown in Figure 6(a), and that the amplitudes of the black painted areas are roughly 5 times larger than those in Figure 6(a). Further, it is seen that the output waveform of the amplifier has rapid changes in output voltage from T_1 to T_2, and that the rapid change V_c is −25.7 V (see Figure 7(b)). The simulation result in the case of using the SCF at the output of the amplifier is shown in Figure 7(c). The rise time of the output waveform of the SCF is also 20.0 μs. It is obvious from Figure 7(c) that the amplitude of the output voltage of the amplifier becomes 1 V, and that the output waveform has the offset voltage of 2.04 V. Comparing Figure 7 with Figure 6, it is thought that a clock feed through generated by totally different parasitic capacitive components leads to a generation of the offset and the rapid changes in output voltage of the very low level dc current amplifier using SCNF. From other simulation results, it was found that using different parasitic capacitances of C_{dg-n} and C_{dg-p}, and those of C_{gs-n} and C_{gs-p} resulted in the generation of offset voltage of the amplifier, and that using larger parasitic capacitances of C_{dsub-n}, C_{dsub-p}, C_{ssub-n} and C_{ssub-p} led to an error of the equivalent resistance R_{feq}. On the other hand, parasitic capacitances of C_{gsub-n} and C_{gsub-p} did not have an effect on the output voltage of the amplifier.

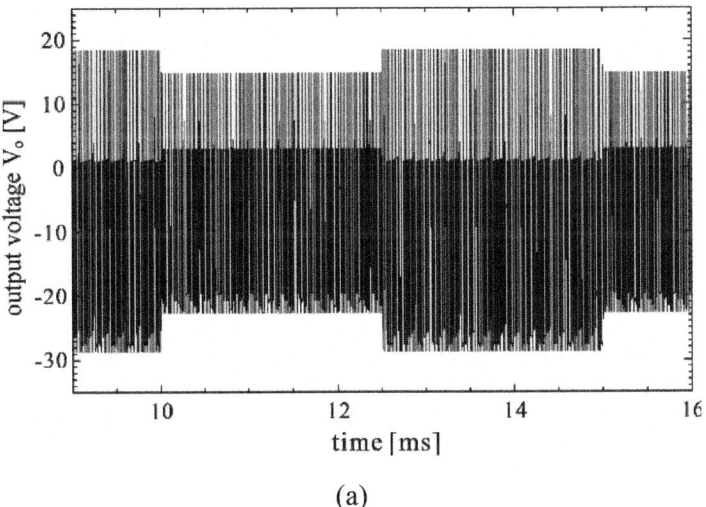

(a)

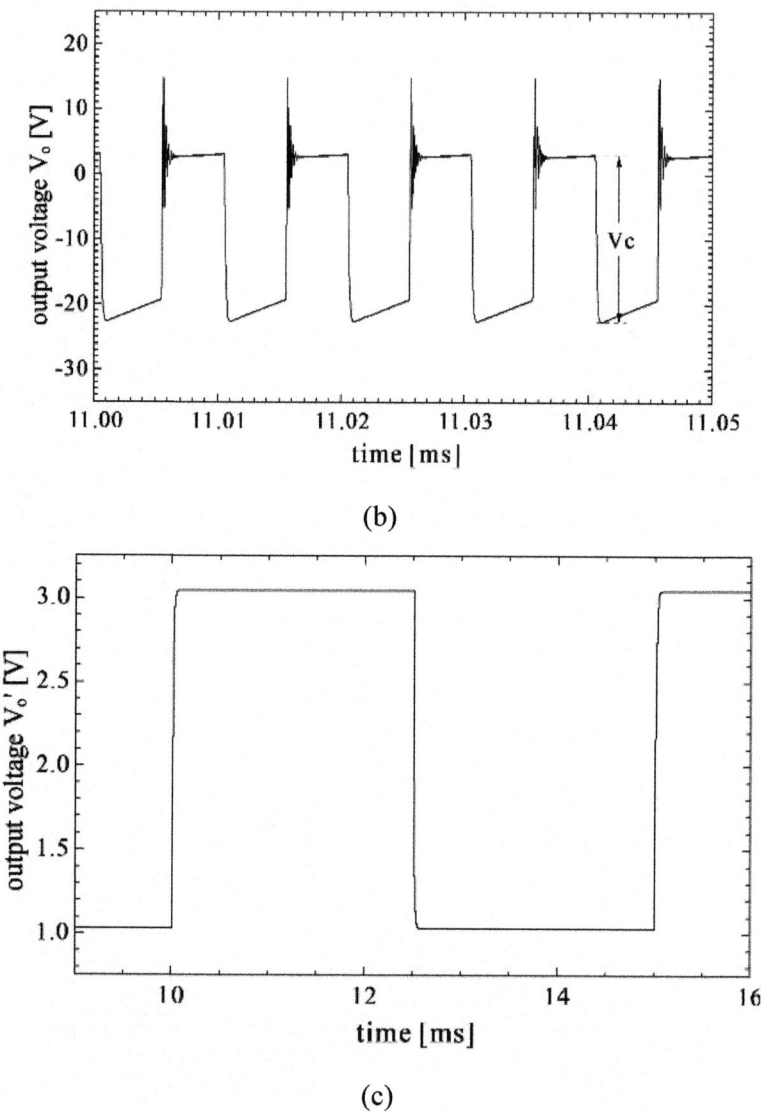

Figure 7: Simulation results in the case of using parasitic capacitances shown in Table 2. (a) Output waveform of very low level dc current amplifier using SCNF, (b) its enlarged waveform at a positive final steady-state, and (c) output waveform of SCF. The rise time of output waveform in Figure 7(c) is 20.0 μs.

EXPERIMENTAL RESULTS AND DISCUSSION

First, the very low level dc current amplifier using SCNF shown in Figure 1 was made and its output responses were observed. The amplifier having high input resistance was composed of two differential amplification stages and an emitter follower stage, and the first stage had two JFETs to obtain high input resistance of the amplifier. The amplification factor K of the amplifier was 62 dB (From DC to 1.2 MHz), and output voltage waveform was observed using an oscilloscope. Since the triangular wave voltage, which had a time period of 5 ms and an amplitude of 10 V, was differentiated by the differentiating capacitor C_s of 1.25 pF, a square wave current with a time period of 5 ms and an amplitude of 10 nA was obtained as an input current I_s to the amplifier. As switches of the SC circuit, we used CMOS analog switches (MAX326, MAXIM Integrated Products, Inc.) having the maximum leakage current of 10 pA. Further, variable capacitors C_1 and C_3 were utilized. Parasitic capacitances of analog switches have a little effect on equivalent resistance of the SCNF R_{feq}, which causes errors in R_{feq} of the amplifier. Thus, the equivalent resistance R_{sc} of the SC circuit with the clock frequency f_s of 100 kHz, was set to 1 MΩ by adjusting capacitance of C_1, and then the attenuation factor of the feedback rate attenuator X was set to 1/100 by adjusting capacitance of C_3. Referring to Equation (7), the total equivalent resistance of the SCNF became 100 MΩ. An offset voltage controller, which was connected to the input of the amplifier and had a gain of unity, was also used to cancel the offset voltage in the experiment. The input stage of the offset controller was composed of a JFET which had much higher input impedance than the negative feedbackcircuit had, and its voltage drift was very small (several μV). Therefore, the offset controller did not have much effect on the current detection sensitivity of the amplifier.

Figures 8(a) and (b) show the output voltage waveform of the very low level dc current amplifier using

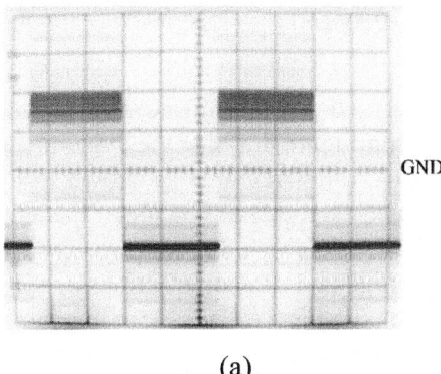

(a)

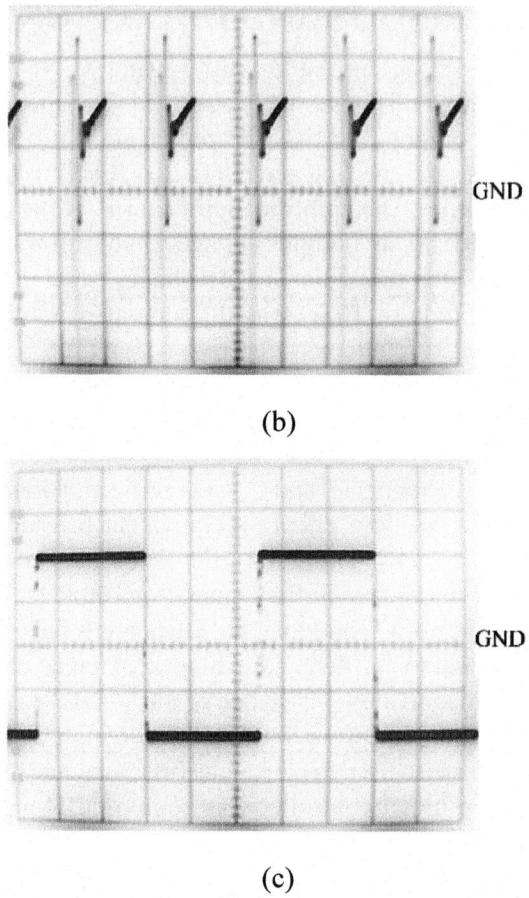

Figure 8: Experimental results. (a) Output waveform (scale H: 1 ms/div, V: 0.5 V/div), (b) its enlarged waveform (scale H: 5 µs/div, V: 0.5 V/div) of very low level dc current amplifier using SCNF, and (c) output waveform (scale H: 1 ms/div, V: 0.5 V/div) of SCF, consisting of C_t of 470 pF and C_h of 220 pF. The rise time of output waveform in Figure 8 (c) is 23.8 µs.

SCNF and its enlarged waveform at a positive final steady-state, respectively. It is seen fromFigure 8(a) that the experimental output waveform does not extensively have the black area as shown in Figure 6(a) and 7(a). This is because rapid changes in output voltage of the amplifier do not appear on a CRT display of the oscilloscope due to its resolution, and as a matter of fact vibrations due to charge and discharge actions of the SCNF can be found at any enlarged waveforms in Figure 8(b). From this figure, it is clear that the output waveform of the amplifier has vibrations and rapid changes as observed in Figure 7(b). Because the amplifier in the PSpice simulation does have an amplitude limiter, in the case of simulation result in Figure 7(b), the minimum output voltage of the amplifier during T_2 was indicated −22.6 V, while in the case of experimental result, the output voltage of the amplifier during T_2 was saturated at −13 V due to the power supply of the negative voltage of the amplifier.

Peak values of the output voltage during T_1 were sampled by the SCF. The experimental result is shown in Figure 8(c). It is clear from Figure 8(c) that using the SCF drastically reduces vibrations of the output of the amplifier, and that the theoretical input current I_s can be obtained by measuring the amplitude value of its output voltage. Thus it is possible to measure an input current of the very low level dc current amplifier using SCNF with the SCF. With respect to the Figure 8(c), the rise time of the output waveform of the SCF is 23.8 μs. On the other hand, that of the output waveform of the very low level dc current amplifier using conventionally used high-ohmage resistor is 92.0 μs. We are easily available to miniaturize SC circuits using IC-compatible techniques. Therefore, using the SCNF is an effective way to obtain both a faster output response and miniaturization of a very low level dc current amplifier.

Finally, a relationship between the equivalent resistance of the SCNF R_{feq} and the clock frequency f_s was investigated. It is clear from Equation (7) that R_{feq} can be set using the clock frequency and capacitances of C_1, C_2, and C_3. In this experiment, we first set to f_s of 100 kHz and C_2 of 1000 pF, adjusted both capacitances of C_1 and C_3 to precisely obtain R_{feq} of 100 MΩ, and then changed f_s ranging from 50 kHz to 200 kHz. The equivalent resistance of SCNF was obtained by measuring output voltage of the SCF. The relationship between R_{feq} and f_s and error rate of R_{feq}, in the case of X of 1/100, are shown in Figure 9. Theoretical values of R_{feq} in Figure 9(a) were calculated using Equation (7) with C_1 of 10.0 pF, C_2 of 1000 pF, and C_3 of 9.32 pF. It is obvious from Figures 9(a) and (b) that the experimental curve agrees well with the theoretical one, and that the error rate of R_{feq} is within 0.86 % for the theoretical values. From this experimental result, it is thought that R_{feq} ranging from 1.1 MΩ with X of 9/10 to 1.0 GΩ with X of 1/500 will be settable.

CONCLUSION

A theoretical output voltage of a very low level dc current amplifier using SCNF with SCF was derived from circuit analysis. It was experimentally demonstrated that a small current of 10 nA could be measured using the amplifier, and that response speed of the amplifier using SCNF was faster than that of the amplifier using conventionally used high-ohmage resistor. A consideration of a very low level dc current amplifier using SCNF with SCF will be needed for practical applications as our future work.

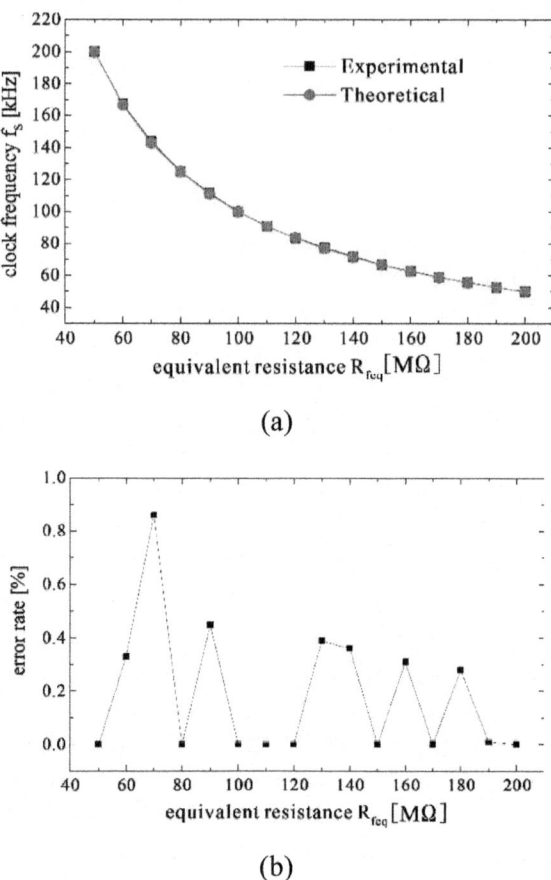

Figure 9: Experimental result: (a) relationship between R_{feq} and f_s and (b) error rate of R_{feq}.

REFERENCES

1. S. P. Presley, "Fast Response Picoammeter," Review of Scientific Instruments, Vol. 37, No. 5, 1966, pp. 643-648. doi:10.1063/1.1720272
2. K. Goto and K. Ishikawa, "Design and Construction of High Speed Pico-Ammeter," The Journal of the Vacuum Society of Japan, Vol. 22, No. 6, 1979, pp. 235-246. doi:10.3131/jvsj.22.235
3. I. Nakamura and T. Kano, "Noise and Fast Response of a Very Small dc Current Amplifier," Oyo Buturi, Vol. 52, No. 4, 1983, pp. 330-338.
4. F. Galliana and P. P. Capra, "Hamon 10 × 100 MΩ Resistor Based Traceable Source for Calibration of Picoammeters in the Range 100 pA -100 nA,"

Measurement, Vol. 43, No. 9, 2010, pp. 1277-1281. doi:10.1016/j.measurement.2010.07.006

5. F. Galliana, P. P. Capra and E. Gasparotto, "Metrological Management of the High dc Resistance Scale at INRIM," Measurement, Vol. 42, No. 2, 2009, pp. 314-321.doi:10.1016/j.measurement.2008.07.002

6. I. Nakamura and M. Takemura, "Fast Respoinse of a Very Low-Level dc Current Amplifier Using Improvement of Feedback Resistor Shielding Structure," IEICE Transactions on Electronics, Vol. J72-C-II, No. 10, 1989, pp. 885-892.

7. I. Nakamura and T. Kano, "High Speed Response of a Very Low Level dc Current Amplifier Using Positive Feedback Loop," Oyo Buturi, Vol. 54, No. 9, 1985, pp. 945-951.

8. A. B. Grebene, "Bipolar and MOS Analog Integrated Circuit Design," John Wiley & Sons Inc., Hoboken, 1984.

9. H. Higa, R. Onaga, N. Nakamura and I. Nakamura, "A Basic Study on a Very Low-Level dc Current Amplifier Using a Switched-Capacitor Circuit: Comparison between Simulation and Experimental Results," Proceedings of ITC-CSCC2005, Jeju, 7 July 2005, pp. 1133-1134.

10. N. H. E. Weste and K. Eshraghian "Principle of CMOS VLSI Design: A Systems Perspective," T. Tomisawa and Y. Matsuyama, Translation, Maruzen, 1999.

Chapter 7

THE ANALYSIS OF THE PERFORMANCE OF MULTI-BEAM FORMING IN MEMORY NONLINEAR POWER AMPLIFIER

Huiyong Li, Xun Li and Chen Wei

School of Electronic Engineering, University of Electronic Science and Technology of China

ABSTRACT

With the increasingly diverse and complex requirements of radar systems and communication systems, the application of multifunction-phased array radar has become a trend, and the digital multi-beamforming technology plays a crucial role in it. In practice, power amplifier (PA) is an essential component in radar systems and communication systems. Unfortunately, it is always nonlinear to provide a high output power. With the purpose of a high output power and efficiency, it is necessary to study the influence of PA nonlinear characteristics on the digital multi-beamforming. In this paper, a form of the multi-beamforming signal and a nonlinear model with memory for PA are given. The output signal *via* the PA model has been analyzed subsequently. As the result of analysis, it can be found that the output signal is divided into the original signal and the interferential signal. The power ratio of original signal to interference signal can reflect the influence of PA nonlinear characteristics on the digital multi-beamforming. Finally, according to the ratio, the results of computer simulation show that the memory effect plays a key role for the small power signal, while the nonlinearity plays an important role for the large power signal.

INTRODUCTION

In recent years, with the development of the military radar technology and the increasing demand for information processing of communications, the application of integrated electronic information system with multifunction-phased array radar has become a trend, and the digital multi-beamforming technology is crucial to the implementation of this system. At present, in contrast to the mature receiving digital multi-beamforming technology, the transmitted digital multi-beamforming technology is still under development. Transmitted multi-beamforming has many advantages. For example, when the radar array intends to search and track objects, the working mode of the common beam is time division, which means that only one job, tracking or searching, can be done at the same time. If the transmitted simultaneous multi-beamforming is used, the two work can be carried out at the same time by using two beams, which are added and synthesized in the digital side and can be transmitted simultaneously. However, because the transmitted signal is the sum of signals, the major bottleneck of the realization of the transmitted simultaneous multi-beamforming is that the transmitted signal envelope is not constant. Therefore, it has a higher requirement on linearity of the transmitter power amplifier.

In order to design an optimal PA, it is necessary to analyze the effect of the PA nonlinearities on transmitted multi-beamforming firstly. The nonlinear distortion of PA has always been a hot research area. A memory PA model is proposed in [1] and the performance of behavioral models is analyzed in [2]. H Ku has analyzed behavioral modeling of nonlinear RF power amplifiers with memory effects in [3]. The major consequence of memory effects has been introduced by W Bosch in [4]. The papers [5, 6] present that memory effects of PA in the wideband system are obvious. However, here the definition of wideband is the high ratio of the transmitted signal bandwidth to the device bandwidth, rather than the signal bandwidth to the carrier frequency. For a transmitted narrowband signal, the memory effects of the PA are evident when the signal bandwidth matches the device bandwidth [7]. The constant envelope signal in the nonlinear PA is studied in [8]. Kohls [9] has used a Bessel series fit to the measured amplifier transfer functions with a computer model to predict third-order intermodulation product beam patterns. In [10], Hemmi describes the nonlinear PA response of an active linear array by a third-order polynomial and develops equations for the beam-pointing angle of the harmonics and intermodulation products. An active phased array multibeam antenna model, including the nonlinear Shimbo model of the amplifiers, has been developed and validated experimentally in [11], which is useful for solid state power amplifiers (SSPAs). In [12], only the AM-AM varieties are considered to affect the beamforming, regardless of the AM-PM varieties and memory effects.

In this paper, the principle of transmitted multi-beamforming is presented in section 2. Then, section 3.1 presents a model based on the Hammerstein model to describe the AM-AM varieties, AM-PM varieties and memory effects.

The output signal model from PA is analyzed in section 3.2. The simulation results are shown in section 4. Finally, section 5 concludes the study. Assumed that each channel and array elements are ideal and each PA is identical. A narrowband signal is considered and the attention will be focused on the effects of the amplitude and phase nonlinearities.

THE PRINCIPLE OF TRANSMITTING MULTI-BEAM-FORMING

As shown in Figure 1, consider an L-element array with elements uniformly spaced on the line of distance equal to d. The transmitted signals are $s_1(n)$, ..., $s_N(n)$ and the weight vectors are $\mathbf{w}_1$, ..., $\mathbf{w}_N$.

$$\mathbf{w}_i = \left[1, e^{j\varphi_i}, \cdots, e^{j(L-1)\varphi_i}\right]^T, i = 1, \cdots, N$$
$$\varphi_i = \frac{2\pi d \sin\theta_i}{\lambda} \tag{1}$$

where φ_i is the space phase of the i th signal, λ is the wavelength and θ_i is the transmitted angle of the i th signal.

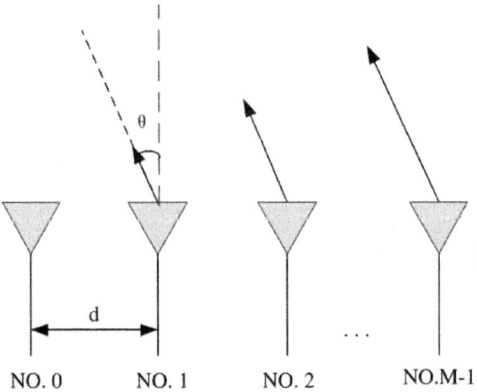

Figure 1: The model of transmitting array.

The space-power-function of transmitted multi-beam is given by

$$f(\theta) = \left|\sum_{i=1}^{N} \mathbf{w}_i^H \mathbf{a}(\theta) s_i(n)\right|^2 \tag{2}$$

where $a(\theta) = [1, e^{j\varphi}, \cdots, e^{j(L-1)\varphi}]^T$ is the steering vector of the scanning beam. When the transmitted signals are orthogonal and the powers of them are equal to each other, the multi-beam pattern can be written as

$$E(\theta) = \frac{f(\theta)}{P_s} = \sum_{i=1}^{N} |\mathbf{w}_i^H \mathbf{a}(\theta)|^2 \qquad (3)$$

where P_s is the power of transmitted signal.

THE ANALYSIS OF TRANSMITTED MULTI-BEAM FORMING BASED ON NONLINEAR MEMORY POWER AMPLIFIER

The Model of PA

A behavior model for PA can be divided into two types: the band-pass PA model for the RF signal processing and the baseband PA model for the envelope information processing. In practical engineering, the input signal of PA is a real signal with radio frequency. Therefore, the band-pass PA model can be very accurate to analyze a variety of components in the nonlinear devices, including harmonics, intermodulation, etc. However, in fact, the processed harmonics will not be transmitted. As the result, it is not convenient to analyze the signal of the band-pass PA model. As we all know, the useful information is carried only by the signal envelope. Moreover, nonlinear characteristics reflected by the behavioral model are only related to the signal amplitude, rather than the frequency. Therefore, a baseband PA model can be used for the signal analysis.

In this paper, the Hammerstein model (Figure 2), a modification of Volterra, has been used, which is given by:

$$y(n) = \sum_{k=1}^{P} a_k x^k(n)$$
$$z(n) = \sum_{i=0}^{M-1} h_i y(n-i) \qquad (4)$$

where $x(n)$, $y(n)$, and $z(n)$ denote the input signal of the memoryless nonlinear PA model, memory linear PA and Hammerstein, respectively. M is the memory depth and P is the order of nonlinear system.

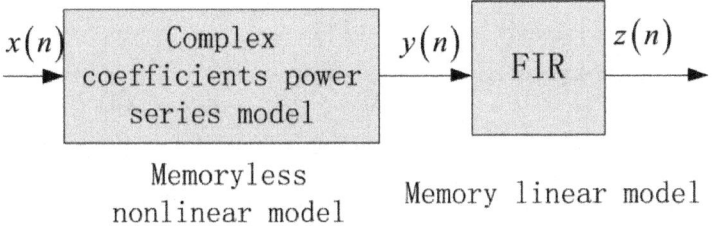

Figure 2: The model of memory nonlinear PA.

The memory linear model uses an $M-$order FIR filter as description. In order to reflect its AM-AM and AM-PM characteristics simultaneously, the memoryless nonlinear model employs a complex coefficients power series model. Assume that $a_k = a_{Ik} + ja_{Qk}$, $x(n) = re^j$. So,

$$\begin{aligned}
y(n) &= \sum_{k=1}^{P} a_k x(n)|x(n)|^{k-1} \\
&= re^{j\theta} \sum_{k=1}^{P} a_k r^{k-1} \\
&= \left[\sum_{k=1}^{P} a_{Ik} r^k + j\sum_{k=1}^{P} a_{Qk} r^k\right] e^{j\theta} \\
&= \sqrt{\left(\sum_{k=1}^{P} a_{Ik} r^k\right)^2 + \left(\sum_{k=1}^{P} a_{Qk} r^k\right)^2} e^{j\left(\theta + \arctan\left(\sum_{k=1}^{P} a_{Qk} r^k / \sum_{k=1}^{P} a_{Ik} r^k\right)\right)}
\end{aligned} \qquad (5)$$

The functions of AM-AM and AM-PM can be given by:

$$\begin{aligned}
A(r) &= \sqrt{\left(\sum_{k=1}^{P} a_{Ik} r^k\right)^2 + \left(\sum_{k=1}^{P} a_{Qk} r^k\right)^2} \\
\Phi(r) &= \arctan\left(\sum_{k=1}^{P} a_{Qk} r^k / \sum_{k=1}^{P} a_{Ik} r^k\right)
\end{aligned} \qquad (6)$$

From the above conclusion, power series with complex coefficients can describe the characteristics of AM-AM and AM-PM. If the PA has no memory, the output of the PA is determined only by the odd-component of the input signal. The even-component of the input signal will not affect the output of the PA, which can be expressed as:

$$y(n) = \sum_{k=0}^{(P-1)/2} a_{2k+1} |x(n)|^{2k} x(n) \qquad (7)$$

The analysis of the transmitted multi-beam signal based on the Taylor model

After frequency mixing of two transmitted real signals, the output signals can be represented as:

$$s_1(t) = A_1(t)\cos(\omega_0 t + \varphi_1(t))$$
$$s_2(t) = A_2(t)\cos(\omega_0 t + \varphi_2(t)) \tag{8}$$

The signal input into the nonlinear PA is given by:

$$s_{in}(t) = A_{w_1}A_1(t)\cos(\omega_0 t + \varphi_1(t) - \phi_{w_1})$$
$$+ A_{w_2}A_2(t)\cos(\omega_0 t + \varphi_2(t) - \phi_{w_2}) \tag{9}$$

where A_{w_1}, A_{w_2} and ϕ_{w_1}, ϕ_{w_2} are the amplitudes and phases of the two weight vectors w_1 and w_2, respectively. $A_1(t)$, $A_2(t)$ and $\varphi_1(t)$, $\varphi_2(t)$ are the amplitudes and phases of the two signals, respectively. ω_0 is the carrier frequency.

Even order harmonic component, produced by PA, can be removed by filters. Therefore, the odd-order harmonic component, which cannot be removed by filters, has a great effect on input signals. According to the power series model, after the signal passes the nonlinear PA, the output signal is given by:

$$s_{out}(t) = a_1 s_{in}(t) + a_3 s_{in}(t)^3 \tag{10}$$

where a_1, a_3 is a real number.

Expanding the cubic term and removing the non-fundamental frequency component, we can get:

$$a_3 s_{in}(t)^3 \approx a_3 \left[\left(\frac{3}{4}A_{s_1}(t)^3 + \frac{3}{2}A_{s_1}(t)A_{s_2}(t)^2\right)\cos(\omega_0 t + \varphi_1(t) - \phi_{w_1}) \right.$$
$$+ \left(\frac{3}{4}A_{s_2}(t)^3 + \frac{3}{2}A_{s_2}(t)A_{s_1}(t)^2\right)\cos(\omega_0 t + \varphi_2(t) - \phi_{w_2})$$
$$+ \left(\frac{3}{4}A_{s_2}(t)A_{s_1}(t)^2\right)\cos(w_0 t + 2(\varphi_1(t) - \phi_{w_1}) - \varphi_2(t) - \phi_{w_2})$$
$$\left. + \left(\frac{3}{4}A_{s_1}(t)A_{s_2}(t)^2\right)\cos(w_0 t + 2(\varphi_2(t) - \phi_{w_2}) - \varphi_1(t) - \phi_{w_1}) \right] \tag{11}$$

where $A_{s_1}(t) = A_{w_1}A_1(t), A_{s_2}(t) = A_{w_2}A_2(t)$.

The real signal is a double-sideband signal. To analyze conveniently, we study one side of the spectrum, and the analytic signal of output is given by:

$$S_{\text{out}}(t) = a_1\left(A_{s_1}(t)e^{j(\omega_0 t+\varphi_1(t)-\phi_{w_1})} + A_{s_2}(t)e^{j(\omega_0 t+\varphi_2(t)-\phi_{w_2})}\right)$$
$$+a_3\left[\left(\frac{3}{4}A_{s_1}(t)^3 + \frac{3}{2}A_{s_1}(t)A_{s_2}(t)^2\right)e^{j(\omega_0 t+\varphi_1(t)-\phi_{w_1})}\right.$$
$$+\left(\frac{3}{4}A_{s_2}(t)^3 + \frac{3}{2}A_{s_2}(t)A_{s_1}(t)^2\right)e^{j(\omega_0 t+\varphi_2(t)-\phi_{w_2})}$$
$$+\left(\frac{3}{4}A_{s_2}(t)A_{s_1}(t)^2\right)e^{j(w_0 t+2(\varphi_1(t)-\phi_{w_1})-\varphi_2(t)-\phi_{w_2})}$$
$$\left.+\left(\frac{3}{4}A_{s_1}(t)A_{s_2}(t)^2\right)e^{j(w_0 t+2(\varphi_2(t)-\phi_{w_2})-\varphi_1(t)-\phi_{w_1})}\right] \quad (12)$$

According to expression in (12), whether the nonlinear term is the same as the original signal is related to the input signal. Under amplitude modulation, the desired signal can be represented as:

$$S_{\text{desire}}(t) = a_1\left(A_{s_1}(t)e^{j(\omega_0 t+\varphi_1(t)-\phi_{w_1})} + A_{s_2}(t)e^{j(\omega_0 t+\varphi_2(t)-\phi_{w_2})}\right) \quad (13)$$

All items of a_3 in expression (12), which are different from previous signals, are interference signals.

$$S_{\text{jam}}(t) = a_3\left[\left(\frac{3}{4}A_{s_1}(t)^3 + \frac{3}{2}A_{s_1}(t)A_{s_2}(t)^2\right)e^{j(\omega_0 t+\varphi_1(t)-\phi_{w_1})}\right.$$
$$+\left(\frac{3}{4}A_{s_2}(t)^3 + \frac{3}{2}A_{s_2}(t)A_{s_1}(t)^2\right)e^{j(\omega_0 t+\varphi_2(t)-\phi_{w_2})}$$
$$+\left(\frac{3}{4}A_{s_2}(t)A_{s_1}(t)^2\right)e^{j(\omega_0 t+2(\varphi_1(t)-\phi_{w_1})-\varphi_2(t)-\phi_{w_2})}$$
$$\left.+\left(\frac{3}{4}A_{s_1}(t)A_{s_2}(t)^2\right)e^{j(\omega_0 t+2(\varphi_2(t)-\phi_{w_2})-\varphi_1(t)-\phi_{w_1})}\right] \quad (14)$$

When it is not an amplitude modulation, the cubic term contains a part of desired signal component, and the desired signal can be represented as:

$$S_{\text{desire}}(t) = \left[a_1 A_{s_1} + a_3\left(\frac{3}{4}A_{s_1}^3 + \frac{3}{2}A_{s_1}A_{s_2}^2\right)\right]e^{j(\omega_0 t+\varphi_1(t)-\phi_{w_1})}$$
$$+\left[a_1 A_{s_2} + a_3\left(\frac{3}{4}A_{s_2}^3 + \frac{3}{2}A_{s_2}A_{s_1}^2\right)\right]e^{j(\omega_0 t+\varphi_2(t)-\phi_{w_2})} \quad (15)$$

The interference signal can be represented as:

$$S_{\text{jam}}(t) = a_3\left[\left(\frac{3}{4}A_{s_2}A_{s_1}^2\right)e^{j(\omega_0 t+2(\varphi_1(t)-\phi_{w_1})-\varphi_2(t)-\phi_{w_2})}\right.$$
$$\left.+\left(\frac{3}{4}A_{s_1}A_{s_2}^2\right)e^{j(\omega_0 t+2(\varphi_2(t)-\phi_{w_2})-\varphi_1(t)-\phi_{w_1})}\right] \quad (16)$$

According to expression in (14) and (16), on the condition of amplitude modulation, the power of interference signal mainly consists of the signal which has the same weight vector with the desired signal, and the power of interference signal has the same pointing direction with the desired signal.

However, on the condition of non-amplitude modulation, it is the weight vectors $w_1^{H2}w_2$ and $w_2^{H2}w_1$ that determining the pointing direction of interference signal power.

The analysis of the transmitted multi-beam signal based on the Hammerstein model

A complex envelope of the input signal which is the sum of two narrow-band signals can be expressed by:

$$x(n) = A_{w_1}A_1(n)e^{j\varphi_1(n)-\phi_{w_1}} + A_{w_2}A_2(n)e^{j\varphi_2(n)-\phi_{w_2}}$$
$$= r_1(n)e^{j\varphi_1(n)-\phi_{w_1}} + r_2(n)e^{j\varphi_2(n)-\phi_{w_2}} \qquad (17)$$

where $A_{w_1}, A_{w_2}, \phi_{w_1}$, and ϕ_{w_2} are the amplitudes and phases of the two weight vectors, and $A_1(n)$, $A_2(n)$, $\varphi_1(n)$, and $\varphi_2(n)$ are the amplitudes and phases of the two signals. $r_1(n) = A_{w_1}A_1(n)$ and $r_2(n) = A_{w_2}A_2(n)$.

After the signal passes the memoryless nonlinear model, the output signal is given by

$$y(n) = a_1 x(n) + a_3 x(n)|x(n)|^2$$
$$= (a_1 + a_3 r^2(n))x(n) \qquad (18)$$

where $r^2(n) = r_1(n)^2 + r_2(n)^2 + 2r_1(n)r_2(n)\cos[\varphi_1(n) - \phi_{w_1} - \varphi_2(n) + \phi_{w_2}]$.

If the input signals have constant envelopes, that is $r_1(n)=r_1$ and $r_2(n)=r_2$ then:

$$y(n) = \left[a_1 r_1 + a_3 r_1(r_1^2 + r_2^2)\right]e^{j\varphi_1(n)-\phi_{w_1}}$$
$$+ 2a_3 r_1^2 r_2 \cos\left[\varphi_1(n) - \phi_{w_1} - \varphi_2(n) + \phi_{w_2}\right]e^{j\varphi_1(n)-\phi_{w_1}}$$
$$+ \left[a_1 r_2 + a_3 r_2(r_1^2 + r_2^2)\right]e^{j\varphi_2(n)-\phi_{w_2}}$$
$$+ 2a_3 r_2^2 r_1 \cos\left[\varphi_1(n) - \phi_{w_1} - \varphi_2(n) + \phi_{w_2}\right]e^{j\varphi_2(n)-\phi_{w_2}} \qquad (19)$$

We notice that the first and third coefficients are time-independent complexes, so they can be regarded as original signals. However, the second and fourth coefficients are time-dependent complexes, which means that the information carried by the signal envelope has changed. Hence, they should be considered as interferential signals. The effect of nonlinearity on the transmitted signal can be measured by the power ratio of the original signal to interference.

The output of memoryless PA model is seen as the input of linear FIR filter. Assume that the memory depth is 2. Then, the output signal of memory PA model can be represented as:

$$z(n) = h_0 y(n) + h_1 y(n-1) \qquad (20)$$

According to the concept of phase modulation and frequency modulation, the phase difference of $\varphi(n-1)$ and $\varphi(n)$ is decided by the modulation signal. So, $\varphi(n-1)$ and $\varphi(n)$ are different information. Finally, Equation 10 also can be described as:

$$\begin{aligned}
z(n) = & h_0\left[a_1 r_1 + a_3 r_1(r_1^2 + r_2^2)\right] e^{j\varphi_1(n)-\phi_{w_1}} \\
& + h_0\left[a_1 r_2 + a_3 r_2(r_1^2 + r_2^2)\right] e^{j\varphi_2(n)-\phi_{w_2}} \\
& + h_1\left[a_1 r_1 + a_3 r_1(r_1^2 + r_2^2)\right] e^{j\varphi_1(n-1)-\phi_{w_1}} \\
& + h_1\left[a_1 r_2 + a_3 r_2(r_1^2 + r_2^2)\right] e^{j\varphi_2(n-1)-\phi_{w_2}} \\
& + 2h_0 a_3 r_1^2 r_2 \cos\left[\varphi_1(n)-\phi_{w_1}-\varphi_2(n)+\phi_{w_2}\right] e^{j\varphi_1(n)-\phi_{w_1}} \\
& + 2h_1 a_3 r_1^2 r_2 \cos\left[\varphi_1(n-1)-\phi_{w_1}-\varphi_2(n-1)+\phi_{w_2}\right] e^{j\varphi_1(n-1)-\phi_{w_2}} \\
& + 2h_0 a_3 r_2^2 r_1 \cos\left[\varphi_1(n)-\phi_{w_1}-\varphi_2(n)+\phi_{w_2}\right] e^{j\varphi_2(n)-\phi_{w_2}} \\
& + 2h_1 a_3 r_2^2 r_1 \cos\left[\varphi_1(n-1)-\phi_{w_1}-\varphi_2(n-1)+\phi_{w_2}\right] e^{j\varphi_2(n-1)-\phi_{w_2}}
\end{aligned} \quad (21)$$

According to the expression in (21), it can be seen that the first and second signals are still original signals. However, the third and fourth signals have different phases from the original signals, and the rest of the signals have become new signals which the amplitudes vary with time. That is to say, except for the first and second signals, the others are interferential signals.

SIMULATION

Consider a 16-element array with elements uniformly spaced on the line of distance equal to half wavelength. Two transmitted beams are assumed to be transmitted at angle 30° and −20°. Several null points are set at angle −10°, 10°, and 50°. One thousand sampling points are chosen. The transmitted signals are two linear frequency modulation signals with $s_1(t) = A_{s_1} e^{jKt^2} + n_1(t)$ and $s_2 = A_{s_2} e^{j2\pi Bt + jKt^2} + n_2(t)$, where $K=B/T$. $B=10$ M is the bandwidth of the signal and T is pulse period, $T=NT_s=N/f_s$, $N=1,000$ is the sampling numbers and $f_s = 3.01/2 \times 10^9$ Hz is the sampling frequency. $n_1(t)$ and $n_2(t)$ are system thermal noises. The transmitted signal before it is input to PA has an SNR=90 dB. Although the amplitudes, A_{s1} and A_{s2}, are the same, the two signals which have different carrier frequencies are orthogonal. The first coefficient of PA is $a_1 = 1$. The first tap coefficient of FIR is $h_0 = 1$ and the memory depth is 2. The beam pattern affected by the nonlinear PA is produced by a beamforming technique based on orthogonal projection algorithm. a_1/a_3 is set to measure the memoryless nonlinearity of PA and h_0/h_1 is used to measure the memory effect of PA. The effect of nonlinearity PA on the multi-beamforming at −20° can be shown by changing the two parameters. Figure 3 and Figure 4 show AM-AM and AM-PM variation curves versus different PA coefficients. According to these figures, it can be found that the linearity of $a_3 = -0.01 \times (2+j)$ is less than $a_3 = -0.01 \times (1+j)$.

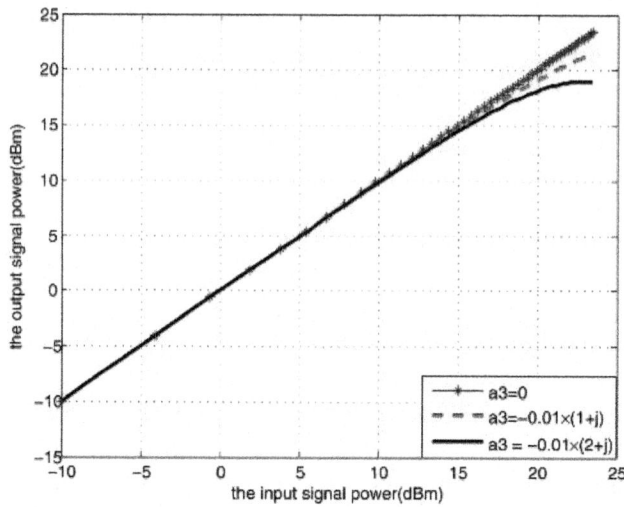

Figure 3: AM-AM conversion.

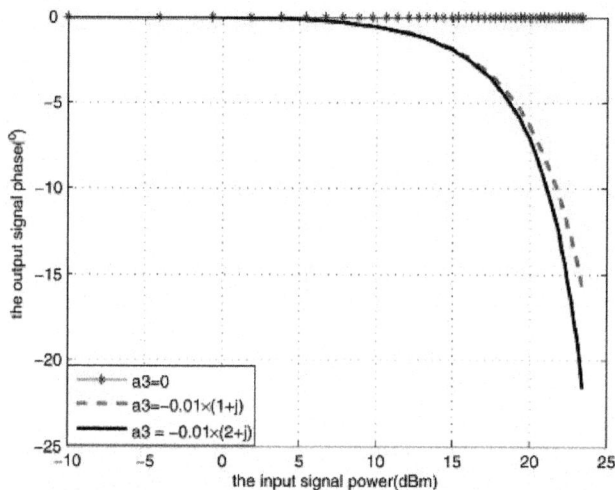

Figure 4: AM-PM conversion.

The power ratio of the original signal to the interference can show the effect of the nonlinearity of PA on the input signal. Figure 5 shows that when $h_1 = 10^{-2}$ and the input signal power is below 7 dBm, the power ratio keeps 40 dB with different nonlinearities. It can be explained that when the input signal power is small, the memory effect plays a dominant role. On the other hand, when the input signal power is 20 dBm, owing to the nonlinearity, the ratio of $a_3 = -0.01 \times (2+j)$ is more than $a_3 = -0.01 \times (1+j)$.

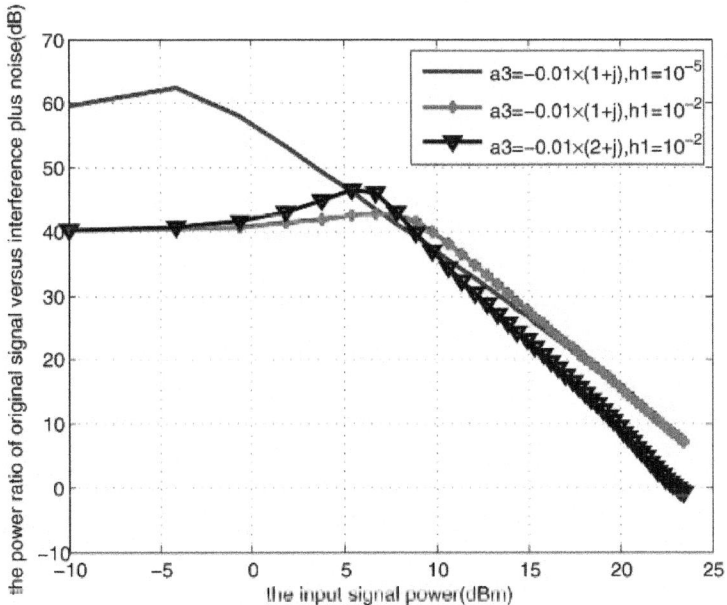

Figure 5: The power ratio of original signal versus interference plus noise of input signal.

Equation 9 shows that the equation value is significant only if a_3 and h_1a_1 have the same dimension, which means we can ignore the effect of h_1a_3 on beamforming. In addition, the nonlinearity represents gain compression characteristics. So, a_3 is negative and h_1 is positive. Under this premise, the total power of interference can be considered as a positive, when the memory effect plays a major role. The power of interference becomes weak as the memory effect is reduced. On the contrary, when nonlinearity plays a key role, the total power of interference is a negative. The power of interference increases as the memory effect is reduced. This is why the power ratio has an upward trend.

To measure the memoryless nonlinearity of PA, we assumed that $a_3 = -0.01 \times (x+j)$, $h_1 = 10^{-5}$ and $A_1 = 19.3$ dBm, where x is an independent variable. Figure 6 shows that when the input signal power A_1 is 19.3 dBm, the power ratio of the different h_1 is almost identical. However, when A_1 is 9.8 dBm, the power ratio of the different h_1 is different. The same conclusion can be found in Figure 6.

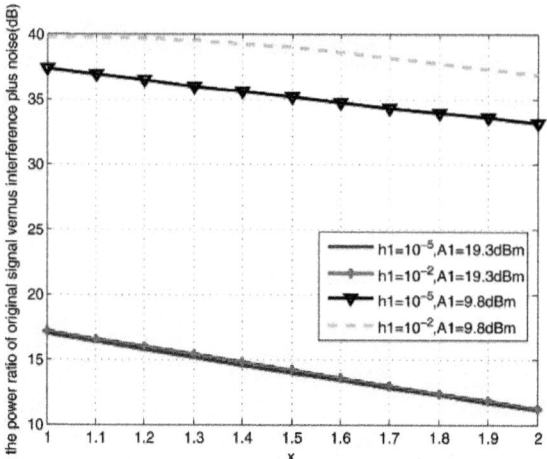

Figure 6: The power ratio of original signal versus interference plus noise with x.

$a_3 = -0.01 \times (1+j)$, $A_1 = 9.8$ dBm, and $A_1 = 19.3$ dBm are assumed to measure the memory effect of PA. From Figure 7, it can be seen that when the input signal power A_1 is 9.8 dBm, h_1 which is more than 10^{-4} will affect the ratio. However, it hardly affects the ratio when h_1 is less than 10^{-5}. On the other hand, when the power of input signal A_1 is 19.3 dBm, it is a turning point that h_1 is 10^{-2}. So, it is concluded that the low power of input signal is sensitive to the memory effect. The reason for the upward trend in Figure 7 has already been explained in Figure 5.

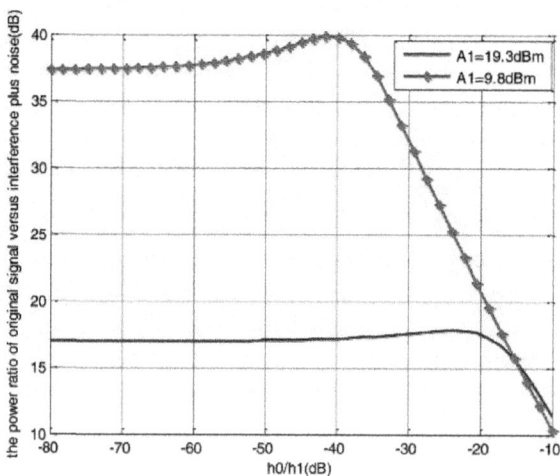

Figure 7: The power ratio of original signal versus interference plus noise with memory effect.

CONCLUSION

In this paper, starting from the signal form, the output signal of PA is obtained *via* the establishment of the PA model. Then the influence of the memory effect and nonlinearity of PA is presented by analyzing the composition of the final output signal. Lastly, computer simulations verify that the memory effect plays a key role, in the presence of the small power signal, while the nonlinearity plays an important role in the presence of the large power signal. A detailed theoretical basis and reference to linearize PA is provided in this paper for other researchers. When the transmitted signals are small power signals, the memory effect should gain attention in linearizing PA. However, if large power signals are transmitted, the importance of the nonlinearity of PA should be attached.

ACKNOWLEDGEMENTS

This research was supported by Applied Basic Research Programs of Sichuan Province (No. 2013JY0004), National Natural Science Foundation of China (No. 61371184) and the Fundamental Research Funds for the Central Universities (No. ZYGX2012J018).

REFERENCES

1. Ding L, Zhou GT, Morgan DR, Ma Z, Kenney JS, Kim J, Giardina CR: A robust digital baseband predistorter constructed using memory polynomials. *IEEE Trans. Commun.* 2004, 52(1):159-165. 10.1109/TCOMM.2003.822188

2. Isaksson M, Wisell D, Ronnow D: A comparative analysis of behavioral models for RF power amplifiers. *IEEE Trans. Microw. Theory Tech.* 2006, 54(1):348-359.

3. Ku H, Kenney JS: Behavioral modeling of nonlinear RF power amplifiers considering memory effects. *IEEE Trans. Microw. Theory Tech.* 2003, 51(12):2495-2504. 10.1109/TMTT.2003.820155

4. Bosch W, Gatti G: Measurement and simulation of memory effects in predistortion linearizers. *IEEE Trans. Microw. Theory Tech.* 1989, 37: 1885-1890. 10.1109/22.44098

5. Vuolevi JHK, Rahkonen T, Manninen JPA: Measurement technique for characterizing memory effects in RF power amplifiers. *IEEE Trans. Microw. Theory Tech* 2001, 49: 1383-1388. 10.1109/22.939917

6. Kim J, Konstantinou K: Digital predistortion of wideband signals based on power amplifier model with memory. *Electron. Lett.* 2001,

37(23):1417-1418. 10.1049/el:20010940

7. Ku H, Mckinley M, Kenney JS: Quantifying memory effects in RF power amplifiers. *IEEE Trans. Microw. Theory Tech.* 2002, 50(12):2843-2849. 10.1109/TMTT.2002.805196

8. Wu Y, Liu Y: The analysis of the effect of high power amplifier and bandwidth to the constant envelop guidance signal. *J Telemetry. Tracking. Command.* 2011, 32(3):14-20.

9. Kohls EC, Ekelman EP, Zaghloul AI, Assal FT: Intermodulation and bit-error ratio performance of a Ku-band multibeam high-power phased array. In *Proceedings of the IEEE International Symposium on Antennas and Propagation. Volume 3*. Piscataway: IEEE; 1995:1404-1408.

10. Hemmi C: Pattern characteristics of harmonic and intermodulation products in broadband active transmit arrays. *IEEE Trans. Antennas Propag.* 2002, 50(6):858-865. 10.1109/TAP.2002.1017668

11. Maalouf KJ, Lier E: Theoretical and experimental study of interference in multibeam active phased array transmit antenna for satellite communications. *IEEE Trans. Antennas Propag.* 2004, 52(2):587-592. 10.1109/TAP.2004.823900

12. Real EC, Charette DP: Non-linear amplifier effects in transmit beamforming arrays. In *Proceedings of the IEEE International Conference on Acoustics Speech Signal Processing. Volume 5*. Piscataway: IEEE; 1995:3635-3638.

Chapter 8

A HIGH DYNAMIC RANGE ULTRALOW-CURRENT-MODE AMPLIFIER WITH PICO-AMPERE SENSITIVITY FOR BIOSENSOR APPLICATIONS

Lei Zhang[1], Zhiping Yu[2], Xiangqing He[3]

[1]Graduated with honors from the Department of Electronic Engineering in Tsinghua University, Beijing, China

[2]Institute of Microelectronics, Tsinghua University, Beijing 100084, People's Republic of China

[3]BS degree from Tsinghua Univer- sity, Beijing, China

ABSTRACT

A novel ultralow-current-mode amplifier (ULCA) serving for on-chip biosensor signal pre-amplification in the integrated biosensing system (IBS) has been presented and verified in SMIC 0.18 μm CMOS technology by elaborately considering gain, bandwidth, noise, offset, and mismatch. The proposed ULCA solved the noise, bandwidth, and current headroom dilemma in the reported works, and can completely satisfy the specifications of IBS. It provides a current gain of 20 dB, 3 dB bandwidth of 7.03 kHz and input dynamic range of 20 bit, with only 1 nA of DC quiescent current, while the input offset current and noise current are less than 16.0 pA and 4.67 pArms, respectively.

INTRODUCTION

Biomedical technology emerges since the past century and is believed to be one of the most promising industries in the 21st century together with micro and nano-electronics industries. Recently, DNA molecule based biosensors are being reported by many famous literatures [1–3]. Naturally, the integrated biosensing system (IBS) which monolithically integrates the biomedical sensor arrays and ASICs such as ultralow-current-mode amplifier (ULCA),

ADC, and DSP in a single chip is avidly expected to be realized to greatly reduce the cost of common sensors used in the hospitals and markets. Actually, the proposed sensing schemes implemented on silicon-nanowire and golden surface [2, 3] are inherently compatible with modern CMOS process, however, the fastidious requirements as summarized in Table 1 of ultralow-current-mode operation and sensitivity (nA or sub-nA) for the following stages of ICs make great challenges to analog IC designers. Obviously, conventional transistor-saturation-based current-mode circuits are out of consideration due to the large noise background induced by the DC quiescent current and subthreshold-based current-mode circuits emerge to be the candidates [4, 5].

Table 1: Specifications of integrated biosensor

DNA releasing voltage	0.9 V
DNA modulation voltage	±0.4 V
Current headroom	±100 nA
Current sensitivity	~100 pA
Max. signal bandwidth	6 kHz
Dynamic range	10 bit
Temperature range	10–40°C

It is well known that since the subthreshold current appears an exponential function of the gate voltage in MOSFET, subthreshold circuits suffer from power fluctuations and process fluctuations between die to die seriously. Fortunately, they suffer little fluctuations on the same die [6] which makes it possible to realize ultralow-current-mode circuits by integrating all the circuit modules on a single chip [7, 8].

Some ULCA topologies have been reported in literatures [9–11], as shown in Fig. 1. The circuit in Fig. 1(a) uses a regulated current mirror to achieve current amplification whose quiescent currents are provided by current sources, while the bandwidth is limited by capacitor C0. However, since quiescent current I0 should be low enough to reduce the noise level and meet the requirement of sensitivity, the required current headroom (in both push and pull directions) can hardly be achieved. One can certainly use the complimentary topology in Fig. 1(b) to meet the headroom requirement and increase sensitivity by removing noise background introduced by quiescent current I0, however, at the cost of losing bandwidth on the low input cases.

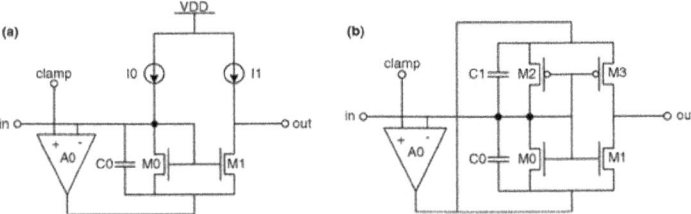

Figure 1: Conventional ULCA topologies.

To meet the specifications in Table 1, in this paper, a ULCA in achieving biosensing pre-amplification purpose has been proposed, characterized and verified by using SMIC 0.18 μm CMOS mixed signal technology. The results show that proposed ULCA can completely satisfy the requirements of IBS application, which makes it a promising candidate for the purpose of pre-amplification of biosensor signals.

CIRCUIT DESCRIPTIONS

The proposed topology of ULCA has been shown in Fig. 2. In this circuit, the current from DNA biosensors is input to a complimentary regulated current mirror composed of N type opamp AN0, P type opamp AP0, and transistor M0, M1, M3, M4, where it is amplified by a factor of 10 (20 dB). Opamp AN1, AP1, and M6–M11 compose a voltage limiter. Initially, due to the "virtual short" mechanism "in" is fixed at "clamp" by AN0 and AP0, and a quiescent current of I_{ref} is constructed in M0 and M3 by AN1 and AP1. When a push input is applied, V_{vn} comes down and so does the output of opamp AN1, thus turns off M2, and the current of M0 and M1 is being sinked by AP0, on the other hand, since V_{vp} is also prone to decrease, current provided by AN0 become smaller, while M5 is turned on by AP1 and compensates the current at node "vp", which limits V_{vp} from decreasing and assures the quiescent current of M0 and M3 exactly equals to the reference current I_{ref} provided by M6–M9 even if the input current I_{in} is much larger than the quiescent current I_{ref}. Similar conclusion can also be made for the pull input cases. It can be seen that M2 and M5 alternatively sustain a quiescent current of I_{ref} for the input current mirror, which in turn keeps a constant bandwidth when input varies between push and pull (positive and negative) directions. Furthermore, I_{ref} can be designed small enough to achieve the required sensitivity and noise level without restricted by the current headroom any more, since the current headroom is no longer depending on the magnitude of quiescent current of the input stage in the proposed ULCA. Therefore, the circuit can provide an extremely high sensitivity and large current headroom at the required bandwidth. Due to the variation and pad leakage issues, I_{ref} is unpractical to be provided off-chip. Therefore, in the design, the biasing stage composed of three steps of current mirroring (each step achieves a conversion factor of 0.1) realized by M12–M16 is introduced, from which a µA off-chip current is down converted by 1,000 times, thus relaxing these unwanted impacts. Capacitors C0 and C1 serve for frequency compensation and bandwidth confinement purposes in the circuit.

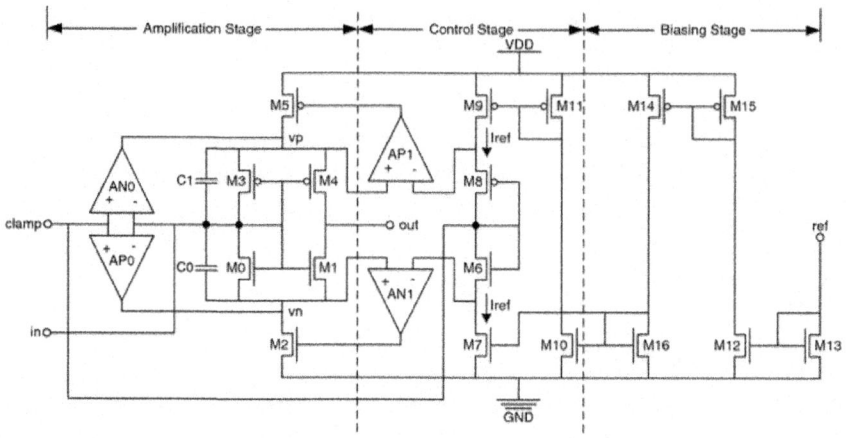

Figure 2: Circuit diagram of proposed ULCA.

The topologies of auxiliary N and P type opamp are shown in Fig. 3(a) and (b). In the circuit, M0, M1, M3, and M4 compose a differential input stage, and are chosen as large dimensions to improve matching and reduce offset, meanwhile, they are biased in their subthreshold regions for the purpose of noise reduction. Transistor M5 and M6 compose a trans-conductance output stage providing current for the following circuits. Capacitor Cn and Cp are serving for frequency compensation in the circuit, thus ensuring the AC stability.

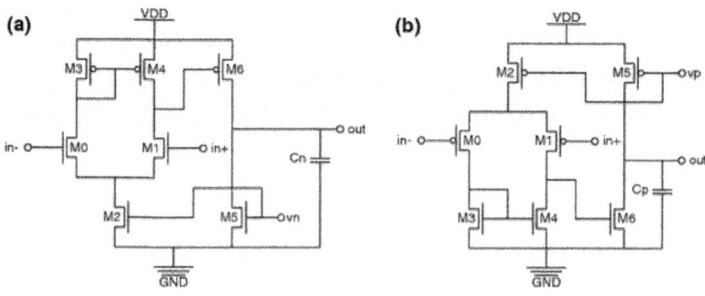

Figure 3: a The auxiliary N type opamp. **b** The auxiliary P type opamp.

CIRCUIT ANALYSIS

As found from Fig. 2, when a push input is applied, the conversion gain is provided by regulated current mirror composed of M0, M1, and opamp AP0, while M3–M5 and opamp AP1 are serving as current sources providing the quiescent current for the stage, which can be simplified to the circuit shown in Fig. 4(a). Complimentary discussion of the pull input case leads to the topology shown in Fig. 4(b).

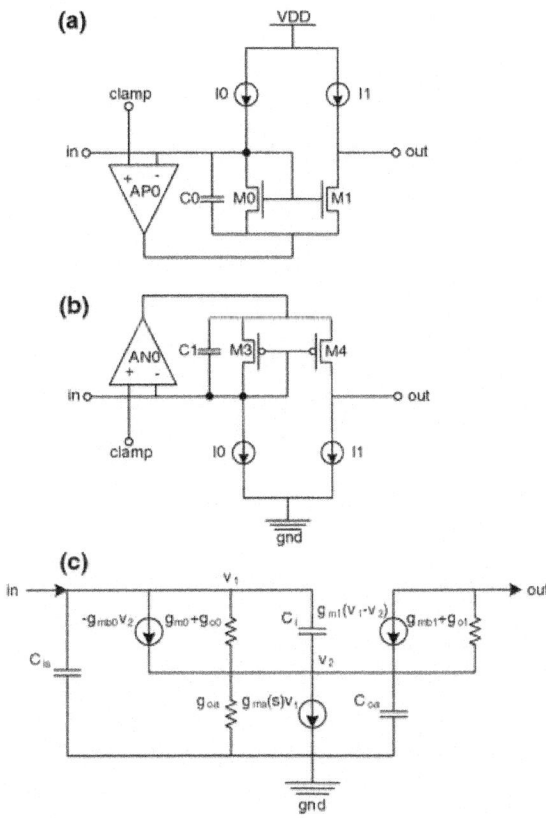

Figure 4: a Simplified circuit of ULCA when a push input is applied. **b** Simplified circuit of ULCA when a pull input is applied. **c** Small signal equivalent circuit of simplified ULCA.

AC Small Signal Analysis

The small signal equivalent circuit for AC analysis is shown in Fig. 4(c). In this circuit, the trans-conductance of opamp is modeled as $g_{ma}(s)$ considering delays introduced by parasitic capacitances of its internal nodes. As suggested by [12], $g_{ma}(s)$ can be written as:

$$g_{ma}(s) = g_{ma}\left(1 - \frac{s}{\omega_a}\right) \quad (1)$$

where g_{ma} is the DC trans-conductance and ω_a models the delay.

Detailed analysis of the equivalent circuit results in the characterization function of $as^2 + bs + c = 0$, where:

$$a = C_{ia}(C_i + C_{oa})\omega_a + C_i\left(C_{oa} - \frac{g_{ma}}{\omega_a}\right)\omega_a \quad (2)$$

$$b \approx C_{ia}(g_{m0} + g_{m1} + g_{oa})\omega_a + (g_{m0} + g_{o0})\left(C_{oa} - \frac{g_{ma}}{\omega_a}\right)\omega_a \quad (3)$$

$$c \approx g_{mb0}(g_{ma} - g_{m1})\omega_a + g_{m0}(g_{mb1} + g_{oa} + g_{ma})\omega_a \quad (4)$$

and the parameters are defined as follows:
- C_{ia}: Input capacitance of opamp.
- C_{oa}: Output capacitance of opamp (Cn or Cp).
- C_i: $C_{gs0} + C_{gs1} + C_0$ (or C_1).
- g_m: Trans-conductance of MOSFET.
- g_{mb}: Body trans-conductance of MOSFET.
- g_o: Output conductance of MOSFET.
- g_{oa}: Output conductance of opamp.

In order to maintain the AC stability, $a > 0$, $b > 0$, and $c > 0$ must be satisfied, resulting in the conditions of $g_{ma} > g_{m1}$ and $C_{oa} > g_{ma}/\omega_a$. Moreover, it can be found that provided the quiescent current I_{ref}, by adjusting the capacitance of C0 and C1, the bandwidth of proposed ULCA can be confined at the expected value.

Noise characterization

Generally, three kinds of noises are considered in CMOS circuit: thermal noise and shot noise, which are white noise, and flicker noise or $1/f$ noise. According to the subthreshold noise characterization in [6, 7] and the discussions in [11], the low frequency noises (flicker noise and some low frequency components of white noise) are being substantially canceled in the input node due to the quiescent current substraction and the symmetrical topology, resulting in that white noise appears more remarkable than flicker noise over the required bandwidth. Furthermore, white noise in the subthreshold MOSFET is basically contributed by shot noise, and the noise power density is given by: $S_I = 2qI$, where I is the DC current and q is the unit charge [6, 7].

The noise performance of proposed ULCA can be characterized by two noise sources, "vn" and "in" with the corresponding power densities of S_{vn} and S_{in}, which can be calculated as usual: by evaluating the output noise current with input open or shorted to ground and dividing by the gain. Meanwhile, considering the practical noises from V_{clamp} and I_{ref} sources, the simplified expressions are reported as:

$$S_{in} = 4qI_{ref}\left(1+\frac{1}{A}\right) + 4kTM^2G_S \tag{5}$$

$$S_{vn} = \frac{4qAI_{ref} + 4kTM^2A^2G_S}{g_{ma}^2} + 2S_{va} + 4kTR_S \tag{6}$$

where A is the current gain, S_{va} is the input–referred noise power density of opamp, $M = 0.001$ is the conversion factor of biasing current mirrors, R_S and G_S represent the source resistance and conductance of the voltage and current sources, k and T are the Boltzman constant and absolute temperature, respectively. In practical implementation, the values of R_S and G_S are 50 Ω and <10^{-7} s, and the noise voltage and current from the sources over the specified the bandwidth are 76 nVrms and 3.4 fArms, much smaller than the typical noises of opamp and transistors in the application, thus being insignificant for the overall noise performance. From (5) and (6), S_{in} and S_{vn} can be reduced by decreasing the input–referred noise of the opamp and the quiescent current I_{ref}, however, trading with the power consumption and the bandwidth of ULCA.

VERIFICATION AND DISCUSSION

The proposed ULCA is verified by using SMIC 0.18 μm CMOS mixed signal technology over the specified temperature range of 10–40°C at all process corners of TT, FF, SS, FS, and SF. Meanwhile, mismatching issues are also analyzed by Monte Carlo simulations over 30 different samples.

As stated in Sect. 3, the DC quiescent current of ULCA comes from the trade-off between noise and bandwidth requirement. From the simulation, the optimized quiescent current is given as $I_{ref} = 1$ nA, transistor channel length is chosen as $L = 1$ μm, and the aspect ratios (W/L) of basic N and P transistor cells are 10 and 20, respectively, while all the capacitances are designed as 1 pF.

Process Corners Analysis

Figure 5(a) shows the gain–bandwidth–product (GBW) and input–referred–noise–current (IRNC) as functions of DC input current I_{in} at all process corners, and the data at $I_{in} = 0$ is summarized in Table 2. It can be seen that within an input range of ±100 pA, GBW and IRNC virtually remain constant at 96.9 dBHz and 4.02 pArms, respectively, while increase monotonously for larger input levels. The minimal bandwidth is 7.03 kHz, a little bit larger than the specified 6 kHz, and the DC gain holds constantly at 20 dB. Meanwhile, GBW and IRNC are suffering little variations from different process corners, and the relative fluctuations are within ±0.3 and ±5.5%, respectively.

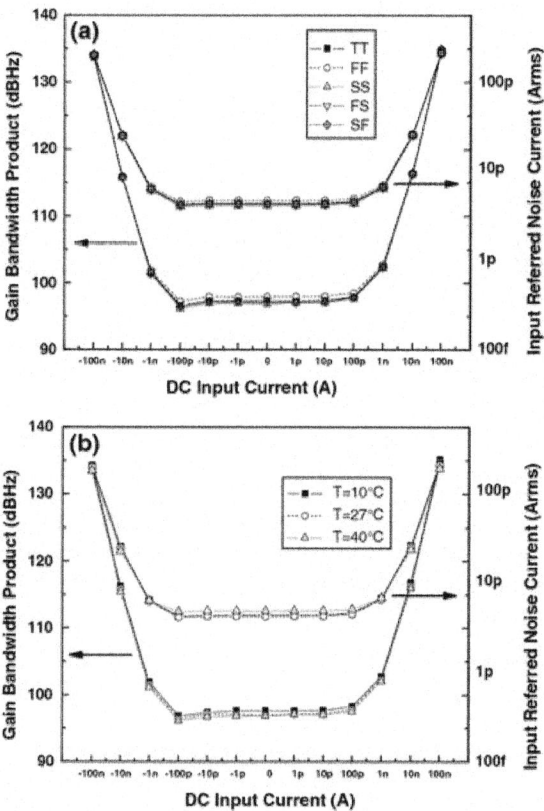

Figure 5: GBW and IRNC versus DC input current varying from −100 to 100 nA. **a** GBW and IRNC at various process corners. **b** Temperature dependencies of GBW and IRNC of proposed ULCA.

Table 2: Process corners data and temperature characteristics of GBW, 3 dB bandwidth, and IRNC at $I_{in} = 0$

	TT	FF	SS	FS	SF	10°C	27°C	40°C
GBW (dBHz)	96.94	97.99	96.85	97.41	97.13	97.60	96.94	96.86
3 dB bandwidth (kHz)	7.03	7.93	6.96	7.42	7.19	7.59	7.03	6.97
IRNC (pArms)	4.02	4.49	3.94	4.19	4.06	4.17	4.02	4.66

Moreover, for $-4.67\text{ pA} \leq I_{in} \leq 4.67\text{ pA}$ the input current is sinked by the noise floor, and the signal-to-noise-ratio (SNR) is less than 0 dB. When $|I_{in}|$ increases up to 100 pA (the sensitivity level of IBS specification) where the noise level still remains constant, SNR linearly ascends with $|I_{in}|$ to 27.5 ± 0.9 dB. For larger input current, the following equation holds until the input current mirror becomes saturated:

$$SNR \propto \frac{I_{in}}{\sqrt{(I_{in} + I_{ref})B}} \propto \frac{I_{in}}{I_{in} + I_{ref}} \rightarrow const. \qquad (7)$$

where B is the bandwidth of the circuit at the input current of I_{in}.

The SNR when the input current mirror becomes saturated can be formulated as follow considering the first order pole of the circuit:

$$SNR \propto \frac{I_{in}}{\sqrt{\sqrt{I_{in}}B}} \propto \sqrt{I_{in}} \qquad (8)$$

Equation (7) suggests that SNR tends to converge before the input current mirror saturates, which indicates that the increase of IRNC does not worsen the SNR. This has been verified by the linear part of IRNC curve in Fig. 5(a) when $|I_{in}| > 1\text{ nA}$. The saturation value of SNR is calculated as 54.6 ± 0.3 dB over the five corners. Along with the increase of I_{in}, the input current mirror becomes saturated, and the SNR should follows (8). However, since the second order conjugate pole gradually becomes dominant and extends the bandwidth as suggested by simulation, the increasing of SNR is slower than (8).

Temperature Dependencies

The temperature dependencies of GBW and IRNC as functions of DC input current I_{in} within the specified temperature range of 10–40°C are illustrated in Fig. 5(b), while the data at $I_{in} = 0$ is also summarized in Table 2. It is concluded from Fig. 5(b) that both the GBW and IRNC exhibit slightly larger dependencies on temperature with respect to process corners, and vary around ± 0.4 and $\pm 7.4\%$, respectively, within the required temperature range of 10–40°C.

Similar conclusion can be made for SNR as analyzed in the previous section, the SNR at $|I_{in}| = 100$ pA is 27.2 ± 0.9 dB, and the saturation value of SNR is 55.0 ± 0.8 dB within 10–40°C.

Mismatch considerations

In reality, the performances of practical circuit also suffers from process fluctuations due to transistor mismatches. To characterize these impacts, Monte–Carlo simulations on gain, 3 dB bandwidth, input–referred–offset (IRO), and IRNC of proposed ULCA are applied over 30 different samples. According to the design kits of SMIC 0.18 μm CMOS mixed signal technology, the standard deviations σ due to threshold voltage V_T and aspect ratio W/L mismatches are 1.85 mV and 0.5% for the applied transistor dimension. In the simulation, W, L, and V_T are randomly selected within the 3σ range at absolute Gaussian distribution, and are applied to the transistors. The results are shown in Fig. 6. It can be seen that the DC gain distributes from 19.8–20.1 dB, and 21 results of 3 dB bandwidth, IRO, and IRNC within the all 30 samples appear good consistency and randomly distribute in the ranges of 7.03–7.27 kHz, 14.3–16.0 pA, and 4.00–4.67 pArms, while other 9 samples are expressing large deviations.

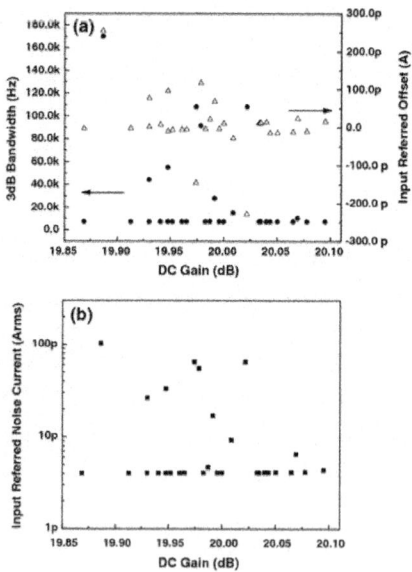

Figure 6: Monte–Carlo simulation results of DC gain, 3 dB Bandwidth, IRO, and IRNC considering device mismatches. **a** Monte–Carlo simulation of 3 dB Bandwidth and IRO as functions of DC gain over 30 samples. **b** Monte–Carlo simulation of IRNC as a function of DC gain over 30 samples.

Specifically, the bandwidth and IRNC deviations are mainly induced by the input–referred offsets V_{off} of opamp AN1 and AP1, and IRO deviation is due to the V_{off} of opamp AN0 and AP0, and mismatches of M0, M1, M3, M4. If the offsets of opamp AN1 and AP1 due to transistor mismatches are as large as hundreds millivolts, the quiescent current of M0 and M3 might considerably differ from I_{ref}, which in turn leads to the large deviations on the 3 dB bandwidth and IRNC. Similarly, if offsets of opamp AN0 and AP0, and mismatches of M0, M1, M3, M4 are considerably large, a large DC output current will be observed in the quiescent state, thus worsening the IRO of the circuit. For further improvement of circuit performances and yield, transistor M0, M1, M3, and M4 in Figs. 2, and 3(a) and (b) have to be carefully laid out or even larger areas are applied to these transistors in the design, however, trading with the bandwidth of the circuit.

The summarized circuit performances from post simulation are listed in Table 3. In the worst case, the input sensitivity is 20.7 pA, which determines the lower rail of input dynamic range, while the higher rail restricted by the maximal input current is 38.4 µA. Therefore, the proposed ULCA is capable of achieving an input dynamic range larger than 20 bit, thus satisfies the IBS specification. The SNR at I_{in} = 38.4 µA is 62.3 dB, roughly 10 dB lower than the value following the square-root law in (8) as a result of the second order conjugate pole of the circuit.

Table 3: Summary of post simulated circuit performances of proposed ULCA

DC gain	20 dB
3 dB bandwidth	7.03 kHz
Phase margin	>90°
Max. output current	0.384 mA
Input referred offset	−14.3 to 16.0 pA
Input referred noise current	4.02 pA
Input dynamic range	>20 bit
DC power dissipation	26.97 µW
Power supply	1.8 V
Layout area	220 µm × 80 µm

CONCLUSION

In this paper, a novel ULCA aiming at the application of signal pre-amplification in the integrated biosensing system has been proposed and verified by using

SMIC 0.18 μm CMOS technology. The proposed ULCA can completely satisfy the prescribed specifications of input current headroom, sensitivity, bandwidth, as well as input dynamic range for IBS applications. The proposed ULCA can also be used for ultralow current amplification in other types of biosensor interfaces, nanoscale device sensing, and optical sensing in the future.

ACKNOWLEDGMENTS

This work was supported by the National Science Foundation of China under Grant 60236020 and Grant 90307016, by a grant from Intel, and in part by a private research grant from Dr. D. Yang. The authors would like to thank Prof. Y. Chen of the Mechanical and Aerospace Engineering Department, University of California, Los Angeles, for the collaboration with his research team.

REFERENCES

1. Gregory Drummond, T., Hill, M. G., & Barton, J. K. (2003). Electrochemical DNA sensors. *Nature Biotechnology, 21*(10), 1192–1199.CrossRef
2. Li, Z., Chen, Y., Li, X., Kamins, T. I., Nauka, K., & Williams, R. S. (2004). Sequence-specific label-free DNA sensors based on silicon nanowires. *NANO Letters, 4*(2), 245–247.CrossRef
3. Guiducci, C., Stagni, C., Zuccheri, G., Bogliolo, A., Beninil, L., Samor, B., & Ricc, B. (2002). A biosensor for direct detection of DNA sequences based on capacitance measurements (pp. 479–482). *ESSDERC 2002*.
4. Mead, C. (1989). *Analog VLSI and neural systems*. Reading, MA: Addison Wesley.MATH
5. Zhang, L., He, X., & Yu, Z. (2007). Design and implementation of ultra low current sensing amplifer with pico-ampere sensitivity aiming at biosensor applications. *Chinese Journal of Electronics, 2*, 247–251.
6. Linares-Barranco, B., & Serrano-Gotarredona, T. (2003). On the design and characterization of femto-ampere current-mode circuits. *IEEE Journal of Solid-State Circuits, 38*(8), 1353–1363.CrossRef
7. Linares-Barranco, B., Serrano-Gotarredona, T., Serrano-Gotarredona, R., & Serrano-Gotarredona, C. (2004). Current mode techniques for sub-pico-ampere circuit design. *Analog Integrated Circuits and Signal Processing, 38*, 103–119.CrossRef
8. O'Halloran, M. O., & Sarpeshkar, R. (2004). A 10-nW 12-bit accurate analog storage cell with 10-aA leakage. *IEEE Journal of Solid-State Circuits, 39*(11), 1985–1996.CrossRef

9. Serrano-Gotarredona, T., Linares-Barranco, B., & Andreou, A. G. (1999). Very wide range tunable CMOS/bipolar current mirrors with voltage clamped input. *IEEE Transactions on Circuits and Systems-I: Fundamental Theory and Applications, 46*(11), 1398–1407.CrossRef
10. Ramirez-Angulo, J., Carvajal, R. G., & Torralba, A. (2004) . Low supply voltage high-performance CMOS current mirror with low input and output voltage requirements. *IEEE Transactions on Circuits and Systems-II: Express Briefs, 51*(3), 124–129.CrossRef
11. Steadman, R., Vogtmeier, G., Kemna, A., Quossai, S. E. I., & Hosticka, B. J. (2006). A high dynamic range current-mode amplifier for computed tomography *IEEE Journal of Solid-State Circuits, 41*(7), 1615–1619. CrossRef
12. Linares-Barranco, B., Rodriguez-Vázquez, A., Huertas, J. L., & Sánchez-Sinencio, E. (1992). On the generation design and tuning of OTA-C high frequency sinusoidal oscillators. *IEE Proceedings of Pt G Circuits, Devices and Systems, 139*(5), 522–528.

Chapter 9

A 12-BIT TRACK AND HOLD AMPLIFIER FOR GIGA-SAMPLE APPLICATIONS

Francesco Cannone and Gianfranco Avitabile

Politecnico di Bari, Bari, Italy

ABSTRACT

The paper presents a track-and-hold amplifier (THA), based on the switched emitter follower topology, suitable for emerging receivers architectures and data acquisition systems. The THA exploits four concurrent techniques, all described in the paper, which allow to reduce the hold-mode feedthrough; to attenuate the differential droop rate; to improve the linearity of the input buffer; and to optimize the third order harmonic distortion for RF sampling operation. The effectiveness of this novel approach is demonstrated using a low cost 0.35 μm SiGe technology. The THA core draws about 145 mA from 3.5 V supply. The THA provides a spurious-free dynamic range (SFDR) better than 72 dB when it is used for sampling an incoming signal of 0.9 V_{pp} centered around 925 MHz at a sampling rate of 0.5 GS/s. The THA allows a max sampling frequency equal to 6 GS/s and a max input frequency equal to 2.5 GHz and provides a SFDR better than 50 dB in all the available working conditions.

INTRODUCTION

In the last decades, the introduction of a number of different telecommunication standards and the corresponding growth of consumer markets strongly pushed forward both technologies and architecture developments for the transceiver analog front end. The cost effectiveness and the design reusability of the transmitter and receiver architecture became a key issue and the approaches based on the software radio (SR) paradigm [1] and RF sampling [2] begun to furnish viable solutions even for consumer applications. The acquisition and successive quantization of the received signal, indeed, allow a numerical manipulation of the signal with evident advantages in terms of re-configurability and adaptability of the receiver to different standards, most of all, when these operations are performable over a wide RF input frequency range covering several telecommunication standards.

The SR exploitation is strongly subject to the availability of high frequency ADCs and DSPs, which have to be still improved in terms of both speed and accuracy to be adequate to the challenge. At the state-of-the-art, the receiver architecture still relies on a preliminary down conversion to match the available ADCs and DSPs performances. As an alternative to the mixer, the sample & hold (S/H) mixer can perform the down conversion operation with comparable performances [2]. The resulting structures are usually referred as software defined radio (SDR) receivers.

An efficient reduction of the conversion rate of the sampling frequency is a key issue in the receiver in that it relaxes the DSP requirements. This goal can be achieved using various techniques, such as analog-decimation and sub-sampling ADCs, and their effectiveness strongly relies on the THA architecture. As a matter of facts, multi-standard operation, resolution and flexibility requirements call for high sampling frequencies and high resolution track and hold amplifiers (THAs). The THAs architectures fall in two main categories, open-loop and closed-loop architectures. Open-loop structures allow high-speed operations while closed-loop structures assure better accuracy. Different technologies lead to different THA solutions.

The use of advanced and, thus, expensive processes furnishes the means to satisfy speed and bandwidth requirements.

Recent literature reports InP THA operating up to 70 GS/s [3, 4]. Usually, a CMOS ADCs requiring a two-chip solution extremely difficult to package while preserving high-speed performance follows the InP THAs. In [5] is reported a THA in an InP BiCMOS process. The SiGe BiCMOS technologies are promising and cheaper alternatives which allow the use of high-f_T HBTs with CMOS devices even if these HBTs operate at lower frequencies than InP HBTs. Moreover, the literature reports high sampling frequency THAs fabricated in advanced RF-CMOS technologies [6].

Concurrently with the improvements achieved by the technologies, THA architectures must be improved in order to guarantee a suitable equivalent number of bits (ENOB). The most popular architectures are based on the diodes-bridges and switched emitter follower (SEF) stages. Usually the diode-bridge architecture requires higher supply voltage than the SEF architecture that, however, normally consumes more power.

The paper introduces a THA architecture whose core is a SEF sampling scheme which is the preferred choice when high sampling rates ($f_s > 0.5$GS/s) in conjunction with medium-to-high resolution performances (≥ 8 bit) are required [7–15]. The proposed THA allows to increase the ENOB at high sampling rates. To achieve this result is mandatory to attenuate the effects of all the non-idealities of the THA. Four techniques are concurrently used to this purpose. The first technique described in [15] is able to minimize the differential droop rate, which is one of the key point in SEF architectures. The second technique aims to minimize the harmonic distortions by using the technique presented in [16]. The third technique gives an improved linearity and is based on an input buffer topology described in [17, 18], while the fourth technique introduced in [18] allows an advanced hold-mode feedthrough cancellation. The process independence of the reported techniques allows the porting of the THA in advanced processes when higher performances are required.

The paper is organized as follows: in Sect. 2 the proposed THA is presented and analyzed and in Sect. 3 the prototype measurements are discussed and compared with other THAs operating at GHz sampling rates and medium-to-high resolution.

PROPOSED THA

The open-loop THA has a pseudo-differential structure composed by two similar branches. The pseudo-differential topology allows to reduce the effect of the even order distortion components coming from all the "common-mode" non-idealities, like even order buffering distortion, common-mode pedestal errors and so on.

As depicted in Fig. 1, the THA is composed by three main blocks: the input buffer, the sampling circuit and the output buffer. The input signal is buffered, sampled by the SEF and, then, fed to the load via the output buffer. Two secondary blocks, the inter-stage buffer and the auxiliary amplifier, allow to minimize the effects of the signal feedthrough on the hold capacitor, C_H, during the hold phase.

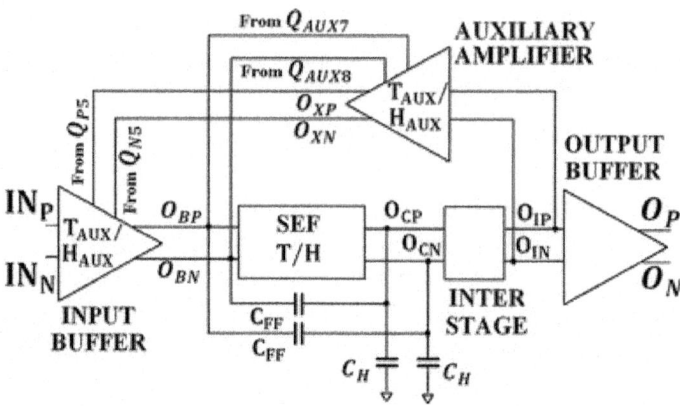

Figure 1: THA architecture.

In Fig. 2 the simplified schematic of the input buffer, the inter-stage buffer and the SEF stage are sketched.

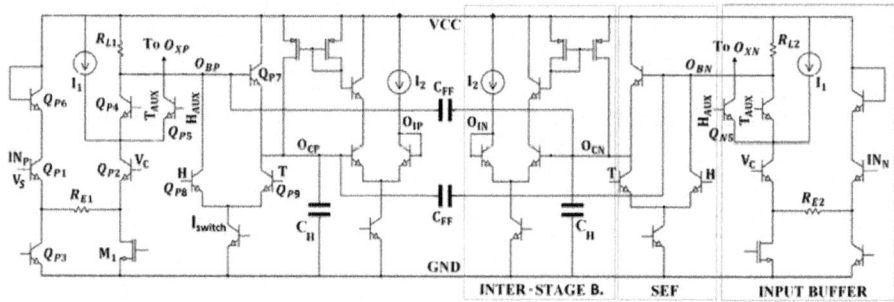

Figure 2: Simplified schematic of Input buffer, inter-stage buffer and SEF stage.

The THA combines the techniques presented in [15–18]. The THAs in [15, 16] rely on the same topology and the same technology. Differently from the proposed THA, these circuits have an emitter degenerated cascode amplifier, shown in Fig. 3(c), as input buffer. Moreover, they do not use the inter-stage and auxiliary amplifier for reducing the hold-mode feedthrough, but a clamping diode which reduces the gain of the input amplifier during the hold mode. These THAs adopt the same output buffer topology and the HD3 reduction technique as the THA described in this paper.

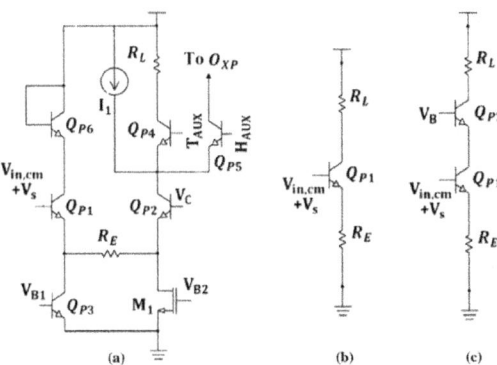

Figure 3: Single ended version for: **a** adopted input buffer, **b** emitter degenerated amplifier, **c** emitter degenerated cascode amplifier.

The THAs in [15, 16] achieve high SFDR but a lower maximum sampling frequency when compared to the present THA, even being implemented in a much more performant and expensive technology. The improvements introduced in the THA presented in this paper depend on the improved control of the SEF operations given by the input buffer depicted in Fig. 3(a) and the auxiliary amplifier, if compared to the emitter degenerated cascode amplifier with the clamping diode of [15, 16].

The THA in [18] adopts the input buffer introduced in [17] and shares with the proposed THA the same technology process, the input buffer and auxiliary amplifier topologies but it uses a different output buffer and also a different sizing of many design parameters. The re-design and the substitution of the output buffer allowed to improve the performances of the THA in terms of SFDR and differential droop rate.

The Input Stage

The input buffer and SEF must be extremely linear during the tracking phase if an overall high linearity has to be achieved. The simplified schematic of the input buffer is depicted in Fig. 2. The input buffer is based on the scheme introduced in [18] which allows to design a degenerated common-emitter cascode amplifier with high linearity. This scheme is a modified version of the input buffer reported in [17]. The change consists in the replacement of the common base transistor of the original structure with the switching pair (Q_{P4}–Q_{P5} in Fig. 2). This modification allows to exploit the hold-mode feedthrough cancellation technique. In track mode the buffer has the same behavior of that in [17].

Normally for applications in the GHz range, the input buffer topologies are based on the degenerated common emitter amplifier. Some common choices are shown in Fig. 3(b), (c) in single-ended arrangements for sake of simplicity.

The buffer depicted in Fig. 3(a) assures distortion performances similar to the cascode (Fig. 3(c)) with the advantage of a voltage supply requirements much lower and similar to the simple degenerated common emitter amplifier (Fig. 3(b)). At high frequencies the simplified analytical model described in [17], confirmed by the simulations result, is able to predicts with a good approximation that the linearity of the circuits in Fig. 3(a), (c) are similar and better than that of Fig. 3(b).

The couple Q_{P1} and R_E creates a trans-conductance amplifier similar to Fig. 3(b), the difference is that the other terminal of R_E is connected to the emitter of Q_{P2} instead of the ground. Since Q_{P2} is strongly current biased in order to work at very high frequencies, its emitter shows a very low impedance thus resulting in a "virtual" ground for the couple Q_{P1}–R_E which, considering only the first order effects, thus, behaves as the circuit in Fig. 3(b). The current dependent by the input signal and generated by the trans-conductor Q_{P1}–R_E reduced by the amount of bias current pulled by M_1 is fed by Q_{P2} to the load R_L through Q_{P4}. The current source I_1 sketched in Fig. 3(a) is used to reduce the supply requirements by removing part of the current through Q_{P4}, and thus through R_L. Since the emitter of Q_{P4} is a very low impedance node, loading it with a PMOS current source does not produce a sensible reduction of both the bandwidth and linearity, given the resulting reduced voltage swing. Hence, Q_{P1}–R_E–Q_{P2}–M_1–Q_{P5} form the highly linear trans-conductance amplifier of an equivalent "conventional" cascode and Q_{P4} the common base transistor. In the proposed circuit, the only nodes subject to a high voltage swing are the base and emitter of Q_{P1} and the collector of Q_{P4}.

It can be noticed that the distortion of the circuit in Fig. 3(b) is significantly higher because it is mainly determined by the Q_{P1} base–collector junction which undergoes a signal whose amplitude is two times that of the transistors subjected to high swing in Fig. 3(a, c).

The circuit in Fig. 3(a) draws more current than the one in Fig. 1(c), resulting in a higher power consumption because the larger drawn current is not completely compensated by the reduction in the supply voltage. This is the main drawback of the structure [17].

The design of the input buffer is brought out using the distortion model reported in [17]. The frequency dependent component of the third order harmonic distortion HD3 can be expressed as:

$$|HD_{3,im}| \simeq j\omega \frac{V_S^2}{4} * \left\{ \frac{C_{\pi,Q_{P1}}}{V_T^2 T^3} \left(\frac{R_{B,Q_{P1}} + R_E}{R_E} \right) \right.$$
$$+ \left(\frac{g_{m,Q_{P1}}}{g_{m,Q_{P2}}} \right)^4 \frac{C_{\pi,Q_{P2}}}{V_T^2 T^3} \left(\frac{R_{B,Q_{P2}} + R_E}{R_E} \right)$$
$$+ \left(\frac{g_{m,Q_{P1}}}{g_{m,Q_{P4}}} \right)^3 \frac{C_{\pi,Q_{P4}}}{V_T^2 T^3}$$
$$\left. + \alpha \left[\frac{R_{B,Q_{P1}}}{R_E} \frac{C_{\mu,Q_{P1}}}{\left(1+\frac{V_{cb,Q_{P1}}}{V_j}\right)^2} + \frac{C_{\mu,Q_{P4}}}{\left(1+\frac{V_{cb,Q_{P4}}}{V_j}\right)^2} \right] \right\}, \quad (1)$$

where $\alpha = m_j(m_j + 1)/V_j^2$, m_j and V_j are process parameters determining the capacitance of the reversed biased base–collector junction, VSVS is the input signal voltage across the node IN_p, R_B is the base resistance, $T = g_{m,Q_{P1}} R_E$ is the loop gain, V_T is the thermal voltage. The load resistors $R_{L1,2}$ and the degeneration resistors $R_{E1,2}$ are 105 and 95 Ω, while the dc voltage at the node $IN_{N,P}$, V_C, T_{AUX} and $O_{BP,N}$ are 1.8, 1.3, 2 and 2.3 V respectively in track mode.

The SEF Stage

The sampling circuit is the classical SEF optimized for high sample rates and high resolution. As shown in Fig. 2, the three transistors Q_{p7}, Q_{p8}, Q_{p9} represent the core of the SEF fed by I_{switch}, and loaded by the hold capacitance, C_H. In track-mode, the signal T is high and Q_{p7} acts as an emitter follower while Q_{p9} is in a common base configuration and the voltage across C_H tracks the input signals. The emitter coupled Q_{p8} and Q_{p9} are used to switch the current, I_{switch}, from Q_{p7} to R_{L1} during the hold phase. The augmented current in R_{L1} during the transition from the track to hold mode pulls down the voltage across the base of Q_{p7}.

The values of R_{L1}, I_{switch} and C_H are the key points to design a high sampling frequency THA. The voltage drop value must be enough for the SEF operation but simultaneously must avoid the complete switching off of Q_{p7} as this would reduce the speed of the THA. To improve the bandwidth, the load resistor R_{L1} of the input amplifier has to be small to maintain small the time constant at the node, but this means that, to produce the required voltage drop, a current higher than the optimum value for sampling rate could be needed. Finally the value of I_{switch} which optimizes the f_{max} of Q_{p7} usually is not compatible with the current necessary to produce the suitable voltage drop.

The values of I_{switch} and C_H are the main parameters which determine the linearity of the THA provided by the SEF stage. The suitable value of I_{switch} determines the compliance with the target distortion performances for a given value of C_H. The proposed THA use the technique described in [16] to further improve the HD3. This solution is based on the existence of a minimum of the HD3 related to the value of C_H as shown in Fig. 4(a).

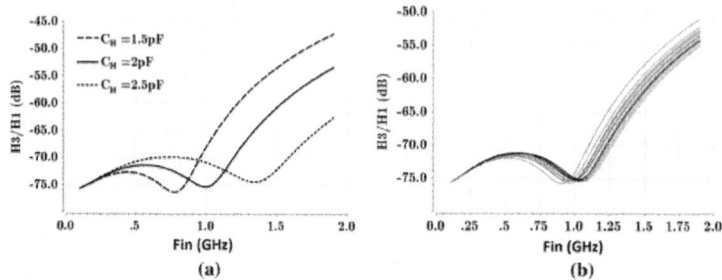

Figure 4: a Third harmonic versus hold capacitor. **b** Third harmonic versus process variations.

This effect is due to a compensation between two different distortion sources which have different slope and opposite sign. One of these sources is related to the distortion component proportional to ω^3 [19]. The imaginary part of the HD3 can be expressed as:

$$\text{IM}(\text{HD}_3) = \left| \left(\frac{\omega}{\omega_1}\right)^\alpha - \left(\frac{\omega}{\omega_2}\right)^3 \right| = 0, \qquad (2)$$

where ω_1, ω_2 and α ($\alpha < 3$) are fitting parameters extracted from the simulations, therefore (2) is a partial aid for the designers and in order to better manage this technique, an approximated analytical expression of the minimum should be useful. At the moment, the study on these distortion sources with different slope and opposite sign has not still provided this kind of equation. The basic idea is to size C_H for placing the minimum of HD3 close to the center band of the input signal. To verify the reliability of this technique over process variation, several Monte carlo simulations have been done. As shown in Fig. 4(b) the technique maintains its efficiency over process variation around the design point (1 GHz in this case).

Actually as mentioned before, the value of C_H directly affects the speed and linearity of the THA, then to exploit this technique a tradeoff is necessary.

The Output Buffer

As depicted in Fig. 5, the output buffer is obtained by the cascade of two voltage followers. These are optimized to reduce the differential droop-rate by using the technique described in [15]. The two cascaded stages have two different values of bias current, I_{ob1} and I_{ob2}. In the hold phase, the differential droop rate is one of the main factors limiting the resolution of the THA. This error has been mitigated by properly biasing the first stage of the output buffer.

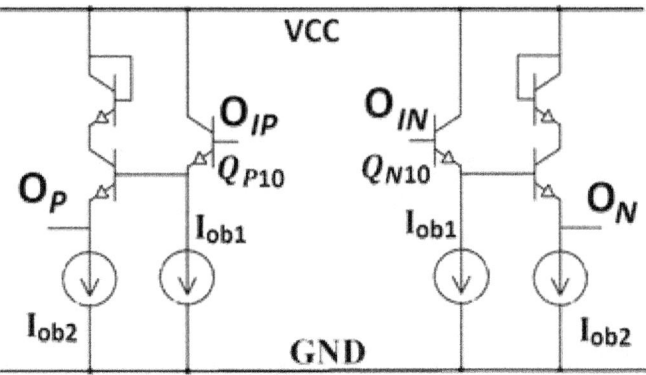

Figure 5: Simplified schematic of output buffer.

It is possible to show that the differential droop rate is mainly due to the mismatch between the base currents of Q_{P10} and Q_{N10} [20] which depends on the sampled value across C_H during the hold phase. Taking into account that the order of magnitude of the current difference is proportional to the bias current, I_{ob1}, this latter must be kept as small as possible. On the other hand, the THA must drive the capacitive load presented by the first stage of the ADC. For this reason, a second buffer biased with a higher value of current, I_{ob2}, has been added to achieve a satisfactory slew rate.

Figure 6 reports the differential droop rate with a single buffer, the second one, and with the configuration with two cascaded buffers which result to be 1.55 and 0.18 mV/ns respectively.

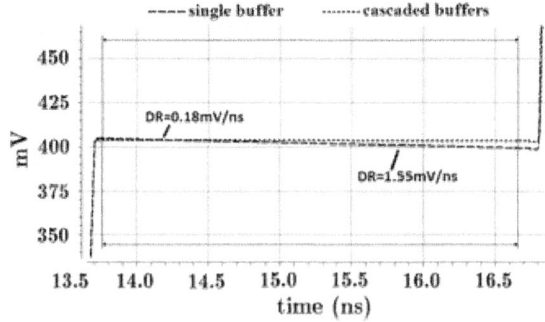

Figure 6: Droop rate in the case of a single buffer and cascaded buffers.

The currents I_{ob1} and I_{ob2} are 1.2 and 12 mA. This technique is simple and effective. Indeed the comparison evidences that the proposed buffer provides a differential droop rate, equal 0.18 mV/ns, almost ten times better than the other one confirming that this error is roughly proportional to the value of the base current.

The relatively low dropped value is also due to the high value of the hold capacitor which is optimized for the specific application.

The Hold-mode Feedthrough Reduction Technique

The signal feedthrough on the hold capacitor C_H during the hold mode is one of the most important source of degradation of the overall performances. To reduce the effect of this non-ideality, usually the feed-forward capacitor C_{FF} [21] is adopted as shown in Fig. 2, generally in conjunction with other circuital solutions.

Different techniques can be used to further reduce this sampling error. The solution used for the THA is based on the solution introduced in [18]. The basic idea for attenuating the feedthrough is to reduce the swing at the entrance of the SEF stages. To this purpose, the same approach described in [7] has been exploited. However, a drawback of that technique is the sensible differential pedestal which derives from the asymmetric common mode drop during the hold phase at the differential input of the SEF stage (O_{BP}, O_{BN} in Fig. 1). If this differential pedestal is not properly managed it introduces a significant sampling error. The proposed THA is capable to reduce the differential pedestal thanks to the use of the auxiliary amplifier depicted in Fig. 7.

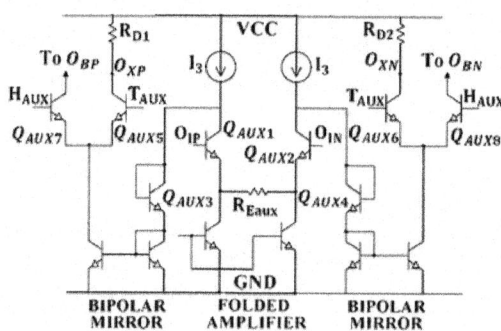

Figure 7: Auxiliary amplifier.

Indeed, to avoid the mentioned common mode drop it is sufficient that a proper copy of the input signal is present at the input of the SEF during the entire hold phase while the sampled value is stored on C_H. This is achieved by using the inter-stage buffer and the auxiliary amplifier. This latter tracks the differential voltage across C_H during the track-mode by means of the inter-stage buffer. During the transition from the track-mode to the hold-mode the input signal just before the sampling instant is restored at the differential input of the SEF stage by the auxiliary amplifier. To this purpose at the transition from track-mode to the hold-mode, the current flowing in Q_{P2} is switched from R_{L1} to R_{D1} through the couple Q_{P4} and Q_{P5} and the current flowing in R_{D1} is switched towards R_{L1} through the couple Q_{AUX5} and Q_{AUX7}. In this way, neglecting the effects of the layout parasitic and substrate coupling, the feedthrough is virtually completely removed. It is important to underline that this mechanism does not affect the SEF operation, indeed the sampling is properly realized switching-off Q_{P7} by steering the bias current from Q_{P7} to R_{L1}, until a fast and sufficient voltage drop at O_{BP} is assured.

EXPERIMENTAL RESULTS

The Prototype and the Measurement Set Up

The THA prototype has been designed using AMS 35 μm SiGe f_T—65 GHz technology. A photograph of the multifunction chip containing the THA is reported in Fig. 8. Its dimensions are about 0.65 × 1.4 mm². A QFN-64 package houses the chip and the test board used for measuring the THA is shown in Fig. 9. The THA absorbs 145 mA from 3.5 V supply voltage. Figure 10 depicts the test bench used to evaluate the THA prototype. A spectrum analyzer measures the output of the THA in both the track and hold phases while the input signal is fed by a signal generator to the DUT using a 1:1 balun mounted on the test board. The clock signal is equally fed in a balanced form with a similar balun. The output pins are loaded by a suitable voltage divider in order to reduce the effects of the artifacts produced by the board and bonding parasitic components. The chosen value for R is 1 kΩ.

Figure 8: Die photograph of the multi-function chip containing the prototype THA.

Figure 9: Photograph of PCB used for testing the THA.

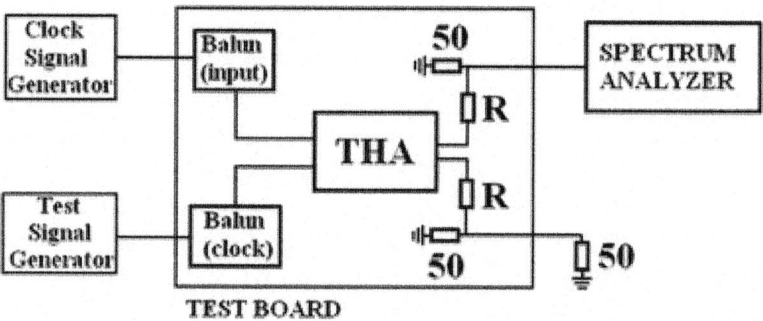

Figure 10: Measurement set up.

The spectrum analyzer (HP-E4407B) has a single ended input and, thus, only one of the DUT differential outputs is measured while the other one is loaded by a 50 Ω termination. The unbalanced measurement suffers from a second harmonic component which is higher than the real residual second harmonic that pseudo-differential THA is able to provide and that would be measured by a true balanced measurement. All the simulations show that the SFDR is dominated by the third harmonics, being the residual second harmonics always lower than the third harmonics. The single-ended measurement does not allow to exactly estimate the value of the residual second harmonics provided by the THA. Taking into account the simulations results and the single ended measurement limits, we have assumed that the real second harmonics are lower than the third harmonics and therefore we have eliminated them from the measurements of the SFDR. The DUT has been tested both in Track and T & H mode with variable clock frequencies and variable input frequency. The resolution bandwidth of the spectrum analyzer was set to its minimum value of 10 Hz in order to obtain the best noise floor that was around −130 dBm.

The Measurements

The THA has been designed to maximize the linearity for sampling a RF incoming signal center around 1 GHz in the band from 900 MHz to 1.1 GHz with the minimum distortion around 1 GHz under sub-sampling condition at a sampling rate of 0.5 GS/s. The effect of the distortion cancellation discussed in Sect. 2 and reported in [16] theoretically guarantees extremely linear operations for the THA. The measurements showed that the minimum is at 925 MHz closely matching the predicted value. This shift is mainly due to the process variations, indeed the value of this shift (75 MHz) is compatible with the Montecarlo simulations reported in Fig. 4. It is worth to note that the proposed technique even in presence of this shift is still effective since at 1 GHz the SFDR reaches the very good value of 70 dB.

Figure 11 reports the fundamental and 3rd harmonic at this frequency. As mentioned before, being the measured signal single ended, the second harmonic exists but it is has been not considered in the measurements since, as simulations show, the SFDR is set by the third harmonic thanks to the pseudo-differential topology of the THA. As can be clearly seen, the instrument noise floor masks the 3rd harmonic and the suppression is better than 74 dB, compatible with a 12 bit A to D quantization.

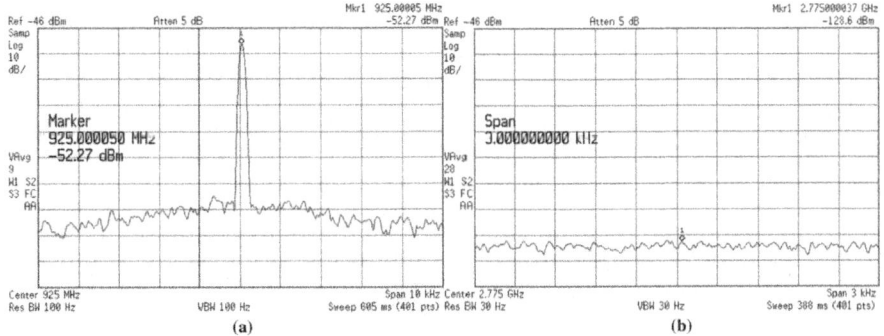

Figure 11: First (a) and third (b) harmonic @ f_{IN} = 925 MHz and f_s = 500 MS/s.

The THA has been tested with a set of sinusoidal input signals at the full scale range level (0.9 Vpp) in the 800 MHz to 1.2 GHz frequency range in 25 MHz steps. The THA outputs have been evaluated by the spectrum both in track and in track and hold mode with f_s equal to 0.5 GS/s.

In Fig. 12 the measured SFDR in this band is shown, the values have been reported in correspondence of the input frequencies. It is possible to note how the minimum of the distortion is almost centered around 1 GHz and there is only a small shift toward the lower frequencies. In the same Fig. 9 for comparison reasons also the simulated SFDR is reported.

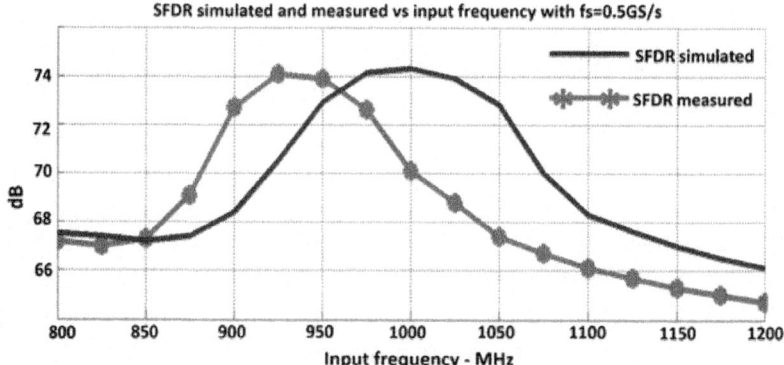

Figure 12: Measured and simulated SFDR for f_{IN} from 800 to 1200 MHz with f_s equal to 500 MS/s.

The measured Full Power Bandwidth is equal to 2.5 GHz and the measured differential droop rate is equal to 0.18 mV/ns.

The THA has been tested for different values of the input frequency from 50 MHz to the max input frequency equal to 2.5 GHz and different values of the sampling frequency from 0.5 GS/s to the max sampling frequency equal to 6 GS/s.

The linearity in track mode, HD3, is the parameter which mainly determines the resolution of the THA for sampling frequency close to 1 GS/s, while, for higher sampling frequency, the resolution is not still dominated only by the linearity in track mode but by all the other non-idealities. As a result, in the spectra of the sampled outputs it is no more present the maximum of the SFDR in correspondence of the minimum of HD3.

The SFDR observed at the output versus the sampling frequency is reported in Fig. 13 with an input frequency of 925 MHz and a sampling frequency ranging from 2 to 6 GS/s.

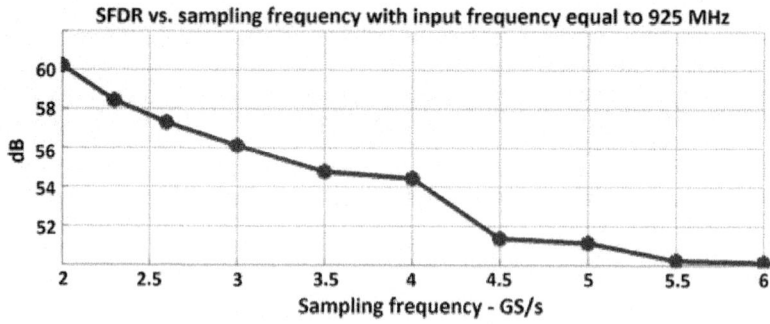

Figure 13: Measured SFDR for f_{IN} equal to 925 MHz versus f_s from 2 to 6 GS/s.

In Table 1 the comparison among THAs operating at GHz sampling rates and medium-to-high resolution reported in literature using different technologies is summarized.

Table 1: Comparison results of THAs operating at GHz sampling rates and medium-to-high resolution

References	FSR V_{pp}	Fin (GHz)	Fs (GHz)	SFDR (dB)	Max Fs (GHz)	Max Fin (GHZ)	Power dissipation	Technology	FOM (pJ/conv)
[7]	0.4	0.55	1.1	62	1.6	0.8	258 mW	0.18 µm—CMOS	5.01
[8]	2	1	10	49	10	3	800 mW	0.25 µm—SiGe	0.35
[9]	0.9	1.1	4	51	8	1.1	590 mW	0.5 µm—SiGe	0.25
[10]	0.8	0.4	2	54	2	0.9	550 mW	0.35 µm—SiGe	0.67
[11]	0.3	0.5	1	62	1	0.5	25 mW	0.8 µm—SiGe	0.02
[12]	0.3	0.5	2.5	45.8	2.5	0.5	n.a.	0.13 µm—CMOS	n.a.
[13]	0.7	2	1	45.1	1	15	n.a.	InP-DHBT	n.a.
[14]	0.35	0.1	0.4	62	0.5	1	164 mW	0.5 µm—SiGe	0.32
[15]	1	1	0.5	73	1.5	1.2	480 mW	0.25 µm—SiGe	0.09
This work	0.9	0.92	0.5	74	6	2.5	507 mW	0.35 µm—SiGe	0.05

For better comparing the reported THAs, the following figure of merit has been adopted, FOM = Power/(2^{ENOB} * Fs). The ENOB has been estimated starting from the available values of SFDR, in this way this FOM takes into account the power, the linearity and the speed being equal to the following ratio, FOM = Power/(Linearity * Speed).

It is possible to note that the proposed THA, properly adopting the mentioned techniques, is able to provide a better linearity than the other reported THAs, for the desired RF sampling. In this operating condition, the measured SFDR is similar to that reported in [15] which, however, uses a more expensive technology. Moreover, even when the proposed THA is used in operating conditions different from those corresponding to the optimum (that is f_{IN} from 900 to 1100 MHz and f_S < 1GS/s), it provides better performances than the other THAs in terms of max sampling frequency and max input frequency and similar in terms of SFDR, FSR and power consumption. Only the THA introduced in [11] presents a power consumption really lower than all the other THAs.

Looking at the FOM which represents a more complete characterization of the THAs than a single result like the power consumption, the performances provided by the THAs in [11, 15] and those described in this work are very close and are the best reported ones. As a matter of facts, given the technology, the higher is the sampling frequency the higher is the power consumption. In fact among the compared THAs those in [8, 9] and in that here presented have the higher maximum sampling frequency and the higher power consumption. In [11] a good optimization in terms of power saving has been proposed, nevertheless, the maximum sampling frequency is the lower one and this result in the lower power consumption.

In general, the increased bias currents and the supply voltage improve the speed but, also, the linearity and the power consumption. The proposed THA reaches a very high level of linearity matched only by the THA in [15] with a reasonable power consumption. This is evident looking at the Table 1, indeed aside the THA in [11] and [15], the FOM of the THA described in this work is sensibly better than all the other ones. This makes the proposed THA a suitable choice for high linearity and high speed sampling systems.

CONCLUSIONS

The work presented a THA operating up to 6 GS/s, with a FPBW equal to 2.5 GHz optimized for RF sampling suitable for advanced A to D stage. Measurements reported a SFDR better than 70 dB, with a 900 mVpp input signal in RF sampling operation and a SFDR better than 50 dB in all the other available sampling condition. The comparison highlighted that the THA introduces significant improvements into the state-of-the-art making it a suitable choice for advanced receivers architectures and data acquisition systems.

REFERENCES

1. Mitola, J. (1995). The software radio architecture. *IEEE Communications Magazine, 33*(5), 26–38.
2. Jakonis, D., Folkesson, K., Eriksson, P., & Svensson, C. (2005). A 2.4-GHz RF sampling receiver front-end in 0.18-μm CMOS. *IEEE Journal of Solid State Circuits, 40*(6), 1265–1277.
3. Daneshgar, S., Griffith, Z., Seo, M., & Rodwell, M. J. (2014). Low distortion 50 GSamples/s track-hold and sample-hold amplifiers. *IEEE Journal of Solid State Circuits, 49*(10), 2114–2126.

4. Deza, J., Ouslimani, A., Konczykowska, A., Kasbari, A., Godin, J., & Pailler, G. (2013). 70 GSa/s and 51 GHz bandwidth track-and-hold amplifier in InP DHBT process. *Electronic Letters, 49*(6), 388–389.

5. Gathman, T. D., Madsen, K. N., Li, J. C., Oh, T. C., Buckwalter, J. F. (2014). A 30 GS/s double-switching track-and-hold amplifier with 19 dBm IIP3 in an InP BiCMOS Technology. *Proceedings of IEEE International Solid-State Circuits Conference* (pp. 499–501).

6. Ma, S., Wang, J., Yu, H., Ren, J. (2014). A 32.5-GS/s two-channel time-interleaved CMOS sampler with switched-source follower based track-and-hold amplifier. *Proceedings of IEEE MTT-S International Microwave Symposium* (pp. 1–3).

7. Dinc, H., & Allen, P. E. (2009). A 1.2 GSample/s double-switching CMOS THA with 62 dB THD. *IEEE Journal of Solid State Circuits, 44*, 848–861.

8. Halder, S., Gustat, H., Scheytt, C. (2006). An 8 bit 10 GS/s 2 Vpp track and hold amplifier in SiGe BiCMOS technology. *Proceedings of ESSCIRC* (pp. 416–419).

9. Smola, D., Huijsing, J. H., Makinwa, K. A. A., Ploeg, H. V. D., Vertregt, M., Breems, L. (2006). An 8-bit, 4-Gsample/s track-and-hold in a 67 GHz f_t SiGe BiCMOS technology. *Proceedings of ESSCIRC* (pp. 408–411).

10. Vessal, F., & Salama, C. A. T. (2002). A bipolar 2-GSample/s track-and-hold amplifier (THA) in 0.35 pm SiGe techmology. *Proceedings of ISCAS, 5*, 573–576.

11. Boni, A., Parenti, M., & Vecchi, D. (2006). Low-power GS/s track-and-hold with 10-b resolution at nyquist in SiGe BiCMOS. *IEEE Transactions on Circuits and Systems II, 53*(6), 429–433.

12. Macedo, M., Roberts, G. W., Shih, I. (2012). Track and hold for giga-sample ADC applications using CMOS technology. *Proceedings of ISCAS* (pp. 2725–2728).

13. Bouvier, Y., Ouslimani, A., Konczykowska, A., & Godin, J. (2009). A 1-GSample/s, 15-GHz input bandwidth master-slave track-and-hold amplifier in InP DHBT technology. *IEEE Transactions on Microwave Theory and Techniques, 57*(12), 3181–3187.

14 Razzaghi, A., Chang, M. F. (2003). A 10-b, 1-GSample/s track-and-hold amplifier using SiGe BiCMOS technology. *Proceedings of CICC* (pp. 433–436).

15. Cannone, F., Avitabile, G., Coviello, G. (2013). A 11-bit track

and hold amplifier in 0.25 μm SiGe BiCMOS for RF sampling receivers. *Proceedings of MWSCAS* (pp. 792–795).

16. Cannone, F., Cascella, D., Avitabile, G., Coviello, G. (2012). A high bandwidth 11-bit 1.5 GS/s track and hold amplifier in 0.25 μm SiGe BiCMOS. *Proceedings of SMACD* (pp. 459–452).

17. Avitabile, G., Cascella, D., Cannone, F., & Coviello, G. (2012). Low distortion input buffer for high resolution GS/s rate track-and-hold amplifiers. *Electronic Letters, 48*(13), 755–757.

18. Cascella, D., Cannone, F., Avitabile, G., Coviello, G. (2012). A 2.5-GS/s 62 dB THD SiGe track-and-hold amplifier with feedthrough cancellation technique. *Proceedings of ICECS* (pp. 109–112).

19. Hoskins, M. J., Williams, D. R. (2003). High-speed SiGe HBT track-and-hold. *Proceedings of Instrumentation and Measurement Technology Conference* (pp. 1448–1453).

20. Baumheinrich, T., Pregardier, B., & Langmann, U. (1997). A 1-GSample/s 10-b full Nyquist silicon bipolar track&hold IC. *IEEE Journal of Solid State Circuits, 32*, 1951–1960.

21. Vorenkamp, P., & Verdaasdonk, J. P. M. (1992). Fully bipolar, 120-Msample/s 10-b track-and-hold circuit. *IEEE Journal of Solid State Circuits, 27*(7), 988–992.

Chapter 10

SYMBOLIC ANALYSIS OF ANALOG CIRCUITS CONTAINING VOLTAGE MIRRORS AND CURRENT MIRRORS

E. Tlelo-Cuautle[1], C. Sánchez-López[2], E. Martínez-Romero[3], Sheldon X.-D. Tan[4]

[1]INAOE, Tonantzintla, Mexico
[2]UAT, Apizaco, Tlaxcala, Mexico
[3]IMSE-CSIC, Sevilla, Spain
[4]University of California, Riverside, CA, USA

ABSTRACT

The pathological elements voltage mirror (VM) and current mirror (CM) have shown advantages in analog behavioral modeling and circuit synthesis, where many nullor-mirror equivalences have been explored to design and to transform voltage-mode circuits to current-mode ones and viceversa. However, both the VM and CM have not equivalents to perform automatic symbolic circuit analysis. In this manner, we introduce nullor-equivalents for these pathological elements allowing to include parasitics and to perform only symbolic nodal analysis. The nullor-equivalent of the CM is extended to provide multiple-outpus (MO-CM). Finally, two active filters containing VMs, CMs and MO-CMs are analysed to show the usefulness of the models.

E. Tlelo-Cuautle was Visiting Researcher at University of California Riverside during 2009–2010 under a CONACT sabbatical leave grant

INTRODUCTION

The nullator and norator elements are quite useful in analog design automation (ADA). For instance, the ideal behavior of the voltage follower (VF) can be modeled using nullators, with the aim to synthesize different VF topologies. Further, the VF can be evolved to synthesize the voltage mirror (VM) [35]. The norator is useful to model the behavior of the current follower (CF), and it can be superimposed with the VF to design more complex devices, e.g. current conveyors [38].

The VF, VM, CF and the current mirror (CM) form the four unity-gain cells (UGCs) [34]. Among them, the CM could be the most useful cell covering a wide range of applications [9, 13, 31, 33, 43], and also it can provide multiple-outputs (MO-CM). The four UGCs can be combined to model the behavior of already known and new active devices [3]. For example, the inverting properties of the VM and CM leads us to design inverting and positive-type current conveyors [1, 25, 26, 27], which properties can improve the ones provided by the four terminal floating nullor (FTFN) [12].

In symbolic analysis of analog and VLSI circuits, several methods are described in [7, 17]. In particular, the nullator and the norator are quite useful to perform symbolic analysis by only applying nodal analysis (NA) [11, 22, 39]. Besides, the symbolic NA method requires that all active devices be modeled using nullors [14, 22, 39].

The modeling of active devices possessing inverting characteristics can be done by using the VM and CM, as pathological elements [2, 20, 30, 42]. An important thing is that an analog circuit containing VMs and CMs also allows to transform circuit topologies or voltage-mode to current-mode circuits and vice-versa [5, 8, 10, 15, 28, 29, 41]. However, both the VM and CM can not be used in symbolic NA because their inverting characteristics imposse addition/subtraction limitations in the formulation process, i.e. there is not way to perform addition or deletion of columns or rows to preserve the inverting characteristics of either or both the VM and CM, as it is done in a nullor network [37]. In this manner, we are introducing nullor-equivalents for the VM, CM and MO-CM to take advantage of the symbolic NA of nullor networks [7, 37, 39].

SYMBOLIC NA METHOD

As already shown in [22, 37, 39], the main advantage of transforming an analog circuit to a nullor network is to apply only NA in the formulation process, and to obtain a reduced system of equations for operational amplifier based circuits [7, 37], and in general for circuits containing non-NA compatible elements.

The first step of the NA-formulation consists to model all circuit elements, such as: active devices, controlled sources and independent voltage sources using nullors [5, 14, 39]. The modeling process must include grounded admittances as much as possible, because they have only one entry in the NA formulation [37], while floating ones may have up to four entries requiring more computational work. The symbolic NA formulation method ($i = Yv$) can be summarized as follows:

Step 1: Describe the interconnection relationships of norators P_j, nullators O_j, and admittances by generating tables including names and nodes (m, n).

Step 2: Calculate the indexes associated to set row and set column, and group grounded and floating admittances:

   a. ROW: Contains all nodes ordered by applying the norator property which nodes (m, n) are virtually short-circuited. These indexes are used to fill vector i and the admittance matrix Y.

   b. COL: Contains all nodes ordered by applying the nullator property which nodes (m, n) are virtually short-circuited. These indexes are used to fill vector v and the admittance matrix Y.

   c. Admittances: They are grouped into two tables: Table A includes all nodes (ordered), and in each node is the sum of all admittances connected to it. Table B includes all floating admittances and its nodes (m, n).

Step 3: Use sets ROW and COL to fill vectors i and v, respectively. To fill Y: if in Table A a node is included in ROW and COL, introduce that admittance(s) in Y at position (ROW index, COL index). For each admittance in Table B, search node m in ROW and n in COL (do the same but search n in ROW and m in COL), if both nodes exist the admittance is introduced in Y at position (ROW index, COL index), and it is negative.

The solution of the formulation can be obtained by boolean logic operations [32], or by determinant decision diagrams [23, 40]. Elsewhere, we can formulate a much reduced system of equations in a nullor network, because it also allows us to apply circuit reduction methods [6, 16, 21, 24].

VM AND CM NULLOR-EQUIVALENTS

The pathological elements VM and CM are shown in Fig. 1 [2]. These representations are useful in circuit modeling [2, 20, 30, 42], circuit transformation [28, 29, 41], and circuit synthesis [18, 19] However, these representations do not allow to include parasitics and they can not be used within the symbolic NA method, because their inverting characteristics do not allow to perform operations as it is done for networks containing nullators and norators [7, 22, 37, 39]. In this manner, we introduce nullor-equivalents for the VM, CM and MO-CM to solve these problems.

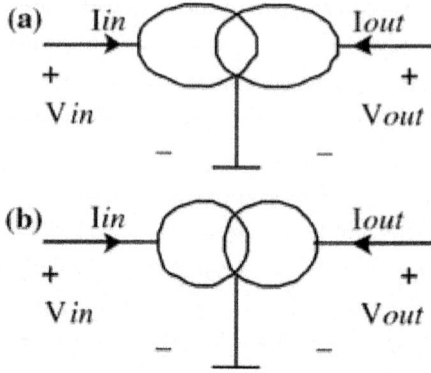

Figure 1: Representations of **a** VM and **b** CM.

Nullor-equivalent of the VM

Among all the possible combinations to generate the nullor-equivalent of the VM, we propose to use the one shown in Fig. 2. In this description Z_{in} and Z_{out} are connected in parallel to the input-port and in series to the output-port, respectively, and they model the input and output parasitics. The resistor of value $1/A_v$ is very useful in the NA formulation because the admittance becomes A_v, and it models the gain or tracking error of the VM. In the ideal case, $Z_{in} = \infty$, $Z_{out} = 0$, and $A_v =$ unity.

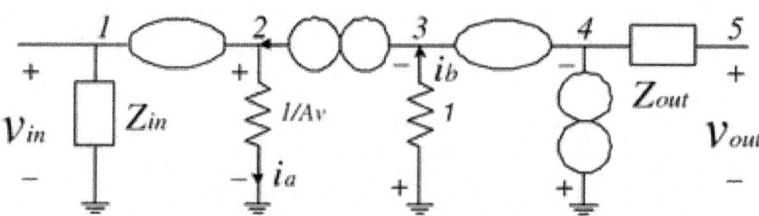

Figure 2: Nullor-equivalent of the VM including gain, input and output impedances.

From Fig. 2, $v_1 = v_{in}$, and since the voltage across a nullator is zero, $v_2 = v_1 = v_{in}$. Because the current through a nullator is zero, v_2 generates a loop-current through nodes 2, 3 and ground. That way, $ia = v2 1/Av = Avvin$ and $i_b = i_a = A_v v_{in}$. Now $v_3 = -1 \times i_b$ and $v_4 = v_3 = -A_v v_{in}$. Finally, we obtain (1). In an ideal VM $v_{out} = -v_{in}$.

$$V_{out}|_{open-circuit} = -A_v V_{in} \qquad (1)$$

Nullor-Equivalent of The Cm

Among all the possible combinations to generate the nullor-equivalent of the CM, we propose to use the one shown in Fig. 3. Indeed, by applying the adjoint network theorem to transform a voltage-mode circuit to a current-mode one [5, 8, 10, 15, 28, 29, 41], this topology is really the adjoint of the VM shown in Fig. 2. In this representation Z_{in} and Z_{out} are connected in serie to the input-port and in parallel to the output-port, respectively, and they model the input and output parasitics. Again, the resistor of value $1/A_i$ is very useful in the NA formulation because the admittance becomes A_i, and it models the gain or tracking error of the CM. In the ideal case, $Z_{in} = 0$, $Z_{out} = \infty$, and $A_i =$ unity.

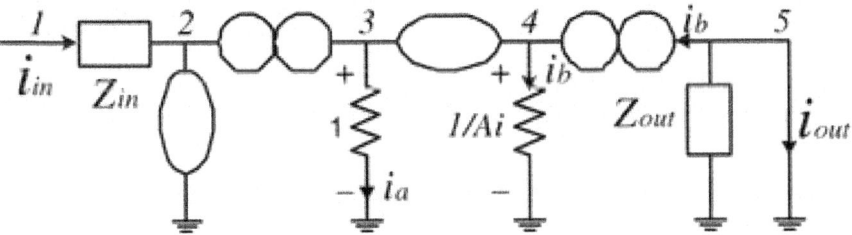

Figure 3: Nullor-equivalent of the CM including gain, input and output impedances.

From Fig. 3, the nullator connected at node 2 does not allow current to flow, so that a loop-current is formed through nodes 1, 2, 3 and ground. In this manner $i_a = i_{in}$, and $v_3 = 1 \times i_a = i_{in}$. The voltage across the nullator is zero so that $v_4 = v_3 = i_{in}$, this generates ib=v41/Ai=Aiiin. By applying Kirchhoff's current law: iout=−(ib+iZout). Finally, we obtain (2). In an ideal CM $i_{out} = -i_{in}$.

$$i_{out}|_{short-circuit} = -A_i i_{in} \qquad (2)$$

Nullor-equivalent of the MO-CM

An extention of Fig. 3, leads us to generate the nullor-equivalent of the MO-CM, as shown in Fig. 4. In this representation we are allowed to include independent gain and output impedance for each output n.

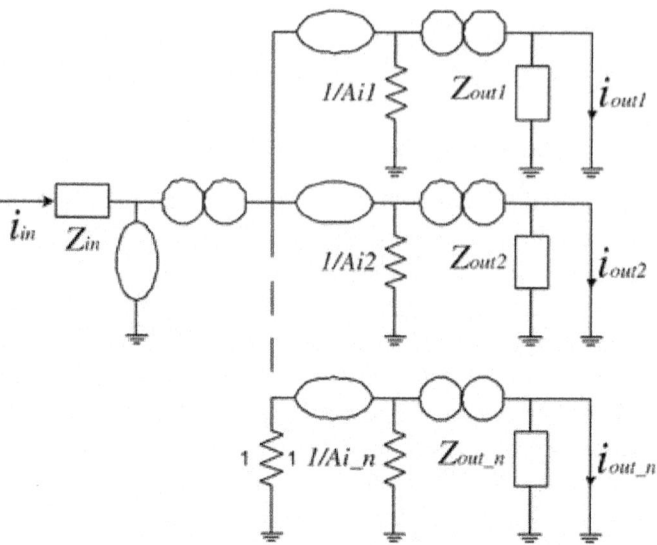

Figure 4: Nullor-equivalent of the MO-CM including input impedance and independent gain and output impedance for each output (n).

SYMBOLIC ANALYSIS OF ANALOG CIRCUITS CONTAINING VMS, CMS AND/OR MO-CMS

The nullor-equivalents of the VM, CM and MO-CMs can be used directly in symbolic NA of analog circuits, where the output of the VM is not connected to the input of the CM or MO-CM. Furthermore, when the circuit to be analysed contains a VM which output is connected to a norator or the input of a CM or MO-CM, we need to apply the superimposing method given in [38], to generate a nullor network containing the same number of nullators and norators, in order to apply the symbolic NA formulation.

Let's us consider the non-inverting and inverting low-pass filter using an inverting positive-type second generation current conveyor (ICCII+) [27]: Its representation using the VM and CM is shown in Fig. 5. As one sees, the output of the VM is connected to the input of the CM. By applying the superimposing method from [38], we obtain the nullor-equivalent network shown in Fig. 6. Basically, the impedance, nullator and norator connected at node 4 in Fig. 2, are superimposed with the impedance, nullator and norator connected at node 2 in Fig. 3. Other nullor-equivalents of inverting and non-inverting current conveyors can be found in [39], which also include parasitic impedances.

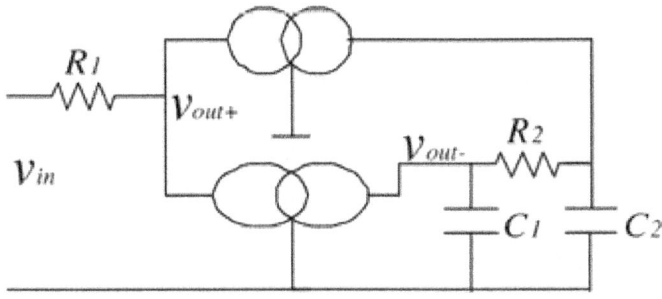

Figure 5: Non-inverting and inverting low-pass filter taken from [27].

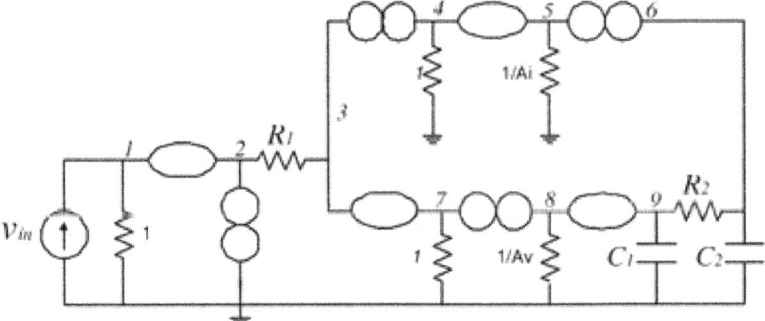

Figure. 6: Nullor-equivalent of Fig. 5.

Were we applying the modified nodal analysis (MNA) formulation to Fig. 6, the system of equations becomes order 6, because we need to introduce two stamps: one for the VM with a voltage-controlled voltage source and other for the independent voltage source. On the other hand, by applying the symbolic NA formulation, the order is reduced to 5, because the independent voltage source has been transformed to a current source [37], and the order becomes to be the number of nodes minus the number of nullator-norator pairs [7]. As a result, for large networks containing many VMs, the symbolic NA formulation is better than by applying the MNA method. For instance, by performing the NA formulation from Fig. 6, the following sets are obtained:

$ROW = \{(1), (3,4), (5,6), (7,8), (9)\}$
$COL = \{(1,2), (3,7), (4,5), (6), (8,9)\}$

The summation of the admittances at each node is presented in Table 1, while the floating admittances are given in Table 2. The NA formulation is given by (3).

$$\begin{bmatrix} v_{in} \\ 0 \\ 0 \\ 0 \\ 0 \end{bmatrix} = \begin{bmatrix} 1 & 0 & 0 & 0 & 0 \\ -G_1 & G_1 & 1 & 0 & 0 \\ 0 & 0 & A_i & G_2+sC_2 & -G_2 \\ 0 & 1 & 0 & 0 & A_v \\ 0 & 0 & 0 & -G_2 & G_2+sC_1 \end{bmatrix} \begin{bmatrix} v_{1,2} \\ v_{3,7} \\ v_{4,5} \\ v_6 \\ v_{8,9} \end{bmatrix} \quad (3)$$

Table 1: Admittances at each node

Node	Admittances
1	1
2	G1
3	G1
4	1
5	Ai
6	G2+sC2
7	1
8	Av
9	G2+sC1

Table 2: Floating admittances

Nodes	Admittance
(2,3)	G1
(9,6)	G2

The solution of this system taken node 9 as the inverting-output is given by (4). However, in the ideal case $A_v = A_i = 1$, and then (4) is reduced to the symbolic transfer function already provided in (5) [27].

$$v_9 = -\frac{A_i v_{in}}{A_v A_i + sR_1C_1 + sR_1C_2 + s^2R_1C_1R_2C_2} \quad (4)$$

$$H(s) = -\frac{1}{1 + sR_1(C_1+C_2) + s^2R_1C_1R_2C_2} \quad (5)$$

From this example, we can conclude on the usefulness of the proposed nullor-equivalents to perform symbolic NA which can be very suitable to enhance the synthesis procedures already introduced in [4,18, 19, 34, 36].

A second example is provided herein by analyzing a universal biquadratic filter using only dual-output CMs and grounded capacitors, it is taken from [33]. Its nullor-equivalent is shown in Fig. 7. More complex active filters based on CM arrays can be found in [31].

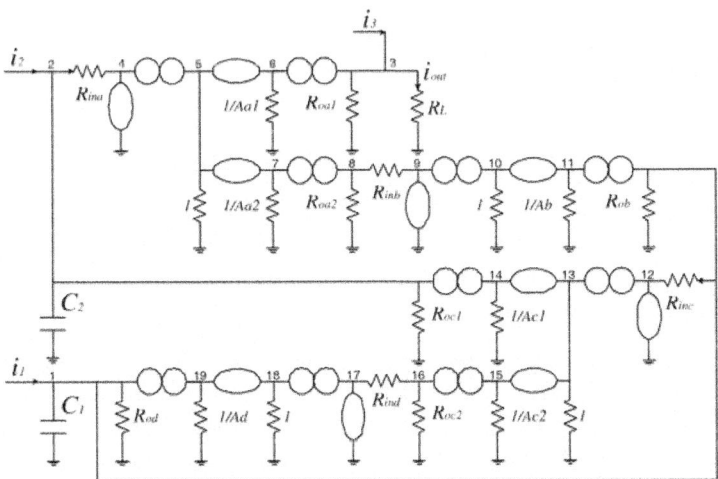

Figure 7: Nullor-equivalent of the current-mode filter taken from [33]

From the integrated circuit (IC) design point of view, a CM can provide multiple-outputs (MO-CM), but all of them with the same sign. Besides, when an output with an opposite sign is needed, another CM must be connected, as it is done in this example.

In this manner, the representation in Fig. 7 shows two dual-ouput CMs (DO-CM) labeled by letters a and c, they include the input resistance R_{ina} and R_{inc}, their gains A_{a1}, A_{a2} and A_{c1}, A_{c2}, and their output resistances R_{oa1}, R_{oa2} and R_{oc1}, R_{oc2}.

To invert the sign in one output of the DO-CMs, two CMs are also included labeled by letters b and d, they include the input resistance R_{inb} and R_{ind}, their gains A_b, A_d, and their output resistances R_{ob}, R_{od}.

By applying the symbolic NA method, the formulation generates a system of order 9 (19 nodes minus 10 nullors) [7, 22, 37]. The sets for the rows and cols become:

$ROW = \{(1, 11, 19), (2, 14), (3, 6), (4, 5), (7, 8), (9, 10),$
$\qquad (12, 13), (15, 16), (17, 18)\}$
$COL = \{(1), (2), (3), (5, 6, 7), (8), (10, 11), (13, 14, 15),$
$\qquad (16), (18, 19)\}$

The four floating resistances are associated to the input-resistance of the CMs, there are six gains and six output-resistances.

The exact symbolic expression has many symbolic product-of-terms. However, when the output-resistances equal to ∞, the reduced symbolic expression is given by (6). As one sees, only the input resistances R_{ina} and R_{inc} are included in the reduced symbolic expression, as already shown in [33]. When the gains are set to unity, (6) becomes (7).

$$I_{out} = I_3 + \frac{-sA_{a1}I_2C_1R_{inc} - A_{a1}I_2 + A_{a1}I_2A_{c2}A_d + A_{c1}A_{a1}I_1}{s^2R_{inc}C_1C_2R_{ina} + s(C_1R_{inc} + C_2R_{ina} - A_{c2}A_dC_2R_{ina}) + A_{c1}A_bA_{a2} - A_{c2}A_d + 1} \quad (6)$$

$$I_{out} = I_3 + \frac{I_1 - sI_2C_1R_{inc}}{s^2C_1C_2R_{ina}R_{inc} + sC_1R_{inc} + 1} \quad (7)$$

CONCLUSION

We introduced new nullor-equivalents to represent the pathological elements possessing inverting characteristics, they were the VM and CM, and we introduced also the nullor-equivalent for the MO-CM. These nullor-equivalents were used in the symbolic NA formulation of two active filters. From the results, it can be appreciated that our proposed nullor-equivalents are quite useful to perform symbolic NA to gain insight on the behavior of the circuits, because they include the input and output parasitics and gain. Furthermore, the generated expressions were the same as the ones already provided in the references, but we calculated first the exact symbolic expressions and then we reduced them by approximating ideal characteristics. In this manner, we conclude that our proposed nullor-equivalents are very suitable to perform symbolic NA which can be used within an ADA environment to enhance circuit modeling and synthesis methods, and to validate circuit-equivalents in the transformation of circuit topologies.

ACKNOWLEDGMENTS

This work is supported by: UC-MEXUS and CONACyT under grants CN-09-310 and 48396-Y; by Promep-Mexico under grant UATLX-PTC-088; by Consejeria de Innovacion, Ciencia y Empresa, Junta de Andalucia-Spain TIC-2532; and by the JAE-Doc program of CSIC co-funded by FSE, Spain.

REFERENCES

1. Awad, I. A., & Soliman, A. M. (1999). Inverting second generation current conveyors: The missing building blocks, CMOS realizations and applications. *International Journal of Electronics* 86(4), 413–432.

2. Awad, I. A., & Soliman, A. M. (2002). On the voltage mirrors and the current mirrors. *Analog Integrated Circuits and Signal Process 32*(1), 79–81.
3. Biolek, D., Senani, R., Biolkova, V., & Kolka, Z. (2008). Active elements for analog signal processing: Classification, review, and new proposals. *Radioengineering 17*(4), 15–32.
4. Cabeza, R., Carlosena, A., & Serrano, L. (1994). Unified approach to the implementation of universal active devices. *Electronics Letters 30*(8), 618–620.
5. Carlosena, A., & Moschytz, G. S. (1993). Nullators and norators in voltage to current-mode transformations. *International Journal of Circuit Theory and Applications 21*(4), 421–424.
6. Daems, W., Gielen, G., & Sansen, W. (2002). Circuit simplification for the symbolic analysis of analog integrated circuits. *IEEE Transactions on Computer-Aided Design of Integrated Circuits and Systems 21*(4), 395–407.
7. Fakhfakh, M., Tlelo-Cuautle, E., & Fernández, F.V. (2010). *Design of analog circuits through symbolic analysis*. United Arab Emirates: Bentham Sciences Publishers.
8. Garcia-Ortega, J. M., Tlelo-Cuautle, E., & Sánchez-López, C. (2007). Design of current-mode gm-c filters from the transformation of opamp-rc filters. *Journal of Applied Sciences 7*(9), 1321–1326.
9. Gupta, M., Aggarwal, P., Singh, P. et al. (2009). Low voltage current mirrors with enhanced bandwidth.*Analog Integrated Circuits and Signal Process 59*(1), 97–103.
10. Herencsar, N., & Vrba, K. (2007). Current conveyors-based circuits using novel transformation method.*IEICE Electronics Express 4*(21), 650–656.
11. Kumar P., & Senani R. (2002). Bibliography on nullors and their applications in circuit analysis, synthesis and design. *Analog Integrated Circuits and Signal Process 33*(1), 65–76.
12. Kumar, P., & Senani, R. (2007). Improved grounded-capacitor SRCO using only a single PFTFN.*Analog Integrated Circuits and Signal Process 50*(2), 147–149.
13. Le, H. B., & Lee, S. G. (2009). A 1-v low power gain boosted self-cascoding current mirror operational transconductance amplifier. *International Journal of Electronics 96*(9–10), 1005–1009.
14. Odess, L., & Ur, H. (1980). Nullor equivalent networks of nonideal operational-amplifiers and voltage controlled sources. *IEEE Transactions*

on *Circuits and Systems 27*(3), 231–235.
15. Palomera-Garcia, R. (2005). Generation of equivalent circuits by FTFN relocation. In*IEEE ISCAS 1-6*, 252–255.
16. Pierzchala, M., & Rodanski, B. (2001). Generation of sequential symbolic network functions for large-scale networks by circuit reduction two-port. *IEEE Transactions on Circuits and Systems I: Fundamental Theory and Applications 48*(7), 906–909.
17. Qin, Z., Tan, S. X. D., & Cheng, C. K. (2005). *Symbolic analysis and reduction of VLSI circuits*. New York, USA: Springer.
18. Saad, R. A., & Soliman, A. M. (2008). Generation, modeling, and analysis of CCII-based gyrators using the generalized symbolic framework for linear active circuits. *International Journal of Circuit Theory and Applications 36*(3), 289–309.MATH
19. Saad, R. A., & Soliman, A. M. (2008). Use of mirror elements in the active device synthesis by admittance matrix expansion. *IEEE Transactions on Circuits and Systems I: Fundamental Theory and Applications 55*(9), 2726–2735.
20. Saad, R. A., & Soliman, A. M. (2010). A new approach for using the pathological mirror elements in the ideal representation of active devices. *International Journal of Circuit Theory and Applications*.
21. Sánchez-López, C., & Tlelo-Cuautle, E. (2008). Novel SBG, SDG and SAG techniques for symbolic analysis of analog integrated circuits. In *SM2ACD*, pp. 17–22.
22. Sánchez-López, C., & Tlelo-Cuautle, E. (2009). Behavioral model generation of current-mode analog circuits. In *IEEE ISCAS*, pp. 2761–2764.
23. Shi, C. J. R., & Tan, S. X. D. (2000). Canonical symbolic analysis of large analog circuits with determinant decision diagrams. *IEEE Transactions on Computer-Aided Design of Integrated Circuits and Systems 19*(1), 1–18.
24. Shi, G. Y., Chen, W., & Shi, C. J. R. (2007). A graph reduction approach analysis. In Proceedings of the Asia South Pacific design automation conference (ASPDAC), pp. 197–202.
25. Sobhy, E. A., & Soliman, A. M. (2007). Novel CMOS realizations of the inverting second-generation current conveyor and applications. *Analog Integrated Circuits and Signal Process 52*, 57–64.

26. Soliman, A. M. (2007). Voltage mode and current mode tow thomas biquadratic filters using inverting CCII. *International Journal of Circuit Theory and Applications 35*(4), 463–467.MathSciNet
27. Soliman A. M. (2008). The inverting second generation current conveyors as universal building blocks. *AEU-International Journal of Electronics and Communications 62*(2), 114–121.
28. Soliman, A. M. (2009). Adjoint network theorem and floating elements in the NAM. *Journal of Circuits Systems and Computers 18*(3), 597–616.
29. Soliman, A. M. (2009). On the DVCC and the BOCCII as adjoint elements. *Journal of Circuits Systems and Computers 18*(6), 1017–1032. MathSciNet
30. Soliman, A. M. & Saad, R.A. (2010). The voltage mirror-current mirror pair as a universal element.*International Journal of Circuit Theory and Applications*. doi:10.1002/cta.596.
31. Souliotis, G., & Haritantis, I. (2008). Current-mode filters based on current mirror arrays. *International Journal of Circuit Theory and Applications 36*(2), 173–183.
32. Tan, S. X. D. (2006). Symbolic analysis of analog circuits by boolean logic operations. *IEEE Transactions on Circuits and Systems II: Analog and Digital Signal Processing 53*(11), 1313–1317.
33. Tangsrirat, W., & Prasertsom, D. (2007). Electronically tunable low-component-count current-mode biquadratic filter using dual-output current followers. *Electrical Engineering 90*, 33–37.
34. Tlelo-Cuautle, E., & Duarte-Villaseñor, M. A. (2008). Evolutionary electronics: Automatic synthesis of analog circuits by GAs chap. In *Success in evolutionary computation series: Studies in computational intelligence*. Springer, Berlin, pp. 165–188.
35. Tlelo-Cuautle, E., Duarte-Villaseñor, M. A., & Guerra-Gómez, I. (2008). Automatic synthesis of VFs and VMs by applying genetic algorithms. *Circuits, Systems and Signal Processing 27*(3), 391–403.
36. Tlelo-Cuautle, E., Guerra-Gómez, I, Reyes-Garcia, C. A., & Duarte-Villaseñor, M. A. (2010). Synthesis of analog circuits by genetic algorithms and their optimization by particle swarm optimization chap. In*Intelligent systems for automated learning and adaptation: Emerging trends and applications*. Information Science Reference: IGI Global, pp. 173–192.

37. Tlelo-Cuautle, E., Martinez-Romero, E., Sánchez-López, C., & Tan, S.X.D. (2009). Symbolic formulation method for mixed-mode analog circuits using nullors. In *IEEE ICECS*.
38. Tlelo-Cuautle, E., Moro-Frias, D., Sánchez-López, C., & Duarte-Villaseñor, M. A. (2008). Synthesis of CCII-s by superimposing VFs and CFs through genetic operations. *IEICE Electronics Express 5*(11), 411–417.
39. Tlelo-Cuautle E., Sánchez-López C., & Moro-Frias D. (2010). Symbolic analysis of (MO)(I)CCI(II)(III)-based analog circuits. *International Journal of Circuit Theory and Applications*
40. Verhaegen, W., & Gielen, G. (2002). Efficient DDD-based symbolic analysis of linear analog circuits.*IEEE Transactions on Circuits and Systems II: Analog and Digital Signal Processing 49*(7), 474–487.
41. Wang, H. Y., Chang, S. H., Jeang, Y. L. et al. (2006). Rearrangement of mirror elements. *Analog Integrated Circuits and Signal Process 49*(1), 87–90.
42. Wang, H. Y., Lee, C. T., & Huang, C. Y. (2005). Characteristic investigation of new pathological. *Analog Integrated Circuits and Signal Process 44*(1), 95–102.
43. Zito D. (2009). A novel low-power receiver topology for rf and microwave applications. *International Journal of Circuit Theory and Applications 37*(9), 1008–1018.

Chapter 11

DESIGN AND STABILITY ANALYSIS OF CFOA-BASED AMPLIFIER CIRCUITS USING BODE CRITERION

Ivailo Pandiev

Department of Electronics, Faculty of Electronic Engineering and Technologies, Technical University – Sofia, Sofia, 1797, Bulgaria

ABSTRACT

In this paper the frequency stability of small-signal high-speed amplifier circuits using Bode criterion is analysed theoretically. In particular, the inverting and non-inverting amplifiers employing current-feedback operational amplifiers are under review. Based on the analysis of the operational principle, the equations for complex transfer functions of both circuits and formulas for the related electrical parameters are obtained. Moreover, using these formulas recommendations for a stable operation are given. As well, design procedure of the amplifiers with resistive and capacitive load is suggested. The efficiency of the proposed procedure and recommendations are verified by simulation modelling and experimental testing of sample electronic circuits.

INTRODUCTION

The small-signal high-speed (with bandwidth > 1 MHz) amplifiers are essential building blocks for video amplifiers, RF/IF amplifiers, high-speed A/D drivers and D/A buffers (Analog Dev., MT-060, 2008; Jung, 2002; Tietze & Schenk, 2008). In the past 10 years current-feedback operational amplifiers (CFOAs) have been basically used as active building blocks for design of high-speed amplifiers. As a kind of the monolithic operational amplifiers (op amps) family, the CFOAs have been realized to overcome the finite gain-bandwidth product of the conventional voltage-feedback operational amplifiers (VFOAs) (Jassim, 2013; Seifart, 2003). However, the CFOA-based amplifier circuits are less understood and documented in comparison with the amplifiers using VFOAs.

Stability analysis is an essential part in the design process of the analogue circuits, especially for the high-speed amplifiers, that use CFOAs. The practical interpretation of the sustainability definition is: 'An amplifier circuit is stable if all voltages and currents are reduced to zero when the input voltages and currents are zero'. Otherwise, in lack of an input signal unexpected oscillations (or self-oscillations) can occur at the output, which is unacceptable. In the theory of electronic circuits, various criteria of stability are designed. For the analysis of analogue circuits, the most commonly used are the criteria of Nyquist (Seifart, 2003) and Bode (Laker & Sansen, 1994; Nagaria, Gopi Krishna, & Singh Rakesh, 2008).

In the analysis of the Bode criteria, the behaviour of the magnitude and phase frequency characteristics or Bode plots are investigated. The Bode plots can be drawn directly from experimental data or computer simulations.

For the amplifier circuits it is assumed that the open-loop transfer function of the CFOAs is stable and their logarithmic a.c. transfer characteristics monotonously decreased with an increase in the frequency ω of the input signal. For this case the Bode criterion says: the closed-loop system containing op amp with negative feedback is stable only if the a.c. transfer characteristic of the open-loop system crosses the x-axis (0 dB) before the linear phase characteristic has reached $-180°$.

A relatively large number of books, publications and company application reports are devoted to the theory and the design of the amplifier circuits employing CFOAs (Jassim, 2013; Jung, 2002; Kamath, 2014; Mancini, 2001, 2002; Palumbo, 1997; Pandiev,2012; Safari & Azhari, 2012; Schmid, 2003; Seifart, 2003; Texas Instruments, 2002; Tietze & Schenk, 2008). The authors' attention is focused on the structure and principle of operation at d.c. and large frequency range (Jung, 2002; Safari & Azhari,2012; Seifart, 2003; Tietze & Schenk, 2008). However, for the analysis of basic amplifier circuits a simplified model of the CFOA is used. The attention of the author in Jassim (2013) is focused on the design technique, employing CMOS CFOA. Some results on the behaviour of the non-inverting amplifier are given, which also confirm the efficiency of the designed CMOS CFOA. In the application report (Schmid, 2003), a method for measuring parasitic components in a prototype or final printed circuit board (PCB) design is discussed. A standard oscilloscope and a low-frequency waveform generator to collect valuable information for SPICE simulations are used there. The authors propose some design recommendations for the selection of the feedback resistor and gain resistor for the basic inverting and non-inverting amplifiers, employing CFOAs (Mancini, 2002; Texas Instruments, 2002). The note on the application (Mancini, 2001) has analysed the frequency stability of the inverting and non-inverting amplifier circuits, taking into account the influence of parasitic capacitance C_N and C_F (where C_N is the capacitance connected to the inverting input and C_F is the capacitance connected in parallel to the feedback resistor of the op amp). In the analysis, a simplified model of the CFOA has been used and at the same time the effect of the load in the loop-gain transfer function is not considered. An analysis of stability and compensation of CFOA is presented (Palumbo, 1997). In the theoretical analysis of the inverting amplifier, integrator and differentiator using a small-signal model of a CFOA, some of the input parasitic capacitances and the effect of the mounting capacitance are not included. Furthermore, the load is considered only as capacitance C_L without taking into account the parallel connection of the load resistance R_L and the load capacitance C_L (i.e. $Z_L = R_L \| (1/sC_L)$). This study is particularly useful for pencil-and-paper design of CFOA and takes into account both the resistive and the capacitive feedback. Recently, compensation techniques for improving amplitude and phase response of CFOA-based inverting amplifier are investigated (Kamath, 2014). The active compensation technique employing composite CFOA consisting of CFOA1 and CFOA2 is used in the place of single CFOA.

In this paper based on theoretical analysis of the high-speed amplifier circuits employing CFOAs and based on the results obtained for the amplifiers using VFOA, the frequency stability is studied by using the Bode criteria (Mancini, 2001). The results of the analysis are used to define the recommendations for improving the stability and to create a modified design procedure based on the procedure proposed (Pandiev, 2012).

The organization of the paper is as follows: in Section 2 the structure and relation between the input and the output voltages and currents for a linear high frequency model of the CFOA used in this work are presented; in Sections 3 and 4 the principles of operation of the inverting and non-inverting amplifiers at low and high frequencies are described; also in Section 4 recommendations for improving the frequency stability are defined, based on the obtained results; the proposed design procedure is given in Section 5; to illustrate the proposed theoretical analyses and the procedure, in Section 6 examples of studying the frequency stability of the inverting and non-inverting amplifiers at several voltage gains and various CFOAs are given. Finally in Section 7, the concluding remarks are given.

CURRENT-FEEDBACK OPERATIONAL AMPLIFIERS

The most common CFOA (or transimpedance amplifier) is equivalent to a (positive second-generation current conveyor) CCII+ plus an output voltage buffer. These op amps have a high impedance non-inverting input y, a low-impedance inverting input x, a current output z and a voltage output o. In some of the CFOAs the port z, between the first stage (CCII+) and the second stage (voltage follower), is defined as an external pin. The port o is the output of the voltage buffer, where the output resistance r_o is very low (several ohms magnitude). The linear model of the CFOA used in this work is presented in Figure 1.

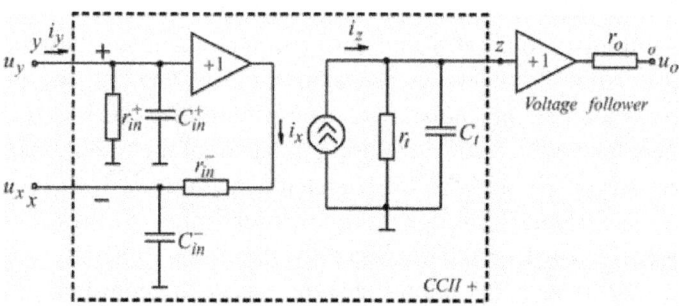

Figure 1: A linear model of the real CFOA.

The model in Figure 1 reflects the small-signal behaviour of the real device. It includes the following elements: input and output buffers (voltage followers); i_x and i_z – input and output current through the current-controlled current source (or current mirror); r_{in}^+ and C_{in}^+ – resistance and capaitance of the non-inverting input; r_{in}^- and C_{in}^- – resistance and capacitance of the inverting input; r_t and C_t – equivalent to Z_t – transmission impedance and r_o – output resistance.

For the linear CFOA, the ideal relations between input and output voltages and currents can be given by the following hybrid matrix:

$$\begin{bmatrix} i_y \\ u_x \\ i_z \\ u_o \end{bmatrix} = \begin{bmatrix} 1/Z_{in}^+ & 0 & 0 \\ 1 & Z_{in}^- & 0 \\ 0 & 1 & 1/Z_t \\ 0 & 0 & 1 \end{bmatrix} \begin{bmatrix} u_y \\ i_x \\ u_z \end{bmatrix},$$

(1)

where $Z_{in}^+ = r_{in}^+ ||(1/pC_{in}^+)$, $Z_{in}^- = r_{in}^-||(1/pC_{in}^-)$ and $Z_t = r_t||(1/pC_t)$.

The matrix representation given in Equation (1) is valid only for ideal input and output voltage buffers with voltage gain equal to one.

In general, the CFOAs have several important applications. For example, the CFOAs are basically used for the design of amplifiers, active filters and oscillators. The objective in this paper is the frequency stability analysis of the basic amplifier circuits, using CFOAs.

PRINCIPLE OF OPERATION AND BASIC ANALYSIS (APPROXIMATELY UP TO 50 MHZ)

The objects of study are the inverting and non-inverting amplifier circuits, employing CFOAs.

An Inverting Amplifier

The schematic structure of the high-speed inverting amplifier, employing CFOA, is shown in Figure 2. In this circuit is introduced a parallel negative feedback through the resistors R_F and R_N. The resistor R_P is used for compensation of the input bias current of the CFOA. If the circuit in Figure 2 is used as a video line driver, the best frequency response can be obtained by the addition of small resistances R_T and R_o at each terminal (with values of 50 or 75 Ω, for example).

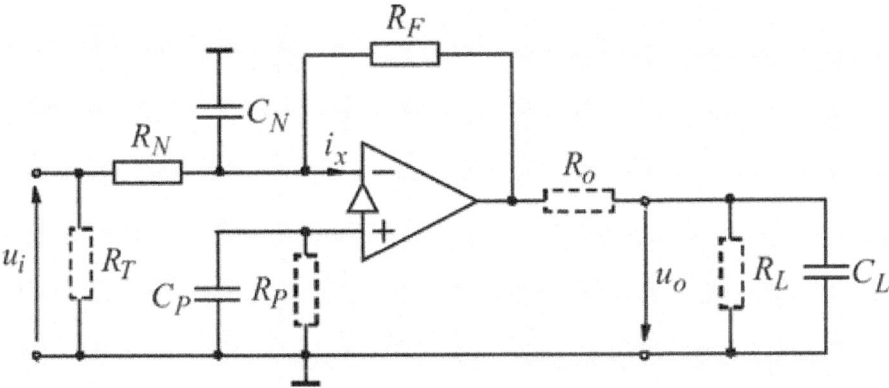

Figure 2: An inverting amplifier circuit using CFOA.

For low frequencies (approximately up to 50 MHz), the influence of the parasitic capacitances C_{in}^+, C_{in}^- and C_L can be neglected, thereby the real linear model of the CFOA (Figure 1) is simplified. Including the external elements, we obtain the a.c. equivalent circuit of the analysed inverting amplifier (Figure 3). The [Y]-matrix of the circuit was composed using the well-known formulas (Boyanov & Shoikova, 1989), and after some transformations (using the condition $r_o \ll r_t$) we obtain the following expression for the transfer function:

$$A_U(s) = \frac{U_o(s)}{U_i(s)} = \frac{-R_F/R_N}{R_F C_t \left(1+\frac{r_o}{R_L}\right)\left(1+\frac{r_{in}^-}{R_N}+\frac{r_{in}^-}{R_F}\right)\left[s+\frac{1+\frac{R_F}{r_t}\left(1+\frac{r_o}{R_L}\right)\left(1+\frac{r_{in}^-}{R_N}+\frac{r_{in}^-}{R_F}\right)}{R_F C_t\left(1+\frac{r_o}{R_L}\right)\left(1+\frac{r_{in}^-}{R_N}+\frac{r_{in}^-}{R_F}\right)}\right]}$$

$$= \frac{H}{s+\omega_p}. \quad (2)$$

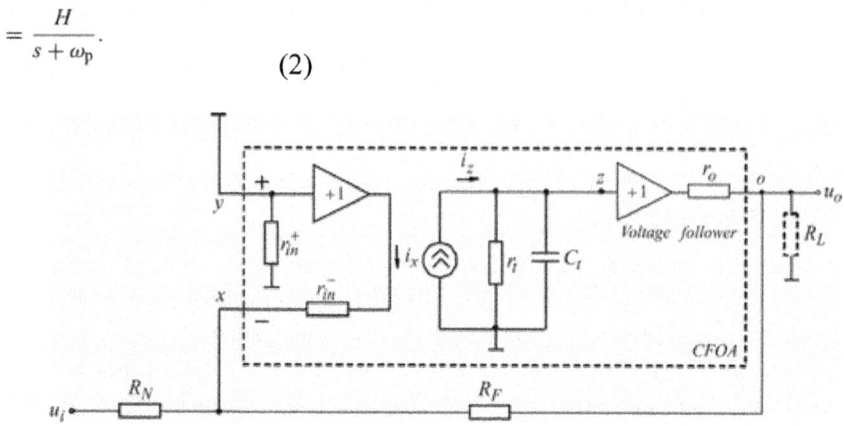

Figure 3: An equivalent circuit of the inverting amplifier using CFOA.

Comparison of the left and right sides of Equation (2) results in the following formulas:

$$H = A_{U0} = \frac{-R_F/R_N}{R_F C_t \left(1 + \frac{r_o}{R_L}\right)\left(1 + \frac{\overline{r_{in}}}{R_N} + \frac{\overline{r_{in}}}{R_F}\right)} \quad (3a)$$

and

$$\omega_p = \frac{1 + \frac{R_F}{r_t}\left(1 + \frac{r_o}{R_L}\right)\left(1 + \frac{\overline{r_{in}}}{R_N} + \frac{\overline{r_{in}}}{R_F}\right)}{R_F C_t \left(1 + \frac{r_o}{R_L}\right)\left(1 + \frac{\overline{r_{in}}}{R_N} + \frac{\overline{r_{in}}}{R_F}\right)}, \quad (3b)$$

where H is the d.c. voltage gain and ω_p is the pole frequency, defining working frequency bandwidth f_{-3dB} of the circuit.

Therefore, for $\overline{r_{in}} \approx 0$, $r_o \ll R_L$ and $R_F \ll r_t$, the working frequency bandwidth $f_{-3dB} \approx 1/(2\pi R_F C_t)$ of the amplifier depends only on the internal capacitance C_t and the external feedback resistor R_F. Moreover, the bandwidth f_{-3dB} does not depend on the resistance R_N, setting the voltage gain of the circuit. The possibility for independent adjustment of gain and bandwidth is one of the main advantages of the amplifiers with CFOAs in comparison to those realized with VFOAs.

Based on the transfer function (2) for the module and the phase is obtained

$$|\dot{A}_U| = \frac{A_{U0}}{\sqrt{1 + (f/f_p)^2}} \text{ and } \varphi_{AU} = 180° - \arctan\left(\frac{f}{f_p}\right).$$

A Non-inverting Amplifier

The electronic circuit of the non-inverting amplifier is shown in Figure 4. In this circuit the CFOA is with serial negative feedback through the resistors R_F and R_N. The resistor R_P is used for compensation of the input bias current of the op amp, and the resistors R_T and R_o are used for the termination of each end of the circuit. The symbol analysis of the circuit in Figure 4 is implemented providing that C_{in}^+, C_{in}^- and C_L are neglected (see Figure 1). The a.c. equivalent circuit of the analysed non-inverting amplifier is presented in Figure 5.

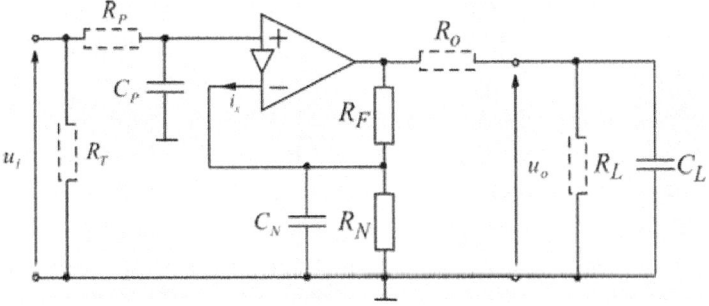

Figure 4: A non-inverting amplifier circuit using CFOA.

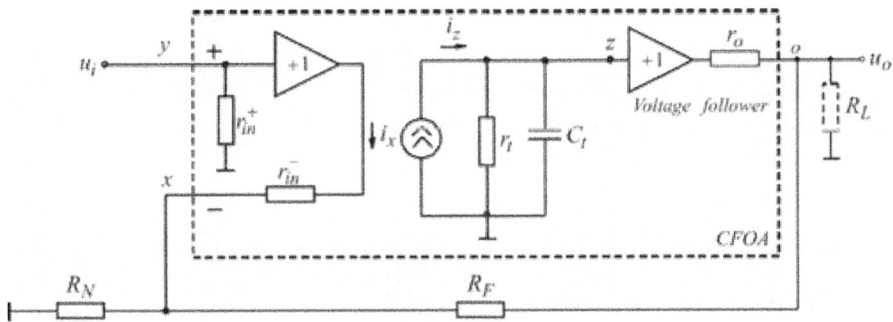

Figure 5: An equivalent circuit of the non-inverting amplifier using CFOA.

In a.c. mode of operation (using the condition $r_o \ll r_t$), the transfer function can be written as

$$A_U(s) = \frac{U_o(s)}{U_i(s)}$$

$$= \frac{1 + R_F/R_N}{R_F C_t \left(1 + \frac{r_o}{R_L}\right)\left(1 + \frac{r_{in}^-}{R_N} + \frac{r_{in}^-}{R_F}\right)\left[s + \frac{1 + \frac{R_F}{r_t}\left(1 + \frac{r_o}{R_L}\right)\left(1 + \frac{r_{in}^-}{R_N} + \frac{r_{in}^-}{R_F}\right)}{R_F C_t \left(1 + \frac{r_o}{R_L}\right)\left(1 + \frac{r_{in}^-}{R_N} + \frac{r_{in}^-}{R_F}\right)}\right]}$$

$$= \frac{H}{s + \omega_p}. \qquad (4)$$

Comparison of the left and right sides of Equation (4) results in the following formula for the d.c. voltage gain:

$$H = A_{U0} = \frac{1 + R_F/R_N}{R_F C_t \left(1 + \frac{r_o}{R_L}\right)\left(1 + \frac{r_{in}^-}{R_N} + \frac{r_{in}^-}{R_F}\right)}. \qquad (5)$$

The formula (3b) is valid also for the non-inverting amplifier, because the denominators of the transfer functions, given in Equations (2) and (4), are in the same form.

According to formula (3b) during the implementation of amplifier circuits with a relatively large gain, the resistor R_N has small values, while the resistor R_F has a relatively large value, determined by the range of the working frequency bandwidth. In these cases, the resistance R_N is close to the resistance r_{in}^-. Therefore, in the design process of amplifier circuits, the value of r_{in}^- and the resulting ratios with R_F and R_N should be always given. The analysis of the datasheets of various CFOAs shows that the resistance r_{in}^- has values from several ohms to several tenths of ohms. For example, the resistance r_{in}^- is equal to 8 Ω for the op amp AD8009 (from Analog Devices), while for the op amp AD844 (from Analog Devices) r_{in}^- is 50 Ω. For relatively large gains and low frequencies (up to 50 MHz) according to formula (3b), the value of the feedback resistor R_F is obtained by $R_F = r_t(f_t/f_{-3dB}) - r_{in}^- A_{U0}$, where $f_t = 1/(2\pi r_t C_t)$ is the transit frequency of the CFOA.

The module and the phase for the non-inverting amplifier are given as

$$|\dot{A}_U| = \frac{A_{U0}}{\sqrt{1 + \left(\frac{f}{f_p}\right)^2}} \text{ and } \varphi_{A_U} = \arctan\left(\frac{f}{f_p}\right).$$

From the analysis of the obtained formulas for the inverting and non-inverting amplifiers at low frequencies (i.e. $f \ll f_p$), the phase shift $\varphi_{\dot{A}_U}$ for the non-inverting amplifier is approximately equal to 0° and for the inverting amplifier is −180°, respectively. For higher frequencies (i.e. $f > f_p$) the gain is decreased to $|\dot{A}_U| = A_{U0} \cdot (f_p/f)$ (the slope of $|\dot{A}_U|$ is approximately equal to −20 dB/dec). Furthermore, $|\varphi_{\dot{A}_U^-}| \approx 90°$, which ensures the stable operation of the circuits (the phase margin is greater than 45°).

The results of these analyses can be used to study amplifier circuits with CFOAs, whose operating frequency bandwidth is up to 50 MHz. In these cases, the frequency response is maximally flat in the pass-band. Sample CFOAs suitable for applications up to 50 MHz are THS6184 (from Texas Instruments), LT1256 (from Linear Technology) and AD844 (from Analog Devices).

ANALYSIS AT HIGH FREQUENCIES OF THE AMPLIFIER CIRCUITS

At higher frequencies (>50 MHz), analyses of the transfer function of the input network, and thus of the overall transfer characteristic of the systems of the amplifiers affect two capacitances C_P and C_N. C_P is the capacitance to non-inverting input, which is formed by the capacitance C_{in}^+ – Figure 2 plus the mounting capacitance (i.e. $C_P = C_{in}^+ + C_M$, where C_M is the parasitic board capacity with values usually up to 3 pF (Texas Instruments, 2002)). C_N is the capacitance to inverting input, which is formed by C_{in}^- and the mounting capacitance (i.e. $C_N = C_{in}^- + C_M$). Also the transfer characteristic of the amplifier circuits is affected by the load capacitance C_L, connected in parallel to the resistance R_L. In the following two subsections of the paper, the effects of C_P, C_N and C_L on the frequency response of the inverting and non-inverting amplifiers are examined separately.

Effect of C_P, C_N and C_L on the Frequency Response of the Non-inverting Amplifier

The analysis of the circuit in Figure 4 is performed according to the method of the nodal voltages. The CFOA is replaced by the linear model, given in Figure 1. The transfer function (using the condition $r_{in}^- \ll r_t$) at $Z_L = R_L \| (1/sC_L)$ (the load impedance is a parallel connection of resistor R_L and parasitic capacitance C_L) and using the condition $r_o \ll r_t$ can be found by

$$A_U(s) \approx \cfrac{\cfrac{1}{(r_{in}^+\|R'_P)C_P}}{s + \cfrac{1}{(r_{in}^+\|R'_P)C_P}}$$

$$\cfrac{\cfrac{r_{in}^+}{r_{in}^+ + R'_P} \cdot \cfrac{C_N R_F + C_t r_o}{C_N C_t R_F r_{in}^-\left(1 + \frac{r_o}{R_L} + \frac{C_L}{C_N}\frac{r_o}{r_{in}^-}\right)}\left(s + \cfrac{1 + \frac{R_F}{R_N}}{C_N R_F + C_t r_o}\right)}{s^2 + s\cfrac{1}{C_N r_{in}^-}\cfrac{\left(1 + \frac{r_{in}^-}{R_N} + \frac{r_{in}^-}{R_F}\right)\left(1 + \frac{r_o}{R_L}\right)}{1 + \frac{r_o}{R_L} + \frac{C_L}{C_N}\frac{r_o}{r_{in}^-}} + \cfrac{1}{C_N C_t R_F r_{in}^-\left(1 + \frac{r_o}{R_L} + \frac{C_L}{C_N}\frac{r_o}{r_{in}^-}\right)}}$$

$$= H \frac{\omega_{p,in}}{s + \omega_{p,in}} \frac{s + \omega_z}{s^2 + \frac{\omega_p}{Q_p}s + \omega_p^2}, \qquad (6)$$

where $R'_P = R_G + R_P$ is the equivalent resistance to the non-inverting input and R_G is the internal resistance of the input voltage source.

The equalization of the left and right sides of Equation (6) results in the following formulas for the basic parameters:

$$H = \frac{r_{in}^+}{r_{in}^+ + R'_P} \frac{C_N R_F + C_t r_o}{R_F r_{in}^- C_t C_N} \frac{1}{\left(1 + \frac{r_o}{R_L} + \frac{C_L}{C_N}\frac{r_o}{r_{in}^-}\right)}$$

$$\approx \frac{r_{in}^+}{r_{in}^+ + R'_P} \frac{1}{r_{in}^- C_t \left(1 + \frac{r_o}{R_L} + \frac{C_L}{C_N}\frac{r_o}{r_{in}^-}\right)} \qquad (7)$$

is the transmission coefficient;

$$\omega_{p,in} = 1/[(r_{in}^+ \| R'_P) C_P] \qquad (8)$$

is the pole angular frequency related to the effect of the resistance R'_P and capacitance C_P to the non-inverting input;

$$\omega_z = \frac{1 + R_F/R_N}{R_F C_N + r_o C_t} \qquad (9)$$

is the zero angular frequency related to the effect of the resistance R_F and capacitance C_N to inverting input of the CFOA;

$$\omega_p = \frac{1}{\sqrt{R_F r_{in}^- C_t C_N \left(1 + \frac{r_o}{R_L} + \frac{C_L}{C_N}\frac{r_o}{r_{in}^-}\right)}} \qquad (10)$$

is the pole angular frequency (self-oscillating frequency or undamped natural frequency) and

$$Q_p = \frac{\sqrt{r_{in}^- C_N \left(1 + \frac{r_o}{R_L} + \frac{C_L}{C_N}\frac{r_o}{r_{in}^-}\right)}}{\sqrt{R_F C_t}\left(1 + \frac{r_{in}^-}{R_N} + \frac{r_{in}^-}{R_F}\right)\left(1 + \frac{r_o}{R_L}\right)}; \qquad (11)$$

is the quality factor of the circuit.

After substitution of $s = j\omega$ in formula (6), based on the obtained general complex function, the module and the phase shift can be found by using

$$|A_U(j\omega)| = A_U(\omega)$$

$$= \frac{\omega_{p,in}}{\sqrt{\omega^2 + \omega_{p,in}^2}} \frac{H\sqrt{\omega_z^2 + \omega^2}}{\sqrt{(\omega_p^2 - \omega^2)^2 + \left(\omega\frac{\omega_p}{Q_p}\right)^2}} \quad \text{and} \quad (12a)$$

$$\phi_{A_U} = \arctan\left(\frac{\omega}{\omega_z}\right) - \arctan\frac{\omega_p \omega}{Q_p(\omega_p^2 - \omega^2)}$$

$$- \arctan\left(\frac{\omega}{\omega_{p,in}}\right). \quad (12b)$$

To compensate the effect of the parasitic capacitance C_P, capacitor C'_P can be placed in parallel with R'_P so that $C'_P \gg C_P$(Stoianov, 2000). The modified transfer function of the input network with capacitor C'_P in parallel with R'_P is

$$T_{in}(s) = \frac{r_{in}^+}{r_{in}^+ + R'_P} \frac{1 + sR'_P C'_P}{1 + s[(r_{in}^+||R'_P)(C_P + C'_P)]}. \quad (13)$$

If $C'_P \gg C_P$ and $R'_P \ll r_{in}^+$, then

$$\frac{1 + sR'_P C'_P}{1 + s[(r_{in}^+||R'_P)(C_P + C'_P)]} \cong 1 \text{ and } u_i \cong u_y.$$

In this compensation method, the working frequency bandwidth is not narrowed and also the spikes are not reduced.

At the compensated effect of the capacitance C_P, the analysis of formulas (6), (12a) and (12b) shows that the transfer function is characterized by a double pole (or a complex pole) with angular frequency equal to ω_P and one real zero with angular frequency $-\omega_z$. However, for $\omega = 0$ the voltage gain has a value $A_U(0) = H(\omega_z/\omega_p) = 1 + (R_F/R_N)$, while for much higher frequencies the gain decreases to zero. Therefore, the transfer characteristic of the non-inverting amplifier circuit is a low-pass type. For $Q_P \geq 0.707$ at a frequency equal to ω_P, the denominator tends to zero and the voltage gain theoretically increases towards infinity. As a result, peaking of the frequency characteristic occurs and the phase shift between the input and output signal increases rapidly to 180°. Moreover, the circuit of the non-inverting amplifier becomes unstable. In the cases where $Q_P < 0.707$ and $\omega_p \ll \omega_z$, the frequency response monotonically decreases and the phase shift decreases to $-180°$ (the phase margin is less than 45°). With increase in the frequency, the first component in the formula (12b) tends to be 90°, while the second component tends to be $-180°$. From the theory of electronic circuits (Seifart, 2003), it is known that for $\omega \geq 10\omega_P$ the phase shift increases to $-180°$. According to formula (12b), the stable operation of the amplifier can be produced at the condition that the difference between ω_p and ω_z does not exceed 10 times. At $\omega_z < \omega_p$ ringing in the output signal occurs, which can cause unstable operation.

Therefore, for the working bandwidth ω_{-3dB}, where $|A_U(j\omega)|$ decreases with 3 dB (or $1/\sqrt{2}$), the following is obtained:

$$\omega_{-3dB} = \frac{\omega_p^2}{\omega_z}\sqrt{1+\left(1-\frac{1}{2Q_p^2}\right)\frac{\omega_z^2}{\omega_p^2}+\sqrt{\left[1+\left(1-\frac{1}{2Q_p^2}\right)\frac{\omega_z^2}{\omega_p^2}\right]^2+\frac{\omega_z^4}{\omega_p^4}}}. \quad (14)$$

At $Q_p = 1/\sqrt{2}$ formula (14) is simplified and yields

$$\omega_{-3dB} = \frac{\omega_p^2}{\omega_z}\sqrt{1+\sqrt{1+\frac{\omega_z^4}{\omega_p^4}}}. \quad (15)$$

For $\omega_z \gg \omega_p$ $\omega_{-3dB} \approx \omega_p$. The value of the module of the general complex function (12a) at $\omega = \omega_p$ is found by the formula

$$A_U(\omega_p) = Q_p H \frac{\sqrt{\omega_z^2+\omega_p^2}}{\omega_p^2}.$$

Based on the analysis of the formulas for ω_p, ω_z and Q_p for the chosen resistance R_F and CFOA at $Q_p < 0.707$, in the frequency response ringing can occur under the condition

$$R_N > \frac{R_F}{k_\omega - 1}, \quad (16)$$

where

$$k_\omega = \sqrt{\frac{R_F C_N}{r_{in}^- C_t}} \frac{1}{\sqrt{1+\frac{r_o}{R_L}+\frac{C_L}{C_N}\frac{r_o}{r_{in}^-}}}$$

is the coefficient of peaking of the output signal.

The magnitude and phase Bode plots for the loop gain of a voltage follower using CFOA are shown in Figure 6. A voltage follower or buffer with op amp is obtained from the non-inverting amplifier by removing the resistor R_N. This circuit is analogous to the emitter (source) follower employing one transistor. The Bode plots are constructed by summing the corresponding logarithmic characteristics of all their sections, realizing zeros and poles:

$$L(\omega) = 20 \lg A_U(\omega) = \sum_{i=1}^{n} L_i(\omega) \text{ and} \tag{17}$$

$$\varphi_{A_U}(\omega) = \sum_{i=1}^{n} \varphi_i(\omega), \tag{18}$$

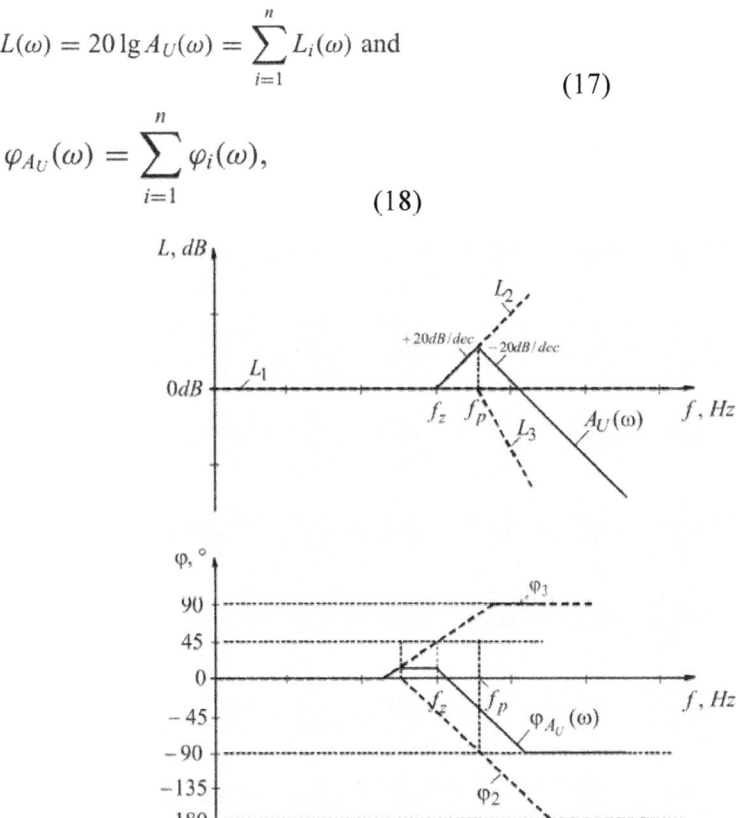

Figure 6: Magnitude and phase Bode plots for the loop gain of a voltage follower employing CFOA.

where $L_i(\omega)$ are the individual logarithmic magnitude characteristics and $\varphi_i(\omega)$ are the individual phase characteristics.

Formulas (12a) and (12b), and the corresponding Bode plots for $\omega_z < \omega_p$, given in Figure 6, show the following:

(1) In the voltage follower operation mode ($A_U(0) = 1$), there is always causes peaking in the frequency response, the coefficient k_ω is greater than unit. By increasing the d.c. voltage gain A_{U0}, the frequency ω_z increases. As a result, the amount of peaks in the frequency response decreases;

(2) The phase margin of the voltage follower and the non-inverting amplifier is greater than $45°$. However, the stability of the circuit is worse. For a given frequency of the input signal, at which the asymptotes L_2 and L_3 intersect, the rate of change in voltage gain is about 40 dB/dec.

The peaks in the magnitude plot lead to an increase in the amplitude of the output signal, which may adversely affect the next stages of the electronic devices and systems.

To compensate the effect of the capacitances C_N and C_L, the following recommendations can be used:

- Reduce the value of C_N by removing ground or power plane around the circuit trace to the inverting input;
- Reduce the value of the feedback resistor R_F. As a result, the operating frequency bandwidth is expanded – formula (14);
- Reduce the value of C_L by minimizing the length of output cables;
- For amplifiers the condition

$$A_{U0} > \sqrt{\frac{R_F C_N}{r_{in}^- C_t}} \frac{1}{\sqrt{1 + \frac{r_o}{R_L} + \frac{C_L}{C_N}\frac{r_o}{r_{in}^-}}}$$

has to be kept, then $\omega_P < \omega_=$. As well it has to maintain the Q_P to be less than $1/\sqrt{2}$, which can be achieved at $(R_F C_t/2) > r_{in}^- C_N + r_o C_L$.

1. Connect a small resistor (10Ω ... 20Ω) between the output of the CFOA and the capacitive load.

For video line drivers (working up to 100 MHz) to isolate the output terminal from the load capacitance, a small resistor R_o, between the output of the op amp and the load, can be connected (Mancini, 2001; THS3091 – datasheet, 2014). Thus, the working bandwidth is narrowed by forming the additional pole frequency of $\omega_{pT,out} = 1/[(R_o \| R_L) C_L]$ and the d.c. transmission coefficient of the output network at $R_o = R_L$ (where R_L is the characteristic impedance of the cable), yielding $T_{out,dc} = R_L/(R_o + R_L)$. Furthermore, the phase margin decreases at higher frequencies.

Effect of C_p, C_N and C_L on the Frequency Response of the Inverting Amplifier

The analysis of the inverting amplifier (Figure 2) is also performed according to the method of the nodal voltages, when the op amp is replaced by the linear model, presented in Figure 1. For the corresponding transfer function (using the condition $r_{in}^- \ll r_t$) is found

$$A_U(s) = \cfrac{\cfrac{1}{C_N C_t R_F r_{in}^-\left(1+\frac{r_o}{R_L}+\frac{C_L}{C_N}\frac{r_o}{r_{in}^-}\right)}\left(-\frac{R_F}{R_N}\right)}{s^2 + s\cfrac{1}{C_N r_{in}^-}\cfrac{\left(1+\frac{r_{in}^-}{R_N}+\frac{r_{in}^-}{R_F}\right)\left(1+\frac{r_o}{R_L}\right)}{1+\frac{r_o}{R_L}+\frac{C_L}{C_N}\frac{r_o}{r_{in}^-}} + \cfrac{1}{C_N C_t R_F r_{in}^-\left(1+\frac{r_o}{R_L}+\frac{C_L}{C_N}\frac{r_o}{r_{in}^-}\right)}}$$

$$= \frac{H\omega_p^2}{s^2 + \frac{\omega_p}{Q_p}s + \omega_p^2}. \qquad (19)$$

The equalization of the left and right sides of Equation (19) results in the following formula for the d.c. transmission coefficient:

$$H = A_{U0} = -\frac{R_F}{R_N}. \qquad (20)$$

The formulas for the angular pole frequency and the quality factor matched with formulas (10) and (11), respectively. Furthermore, the obtained transfer function of the inverting amplifier is not affected by adding the parasitic capacitor C_P to the circuit.

The analysis of formula (19) shows that the transfer function is characterized by one double pole with frequency equal to ω_P. When taking into account the relation r_o/r_t, the transfer function is obtained with one zero with positive real part and one double pole. The value of the zero angular frequency can be found by $\omega_z = (R_F/r_o r_{in}^- C_t)$. The existence of zero with a positive real part in the transfer function makes the amplifier circuit unstable. The comparison of ω_P and ω_z for the inverting amplifier shows that $\omega_P \gg \omega_z$, as it always satisfies the condition

$$\frac{r_o}{R_F}\sqrt{\frac{r_{in}^- C_t}{R_F C_N}}\frac{1}{\sqrt{1+\frac{r_o}{R_L}}} \ll 1.$$

Furthermore, the ω_z is much larger than the unity bandwidth ($f_1 \approx 1/2\pi r_{in}^- C_t$), since $R_F \gg r_o$.

The main disadvantages of the inverting circuit compared to the non-inverting circuit are relatively small input resistance and phase inversion of the input voltage, which can be an unwanted effect for some practical applications.

DESIGN PROCEDURE

As a result of the theoretical analysis, the aforementioned analytical formulas are the bases of the design procedure for amplifiers employing CFOAs. The schematic design for the circuits in Figures 2 and 4 is based on the following sequence:

(1) *Technical specification*. The circuit elements are calculated using predefined: amplitude of the input voltage U_{im} or amplitude of the input voltage source e_G with internal resistance R_G; input resistance R_{iA}; amplitude of the output voltage U_{R_L} with load resistance R_L and capacitance C_L; parasitic board capacitance C_M; output resistance R_{oA}; cut-off frequency f_{-3dB}; relative error ε_{io} [%] defined by the input offset current and voltage, temperature drift $\varepsilon_{\Delta io}$ [%] of the ε_{io} within temperature range ΔT and minimum value of a signal-to-noise (SN) ratio [dB].

(2) *An electronic circuit is selected*. An object of an analysis and design are the amplifier circuits shown in Figures 2 and 4. The inverting circuit (Figure 2) provides smaller input resistance, while the non-inverting circuit (Figure 4) is with greater input resistance. Furthermore, the inverting circuit reduces the influence of the parasitic input capacitance at high frequencies (the inverting input x is a virtual ground – $u_{yx} \approx 0$).

(3) *The op amp is selected*. The main advantages of the CFOAs are the greater slew rate and the wider bandwidth compared to the VFOAs. The higher value of the slew rate is associated with a higher consumption current in a dynamic mode of operation. In order to produce low value of the power dissipation, most of the CFOAs work with supply voltages less than ± 5 V. Then, at high frequencies, without additional output power stage, most CFOAs can get a maximum output current greater than 20 mA. The op amp is selected according to the following conditions:

- Maximum output voltage $U_{om} \geq U_{R_L}$ (U_{om} is the maximum output voltage of the op amp);

- The power supply voltage $V_{CC} = -V_{EE}$ is selected higher than the maximum output voltage U_{om}, as saving the condition $V_{CC\,min} < V_{CC} < V_{CC\,max}$;

- Maximum output current $I_{o,max} > I_L$, where $I_L = U_{R_L}/R_L$;

- Small-signal bandwidth $f_1 > (5\ldots 10)f_{-3dB}$, where f_1 is the cut-off frequency at voltage gain equal to 1;

- Slew rate $SR_{CFOA} > 2\pi f_{-3dB} U_{R_I}$.

(4) *The value of the equivalent quality factor* of the frequency response is obtained:

$$Q_p = \sqrt{\omega_{-3dB}} \frac{\sqrt{r_{in}^- C_N \left(1 + \frac{r_o}{R_L} + \frac{C_L}{C_N} \frac{r_o}{r_{in}^-}\right)}}{1 + \frac{r_o}{R_L}}.$$

(5.1.) *For low-frequency* (up to 50 MHz) *amplifiers*, the value of the feedback resistor R_F is calculated:

- For the inverting amplifier $R_F = r_t(f_t/f_{-3dB}) - (1 + |A_{U0}|)r_{in}^-$, where $f_t = 1/(2\pi r_t C_t)$ is the transit frequency of the CFOA and $|A_{U0}| = U_{R_I}/U_{im}$ is the voltage gain of the circuit;

- For the non-inverting amplifier $R_F = r(f_t/f_{-3dB})_t - r_{in}^- A_{U0}$;

(5.2.) *At higher frequencies* (> 50 MHz) and $\omega_p < \omega_z$, the value of the feedback resistor R_F is calculated:

$$R_F = r_t \frac{f_t f_N}{f_{-3dB}^2} \frac{1}{1 + \frac{r_o}{R_L}} \left[\left(1 - \frac{1}{2Q_p^2}\right) + \sqrt{\left(1 - \frac{1}{2Q_p^2}\right)^2 + 1}\right],$$

where $f_N = 1/(2\pi r_{in}^- C_N)$.

(6) *The value of the gain resistor R_N is calculated*:

- For the inverting amplifier $R_N = R_F/|A_{U0}|$;
- For the non-inverting amplifier $R_N = R_F/(A_{U0} - 1)$.

The calculated values for the resistors R_F and R_N according to the aforementioned formulas have to be consistent with the values from the datasheet of the chosen CFOA.

(7) *The value of the compensation resistor R_P is determined as* $R_P = R_N || R_F$.

(8) *The phase margin* is calculated: $|\varphi_m| = 180° - |\varphi_{A_U}(\omega_1)|$, where $\omega_1 = 2\pi f_1$;

For the inverting amplifier, the phase shift is

$$\varphi_{A_U}(\omega) = 180 - \arctan \frac{\omega_p \omega}{Q_p(\omega_p^2 - \omega^2)}.$$

The obtained value for the phase margin has to be greater than $45°$. Otherwise, other op amps with smaller values of the parasitic capacitances should be chosen.

(9) *The input impedance Z_{iA} is calculated*:

- For the inverting amplifier: $Z_{iA} \approx R_N$;

- For the non-inverting amplifier:

$$Z_{iA} \approx \frac{r_{in}^+ + R'_p}{\sqrt{1 + (f/f_{in}^+)^2}},$$

where $f_{in}^+ = 1/2\pi(r_{in}^+ + R'_p)C_P$ is the cut-off frequency of the input electrical network.

At low frequencies ($f \ll f_{in}^+$) $Z_{iA} \approx r_{in}^+ + R'_P$.

(10) *The output resistance R_{oA} is calculated*: $R_{oA} \approx r_o/\beta A_{d0}$, where $\beta = R_N/(R_N + R_F)$ is the negative feedback coefficient and $A_{d0} \approx r_t/r_{in}^-$ is the d.c. open-loop voltage gain of the chosen CFOA.

(11) *The output offset voltage of the circuit is calculated*. First, the output offset voltage for room temperature – $25°$ (using condition $R_P = R_F \| R_N$) is calculated: $U_{o,err} = (1 + R_F/R_N)U_{io} - R_F I_{io}$, where U_{io} is the input offset voltage and I_{io} is the input offset current of the chosen CFOA. For video drivers the resistor R_P is removed and a resistor R_T in parallel to the non-inverting input of the amplifier is connected. In this case the output offset voltage is $U_{o,err} = (1 + R_F/R_N)[U_{io} - (R_G \| R_T)I_B^+ + (R_F \| R_N)I_B^-]$. Then the relative error $\varepsilon_{io} = (U_{o,err}/U_{R_I})100\%$ is compared to the value given in step No 1.

(12) *The output offset voltage drift is calculated*. First, the output offset voltage drift is calculated: $\Delta U_{o,err} = (1 + R_F/R_N)\Delta U_{io}(T) - R_F \Delta I_{io}(T)$ for the given temperature range ΔT. Then the relative error $\varepsilon_{\Delta io} = (\Delta U_{o,err}/U_{R_I})100\%$ is compared to the value given in step No 1. If the result does not satisfy the specification, a more precise op amp or performing new calculations for the resistances with lower values can be chosen.

13. *The SN ratio is calculated*. First, the resulting noise voltage density at the amplifier's output is calculated: $\bar{s}_{U,out} = \sqrt{\sum_i s_{U_i}^2}$ for $i = 1, 2, \ldots$, where s_{U_i} is the individual noise components.

Then

$$SN = \frac{U_{o,\text{eff}}}{U_{oN}} = \frac{U_{o,\text{eff}}}{\overline{S}_{U,\text{out}}\sqrt{B_{eq}}},$$

where $B_{eq} = 1.57 f_{-3dB}$ is the bandwidth of the circuit multiplied by the correction factor of $\varpi/2 = 1.57$ and $U_{o,\text{eff}}$ is the output effective value. The obtained value for the SN is compared to the value given in step No 1 of the procedure. If the result does not satisfy the specification, other op amps with lower voltage and current noise can be chosen.

VERIFICATION CHECK, EXPERIMENTAL TESTING AND DISCUSSIONS

To verify the theoretical analysis and the proposed design procedure, in this section examples of studying the frequency stability of the inverting and non-inverting amplifiers at several voltage gains are given. A wide bandwidth ($B_1 > 300\,\text{MHz}$) CFOA type AD8011 (AD8011 – datasheet, 2014) and high-current ($I_{o\max} > 200\,\text{mA}$) CFOA type THS3091 (THS3091 – datasheet, 2014) are chosen as active building elements for the investigated electronic circuits. The THS3091 uses an 8-pin SOIC and the 8-pin SOIC with PowerPAD™ (Texas Instruments Incorporated, Dallas, Texas 75243, USA) packages. The package type PowerPAD™ is designed so that the thermal pad is exposed on the bottom side of the integrated circuit (IC). The thermal coefficient for the PowerPAD packages is substantially improved over the basic SOIC.

The verification check of the op amps is performed by Cadence OrCAD® (Cadence Design Systems, Inc., San Jose, CA 95134, USA), using OrCAD PSpice® program with AD8011AN PSpice macro-model (version 1.0) (AD8011 SPICE macro-model, 2014) and THS3091 PSpice macro-model (THS3091 PSpice Model, 2014). The AD8011AN model simulates the input offset voltage and current (offsets will not vary with input common-mode voltage), small-signal closed-loop gain and phase versus frequency, output current limiting and output voltage limiting, slew rate, step response performance (slew rate is based on 10–90% of step response), quiescent current at operating point, noise effects, input impedance and output impedance. The values of the modelled parameters at $V_S = \pm 5\,\text{V}$, $R_L = 1\,\text{k}\Omega$ and $T_A = 27^\circ C$ are as follows: $U_{io} = 2\,\text{mV}$, $I_B^+ = 5.125\,\mu\text{A}$, $I_B^- = 5.125\,\mu\text{A}$, $r_{in}^+ = 517\,\text{k}\Omega$, $C_{in}^+ = 1\,\text{pF}$, $r_{in}^- = 50\,\Omega$, $C_{in}^- = 2.3\,\text{pF}$, $r_t = 1.27\,\text{M}\Omega$, $C_t = 1.52\,\text{pF}$ (or $f_t = 82.5\,\text{kHz}$

), $f_1 \approx 670\,\text{MHz}$ (at $A_{U0} = 1$), $\overline{S_{Ui}} = 7.53\,\text{nV}/\sqrt{\text{Hz}}$ (at $f = 10\,\text{kHz}$), $U_{om} = \pm 4\,\text{V}$, $SR > 1000\,\text{V}/\mu\text{s}$, $I_{o\,\max} = 60\,\text{mA}$ and $r_o = 22\,\Omega$.

The THS3091 model simulates input offset voltage, input bias currents, small-signal closed-loop gain and phase versus frequency (bandwidth is high in gains of +1 V/V and +2 V/V and low at higher gains), output voltage limiting, slew rate, step response performance (slew rate is correct at 2 V step), settling time, quiescent current, noise effects, output impedance and loading effects. The values of the modelled parameters at $V_S = \pm 15\,\text{V}$, $R_L = 100\,\Omega$ and $T_A = 27°C$ are as follows: $U_{io} = 0.9\,\text{mV}$, $I_B^+ = -4.5\,\mu\text{A}$, $I_B^- = -3.5\,\mu\text{A}$ (or $I_{iB} = 4\,\mu\text{A}$ and $I_{io} = 1\,\mu\text{A}$), $r_{in}^+ = 1.1\,\text{M}\Omega$, $C_{in}^+ = 1.2\,\text{pF}$, $r_{in}^- = 32\,\Omega$, $C_{in}^- = 1.4\,\text{pF}$, $r_t = 848\,\text{k}\Omega$, $C_t = 0.8\,\text{pF}$ (or $f_t = 234\,\text{kHz}$), $f_1 \approx 240\,\text{MHz}$ (at $A_{U0} = 1$), $\overline{S_{Ui}} < 6\,\text{nV}/\sqrt{\text{Hz}}$ (at $f > 10\,\text{kHz}$), $U_{om} \approx \pm 3.1\,\text{V}$, $SR > 1000\,\text{V}/\mu\text{s}$ (at $R_F = 1.21\,\text{k}\Omega$ and $A_{U0} = 2$) and $r_o = 100\,\Omega$. The verification check of the models for AD8011AN and THS3091 is performed by comparing the simulation results to the datasheet typical parameters of the real op amps.

In Table 1 the calculated parameters and the simulation results for three values of the working bandwidth – 100, 120 and 150 MHz – are presented. The voltage gain is chosen with value equal to +5 and $R_L || C_L = 50\,\Omega || 20\,\text{pF}$. Based on the proposed design procedure, values for the passive components and values for the basic dynamitic parameters were found. The maximum error between the calculated values of the electrical parameters and the simulation results is not higher than 10%. Moreover, an error of 10% is quite acceptable considering the tolerances of the technological parameters.

Table 1: Comparison between calculated parameters and simulation results for $A_{U0} = +5$, $R_L = 50\,\Omega$ **and** $C_L = 20\,\text{pF}$

Parameter	Calculated results	Simulation results
$A_{U0} = +5, f_{-3dB} = 100\,\text{MHz}, R_F = 1.96\,\text{k}\Omega$ and $R_N = 490\,\Omega$ ($R_F = 1.96\,\text{k}\Omega \pm 1\%$ and $R_N = 487\,\Omega \pm 1\%$), $U_{im} = 100\,\text{mV}$, $R_G = 50\,\Omega$ and $R_T = 50\,\Omega$		
$U_{o,err}$	19.40 mV	19.32 mV
A_{U0}	5	5
Q_p	0.429	0.5
f_{-3dB}	–	100.2 MHz
φ_m	130.5°	126.4°
$A_{U0} = +5, f_{-3dB} = 120\,\text{MHz}, R_F = 1.32\,\text{k}\Omega$ and $R_N = 330\,\Omega$ ($R_F = 1.33\,\text{k}\Omega \pm 1\%$ and $R_N = 332\,\Omega \pm 1\%$), $U_{im} = 100\,\text{mV}$, $R_G = 50\,\Omega$ and $R_T = 50\,\Omega$		
$U_{o,err}$	16.12 mV	16.05 mV
A_{U0}	5	5
Q_p	0.54	0.52
f_{-3dB}	–	121.2 MHz
φ_m	120°	116°
$A_{U0} = +5, f_{-3dB} = 150\,\text{MHz}, R_F = 845\,\Omega$ and $R_N = 211\,\Omega$ ($R_F = 845\,\Omega \pm 1\%$ and $R_N = 210\,\Omega \pm 1\%$), $U_{im} = 100\,\text{mV}$, $R_G = 50\,\Omega$ and $R_T = 50\,\Omega$		
$U_{o,err}$	13.69 mV	13.64 mV
A_{U0}	5	5
Q_p	0.602	0.650
f_{-3dB}	–	140.6 MHz
φ_m	110°	106°

After implementation of a verification check by computer simulations, a combination of simulation and experimental study on various circuits of inverting and non-inverting amplifiers was performed. The tested circuits were implemented on a FR4 PCB laminate with surface-mount device passive components for the resistors and capacitors.

The study of the electronic circuits is performed in two stages. The first stage of computer simulations and experimental study is implemented for the inverting and non-inverting amplifier circuits using CFOA AD8011ARZ, biased with ± 5 V supplies. The values of the chosen passive components for the investigated non-inverting amplifier circuit (Figure 4) are as follows: (1) $R_F = 1\,k\Omega$ and $R_N \to \infty$ at $A_{U0} = 1$; (2) $R_F = 1\,k\Omega$ and $R_N = 1\,k\Omega$ at $A_{U0} = 2$; (3) $R_F = 1\,k\Omega$ and $R_N = 200\,\Omega$ at $A_{U0} = 6$ with tolerances $\pm 1\%$. The resistor R_T is chosen equal to $51\,\Omega$ with tolerance $\pm 1\%$. The coefficient k_ω of ringing of the output signal, according to the formula given in Section 4.1, is 5.53. The a.c. transfer characteristics of the circuits are obtained experimentally by using network analyser HP4195A. To measure the output signal, an active probe type HP41800A with input impedance $100\,k\Omega/3pF$ is used. For the a.c. sweep analysis, the frequency is swept from $100\,kHz$ to $1\,GHz$ by decades, with 100 points per decade. The input voltage source is chosen with amplitude $-10\,dBm$ or $70.8\,mV$ (with initial phase shift equal to zero), to ensure amplitude of the output voltage, not higher than $500\,mV$ at $A_{U0} = 6$. The simulation and experimental results for the module and phase shift at three values of the d.c. voltage gain for the non-inverting amplifier are plotted in Figures 7 and 8. As can be seen for low frequencies approximately up to 10 MHz, the gains (Figure 7) are with constant value and are frequency independent. At A_{U0} equal to 1 and 2 in the form of the frequency response causes peaking and for the voltage follower the amplitude reaches almost 15 dB. Moreover at gains 1 and 2, the module of the transfer function decreases with greater speed, such as for the frequency equal to 500 MHz reaches value equal to -10 dB. The difference between the simulation and the experimental results is due to the influence of the additional parasitic poles determined by the inertial intermediate stages of the CFOA. The simulated values of the zero-pole pair at $A_{U0} = 1$ are $f_z = 30\,MHz$ and $f_p = 250\,MHz$, respectively. For $A_{U0} = 2$, $f_z = 60\,MHz$ and $f_p = 250\,MHz$. At the voltage gain equal to 6, $f_z = 180\,MHz$ and $f_p = 250\,MHz$, the amount of peaking is small, because the frequencies are close and as a result, the transfer characteristic monotonically decreases to unity. The phase margin (Figure 8) for the three gains is greater than $45°$, which means that the amplifier circuits are stable according to the Bode criterion. At gains equal to 1 and 2, the phase shift between the input and the output signal is positive, because $\omega_z < \omega_p$. This additional phase shift has a value less than $50°$, which does not affect the stability of the circuits.

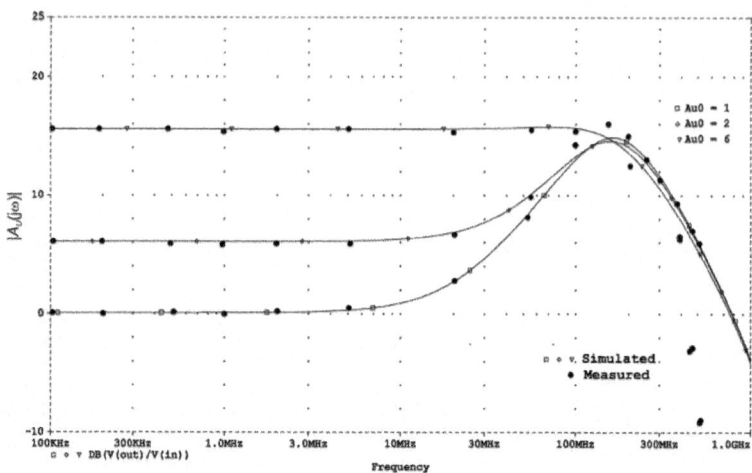

Figure 7: Module of the complex transfer function at $R_F = 1\ \text{k}\Omega$ **and** d.c. voltage gain +1, +2 and +6, respectively.

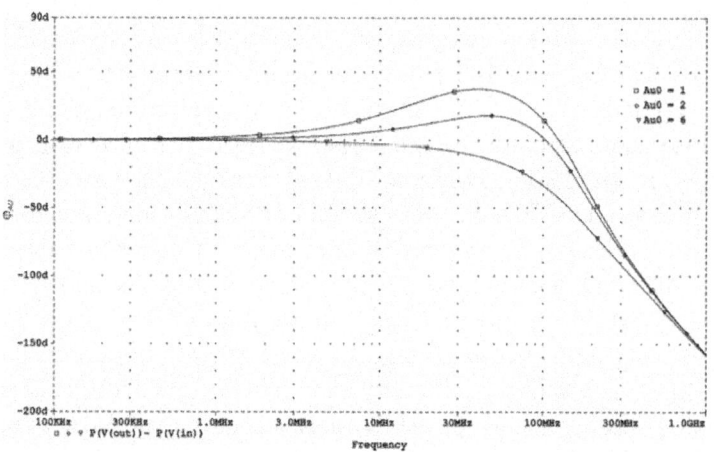

Figure 8: Phase shift of the complex transfer function at $R_F = 1\ \text{k}\Omega$ **and** d.c. voltage gain +1, +2 and +6, respectively.

To verify the results of the theoretical analysis at $R_F = 500\ \Omega$, new non-inverting amplifiers were implemented with the following passive components: (1) $R_F = 500\ \Omega$ and $R_N \to \infty$ at $A_{U0} = 1$; (2) $R_F = 500\ \Omega$ and $R_N = 500\ \Omega$ at $A_{U0} = 2$; (3) $R_F = 500\ \Omega$ and $R_N = 100\ \Omega$ at $A_{U0} = 6$ with tolerances $\pm 1\%$. In this case the coefficient k_ω is equal to 3.91. The simulation results for the transfer characteristics at the three values of the d.c. voltage gain are shown in Figure 9. The comparative analysis of the transfer characteristics in Figure 7 shows that the circuit with the lower value of R_F is with expanded value of working bandwidth and the amplitude of the peaking at voltage gain equal to 1 and 2 is smaller. These simulations and experimental results confirm the correctness of the theoretical analyses and effectiveness of the obtained formulas.

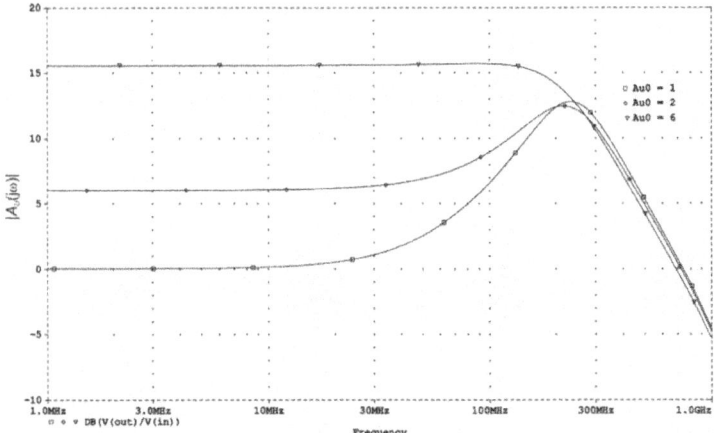

Figure 9: Alternating current transfer characteristics at $R_F = 500\,\Omega$ **and** d.c. voltage gain +1, +2 and +6, respectively.

The inverting amplifier circuit (Figure 2) using CFOA AD8011ARZ is studied for four d.c. voltage gains. The values of the calculated passive components are as follows: (1) $R_F = 1\,\text{k}\Omega$ and $R_N = 1\,\text{k}\Omega$ at $A_{U0} = -1$; (2) $R_F = 1\,\text{k}\Omega$ and $R_N = 499\,\Omega$ at $A_{U0} = -2$; (3) $R_F = 1\,\text{k}\Omega$ and $R_N = 200\,\Omega$ at $A_{U0} = -5$ and (4) $R_F = 499\,\Omega$ and $R_N = 51\,\Omega$ at $A_{U0} = -10$. All resistors were chosen with tolerance $\pm 1\%$. The resistor $R_T = 51\Omega \pm 1\%$ is used for gains -1, -2 and -5. The a.c. transfer characteristic at the four values of the voltage gains is presented in Figure 10. As can be seen, all frequency characteristics are not ringing ($Q_P < 0.707$) and decreased monotonically with increase in the frequency of the input signal. Furthermore, the zero frequency f_z is much greater than the frequency of the double pole. The phase shift between the input and the output signal varies from $+180°$ to $-180°$, as for the frequency approximately equal to 400 MHz the phase shift is $0°$. For gains -1 and -2 the phase margin is greater than $45°$. At gains equal to -5 and -10 for frequencies greater than 240 MHz, the phase margin is less than $45°$ and the circuits are unstable. The phase shift of the output signal varies approximately up to $-90°$. This phase shift is determined by the influence of additional parasitic poles in the frequency response of the op amp.

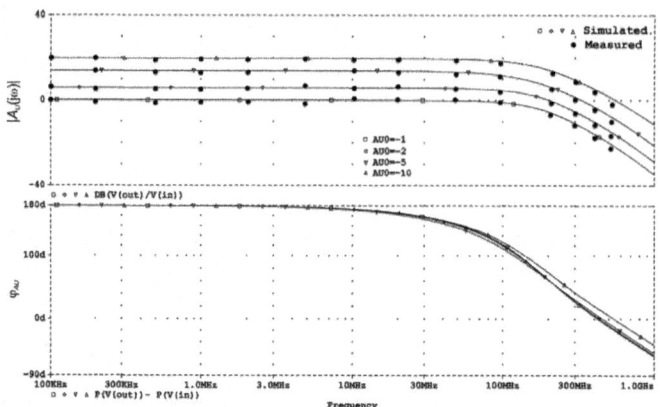

Figure 10: Alternating current transfer characteristics at d.c. voltage gain $-1, -2, -5$ and -10, respectively.

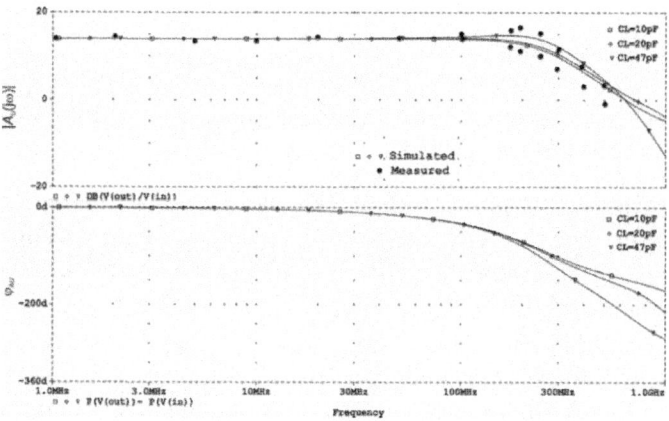

Figure 11: Alternating current transfer characteristics at $R_F = 1000\ \Omega$ and C_L with values $10, 20$ and $47\,\text{pF}$.

The second series of computer simulations and experimental study was performed for the non-inverting amplifier circuits (with $R_F = 1\,\text{k}\Omega$, $R_N = 250\ \Omega$ and $A_{U0} = 5$) at complex (active-capacitive) load employing CFOA THS3091D (using an 8-pin SOIC with PowerPAD™), biased with $\pm 15\,\text{V}$ supplies. The a.c. transfer characteristic at amplitude of the output voltage equal to $4\,\text{V}$ and for three values ($10, 20$ and $47\,\text{pF}$) of the capacitive load of the amplifier is plotted in Figure 11. The R_L was chosen equal to $100\ \Omega$, as the maximum output current is $40\,\text{mA}$. For frequencies up to $30\,\text{MHz}$, the module is with constant value, approximately equal to 5.0 (or $\approx 14\,\text{dB}$). At capacitive load equal to 10 and 20 pF, the quality factor, according to formula (11), is equal to 0.5 and 0.71, respectively. In these cases, the peaks are not observed and the transfer characteristic decreases monotonically with an increase in the frequency of the input signal. At capacitive load equal to 50 pF, the quality factor becomes larger than 0.707 and in the form of transfer characteristic ringing occurs. Furthermore, the phase shift between the input and the output signal increases and at frequency equal to 330 MHz it becomes 135°. At further increase of the frequency, the phase margin becomes smaller than 45°, which decreases the stability of the amplifier.

CONCLUSION

In this paper, a study of the frequency stability analysis for high-speed inverting and non-inverting amplifiers using CFOAs has been presented. Based on the analysis of the principle of operation, equations for the complex transfer functions are obtained, as well as recommendations for improving the stability at complex load and design procedure are defined. The proposed procedure can be useful for the analysis and design of high-speed amplifier circuits employed in various analogue and mixed-signal circuits, such as video amplifiers, line drivers and analogue switches. The results obtained by the theoretical analyses are validated through simulation and experimental testing of sample electronic circuits using monolithic op amps AD8011 and THS3091. The maximum error between calculated values and the simulation results for the basic electrical parameters is not higher than 10%.

ACKNOWLEDGEMENTS

The AD8011 and THS3091 CFOAs, used in this work, were provided by Analog Devices and Texas Instruments, respectively.

REFERENCES

1. AD8011 300 MHz current feedback amplifier – datasheet. (2014). Analog devices. Retrieved from http://www.analog.com/static/imported-files/data_sheets/AD8011.pdf
2. AD8011 SPICE macro-model, rev. A, 9/97. (2014). Analog devices. Retrieved from http://www.analog.com/en/all-operational-amplifiers-op-amps/operational-amplifiers-op-amps/ad8011/products/product.html#product-designtools
3. Boyanov, J., & Shoikova, E. (1989). *Electronic circuits theory*. Sofia: Tehnika.
4. Choosing Between Voltage Feedback (VFB) and Current Feedback (CFB) Op Amps. (2008). Analog devices. Retrieved from http://www.

analog.com/media/en/training-seminars/tutorials/MT-060.pdf.

5. Jassim, H. (2013). A new design technique of CMOS current feedback operational amplifier (CFOA). *Circuits and Systems*, 4(1), 11–15. doi:10.4236/cs.2013.41003 [CrossRef]

6. Jung, W. G. (2002). *Op Amp applications handbook*. Analog devices. Retrieved fromhttp://www.analog.com/library/analogDialogue/archives/39-05/op_amp_applications_handbook.html.

7. Kamath, D. (2014). Bandwidth enhancement of inverting amplifier using composite CFOA block. *International Journal of Innovative Research in Electrical, Electronics, Instrumentation and Control Engineering*, 2(4), 1387–1390.

8. Laker, K., & Sansen, W. (1994). *Design of analog integrated circuits and systems*. New York, NY: McGraw-Hill.

9. Mancini, R. (2001). *Current feedback amplifier analysis and compensation* (Application Report - SLOA021A). Dallas, TX: Texas Instruments Inc.

10. Mancini, R. (2002). *Op amps for everyone. Design guide* (Rev. B). Dallas, TX: Texas Instruments Inc

11. Nagaria, R. K., Gopi Krishna, M., & Singh Rakesh, Kr. (2008). Comparative performance of low voltage CMOS – CFOA suitable for analog VLSI. *WSEAS Transactions on Electronics*, 5(6), 226–231.

12. Palumbo, G. (1997). *Current feedback amplifier: Stability and compensation*. Proceedings of the 40th Midwest Symposium on Circuits and Systems, Sacramento, CA, 249–252.

13. Pandiev, I. (2012). *Analysis and design of CFOA-based high-speed amplifier circuits*. TELFOR, Proceedings of papers, Serbia, Belgrade, 979–982. doi: 10.1109/TELFOR.2012.6419373. [CrossRef]

14. Safari, L., & Azhari, S. (2012). A high performance fully differential pure current mode operational amplifier and its applications. *Journal of Engineering, Science and Technology*, 7(4), 471–486.

15. Schmid, R. (2003). *Measuring board parasitics in high-speed analog design* (Application Report – SBOA094). Dallas, TX: Texas Instruments Inc..

16. Seifart, M. (2003). *Analoge Schaltungen. 6 Auflage*. Berlin: Verlag Technik.

17. Stoianov, I. (2000). *Electronic measurement systems*. Sofia: TU-Sofia.

18. Texas Instruments. (2002). *High speed analog design and application seminar*. Dallas, TX: Texas Instruments Inc.

19. THS3091 current-feedback operational amplifier – datasheet. (2014). Texas Instruments. Retrieved fromhttp://www.ti.com/lit/ds/symlink/ths3091.pdf
20. THS3091 PSpice Model. (2014). Texas Instruments. Retrieved from http://www.ti.com/product/THS3091/toolssoftware
21. Tietze, V., & Schenk, Ch. (2008). *Electronic circuits* (2nd ed.). Berlin: Springer-Verlag

Chapter 12

EQUIVALENT CIRCUIT MODELS FOR OPTICAL AMPLIFIERS

Jau-Ji Jou[1] and Cheng-Kuang Liu[2]
[1]National Kaohsiung University of Applied Sciences Taiwan
[2]National Taiwan University of Science and Technology Taiwan

INTRODUCTION

Electrical equivalent circuit models for optical components are useful as they allow existing, well-developed circuit simulators to be used in design and analysis of optoelectronic devices. A circuit simulator also allows integration with electrical components (package parasitic, laser driver circuit, etc.). Equivalent circuit models were developed and investigated for some optoelectronic circuit elements, including p-i-n diodes, laser diodes, and waveguide modulators (Bononi et al., 1997; Chen et al., 2000; Desai et al., 1993; Jou et al., 2002; Mortazy & Moravvej-Farshi, 2005; Tsou & Pulfrey, 1997). The features of erbium-doped fiber amplifiers (EDFAs) are continuously investigated because of their great importance in optical communication systems. In order to design and analyze the characteristics of EDFAs, it is essential to have an accurate model. A dynamic model of EDFAs is helpful to understand the transient behavior in networks. The EDFA dynamics can also be used to monitor information in optical networks (Murakami et al., 1996; Shimizu et al., 1993). In this chapter, using a new circuit model for EDFAs, the static and dynamic characteristics of EDFAs can be analyzed conveniently through the aid of a SPICE simulator. The dc gain, amplified spontaneous emission (ASE) spectrum, frequency response and transient analysis of EDFAs can be simulated. Semiconductor optical amplifiers (SOAs) are also important components for optical networks. They are very attractive for their wide gain spectrum, and capability of integration with other devices. In the linear regime, they can be used for both booster and in-line amplifiers (O'Mahony, 1988; Settembre et al., 1997; Simon, 1987). Also, much research activities have been done on all-optical signal processing with SOAs (Danielsen et al., 1998; Durhuus et al., 1996). Laser diodes (LDs) are similar devices to SOAs, and they are also the key components for various applications ranging from high-end and high-speed (i.e. fiber communications, and compact-disc players) to low-end and low-speed (i.e. laser pointers, and laser displays) systems. In this chapter, a new unified equivalent circuit model for SOAs and LDs is also presented.

EQUIVALENT CIRCUIT MODEL FOR ERBIUM-DOPED FIBER AMPLIFIERS

Sun et al. (Sun et al., 1996) derived a nonlinear ordinary differential equation to describe EDFA dynamics. Then, Bononi, Rusch, and Tancevski (Bononi et al., 1997) developed an equivalent circuit model to study EDFA dynamics. Based on this equation, Novak and Gieske (Novak & Gieske, 2002) also presented a MATLAB Simulink model of EDFA. However, most EDFA models (Barnard et al., 1994; Freeman & Conradi, 1993; Giles et al., 1989; Novak & Gieske, 2002; Novak & Moesle, 2002) didn't take the ASE into account. Some models or methods of EDFA analysis had been presented with ASE (Araci & Kahraman, 2003; Burgmeier et al., 1998; Ko et al., 1994; Wu & Lowery, 1998), but a complex numerical computation was involved in a model or the ASE was simply taken as an independent light source. Thus, in this section, the Bononi-Rusch-Tancevski model is extended to develop a new equivalent circuit model of EDFAs including ASE. Through the aid of a SPICE simulator, it is convenient to implement the circuit model and to analyze accurately the static and dynamic features of EDFAs.

Circuit Model of EDFA Including ASE

Considering a co-pumped two-level EDFA system, it is assumed that the excited-state absorption and the wavelength dependence of group velocity (v_g) can be ignored. Let the optical beams propagate in z-direction through an EDF of length L. The rate equation and the propagating equations of photon fluxes in time frame can be simplified by transforming to a retarded-time frame moving with v_g. These equations are shown as

$$\left(\frac{1}{\tau}+\frac{\partial}{\partial t}\right)N_2(z,t) = \sum_{k=p,s,A}\left[N_1\sigma_k^a - N_2(z,t)\sigma_k^{ae}\right]\Gamma_k P_k(z,t) \quad (1)$$

$$\frac{\partial P_k(z,t)}{\partial z} = \left[\sigma_k^{ae}N_2(z,t) - \sigma_k^a N_1\right]\Gamma_k P_k(z,t) \quad (2)$$

$$\pm\frac{\partial P_{a,l}^{\pm}(z,t)}{\partial z} = \left[\sigma_l^{ae}N_2(z,t) - \sigma_l^a N_1\right]\Gamma_l P_{a,l}^{\pm}(z,t) + \frac{2\Gamma_l\sigma_l^e\Delta v_{a,l}}{A}N_2(z,t) \quad (3)$$

where $P_k = P_k'/(h\nu_k A)$, P_k' is the power of the kth optical beam, ν_k is the optical frequency, h is Planck›s constant; N_t is the erbium density in the fiber core of effective area A; Γ_k is the overlap factor of the kth beam; τ is the fluorescence lifetime of the metastable level; $\sigma_k^{ae} = \sigma_k^a + \sigma_k^e$, σ_k^a and σ_k^e are respectively the absorption and emission cross sections at the wavelength λ_k; k = p, s, A represent the pump beam (p), signal beam (s), and ASE (A); $P_A(z,t) = P_A^+(z,t) + P_A^-(z,t) = \sum_{l=1}^{m}\left[P_{a,l}^+(z,t) + P_{a,l}^-(z,t)\right]$, m is the number of frequency slots used in the ASE subdivision, $P_{a,l}^+$ and $P_{a,l}^-$ represent the forward and backward ASE fluxes within a frequency slot of width $\Delta\nu_{a,l}$, centered at optical frequency $\nu_{a,l}$ (wavelength $\lambda_{a,l}$). It is noted that s may be replaced by multichannel signals s(1), s(2), ..., and s(M).

By Eqs. (1)-(3), the equations can be obtained

$$\left(\frac{1}{\tau_a} + \frac{d}{dt}\right)\bar{N}_2(t) = \sum_{k=s,p,A}\left[P_k^{in}(t) - P_k^{out}(t)\right] \quad (4)$$

$$P_{s,p}(z=L,t) = P_{s,p}(z=0,t)G_{s,p}(t) \quad (5)$$

$$P_{a,l}^{\pm}(z=\frac{L\pm L}{2},t) \approx P_{a,l}^{\pm}(z=\frac{L\mp L}{2},t)\left[1 + \frac{2\Gamma_l \sigma_l^e \Delta\nu_{a,l}}{A}\int_0^L \frac{N_2(z,t)}{P_{a,l}^{\pm}(z,t)}dz\right]G_l(t) \quad (6)$$

where $\tau_a^{-1} = \tau^{-1} - \sum_{l=1}^{m} 4\Gamma_l \sigma_l^e \Delta\nu_{a,l}/A$, $\bar{N}_2(t) = \int_0^L N_2(z,t)dz$, $P_{s,p}^{in} = P_{s,p}(z=0,t)$, $P_{s,p}^{out} = P_{s,p}(z=L,t)$,

$P_A^{in}(t) = \sum_{l=1}^{m}\left[P_{a,l}^+(z=0,t) + P_{a,l}^-(z=L,t)\right]$, $P_A^{out}(t) = \sum_{l=1}^{m}\left[P_{a,l}^+(z=L,t) + P_{a,l}^-(z=0,t)\right]$, and

$G_k(t) = \exp\left\{\Gamma_k\left[\sigma_k^{ae}\bar{N}_2(t) - \sigma_k^a N_t L\right]\right\}$. For simplicity, let an approximation make in this model: $P_{a,l}^{\pm}(z,t) \approx$ constant in Eq. (6) and write

$$P_{a,l}^{\pm}(z=\frac{L\pm L}{2},t) \approx \left[P_{a,l}^{\pm}(z=\frac{L\mp L}{2},t) + \frac{2\Gamma_l\sigma_l^e\Delta\nu_{a,l}}{A}\bar{N}_2(t)\right]G_l(t) \quad (7)$$

In general, the forward ASE remains constant at moderate pump power if the high-gain EDF length is not too long (around 4m in the case of (Pederson et al., 1990)). The forward ASE grows with pump power if the EDF fiber is long. Moreover, for a long EDF fiber (>10m in this case), the growth (or attenuation) of forward ASE along fiber length can not be ignored if the pump power is large (or small). A subdivision of EDF into small segments is necessary in case of long fiber. A similar conclusion holds for the backward ASE. The validity of the approximation of constant ASE power along the EDF will be shown in next subsection. Subdividing the EDF into n segments with lengths L_i, i = 1, 2, ..., n, an equivalent circuit model of EDFA including ASE contributions is developed for Eqs. (4), (5), and (7), as shown in Fig. 1, where

$V\,N\,(t)$; the subscript i in the $P_{s(M),i}^{in(out)}$, $P_{p,i}^{in(out)}$, $P_{A,i}^{\pm in(out)}$, or $P_{SE,i}^{\pm in(out)}$ represents the number of EDF segments; $I_{total,i}^{in(out)} = \sum \left(P_{s(M),i}^{in(out)} + P_{p,i}^{in(out)} + P_{A,i}^{\pm in(out)} \right)$. M is the number of channels in a multichannel system; $P_{A,i}^{+in} = P_{A,i-1}^{+out} + P_{SE,i}^{+} = \sum_{l=1}^{m} \left(P_{a,l,i-1}^{+out} + P_{se,l,i}^{+} \right)$ with $P_{A,0}^{+out} = 0$; $P_{A,i}^{-in} = P_{A,i+1}^{-out} + P_{SE,i}^{-} = \sum_{l=1}^{m} \left(P_{a,l,i+1}^{-out} + P_{se,l,i}^{-} \right)$ with $P_{A,n+1}^{-out} = 0$; and $P_{se,l,i}^{\pm} = 2\Gamma_l \sigma_l^e \Delta v_{a,l} \bar{N}_{2,i}/A$.

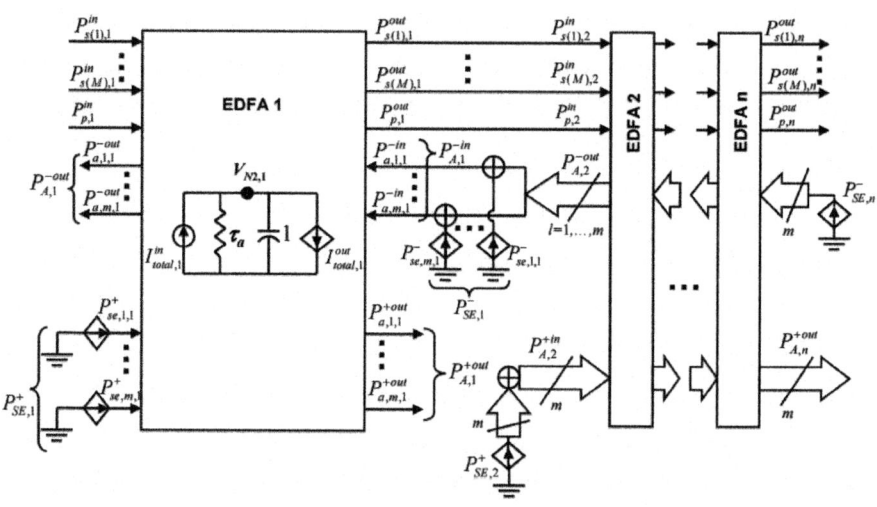

Figure 1: Equivalent circuit model of EDFA including ASE

Static Gain of EDFA

A forward 980nm-pump EDFA with 12m EDF length and 50μW input signal power at 1558nm are considered. Other parameters used in these simulations are: $N_T = 7.7 \times 10^{24} m^{-3}$, $\tau = 10ms$, $A = 2.5 \times 10^{-11} m^2$, $\sigma_s^a = 2.4 \times 10^{-25} m^2$, $\sigma_s^e = 3.8 \times 10^{-25} m^2$, $\sigma_p^a = 2.0 \times 10^{-25} m^2$, and NA = 0.18. These parameters are obtained from manufacturing data or the fitting of experimental gain (Jou et al., 2000; Lai et al., 1999). The dc gain is shown in Fig. 2, as a function of pump power. The square keys represent experimental data (Lai et al., 1999). As a comparison, the numerical calculation is from Eqs. (1)-(3). The numerical results calculated with ASE are shown as the filled circle keys. The EDF length is subdivided into 1200 segments in this numerical computation. The dash curves represent the circuit-model simulation without ASE (Bononi et al., 1997). The solid curve represents the simulation result using this equivalent circuit model. In the circuit models, the parameters used are: the EDF segment number n = 5, the centered wavelength of ASE light λ_a

= 1540nm, the bandwidth Δv_o = 5.06 THz (from 1520nm to 1560nm), and the average cross-section σ_a^e = 5.6×10^{-25}m^2 and σ_a^a = 4.8×10^{-25}m^2.

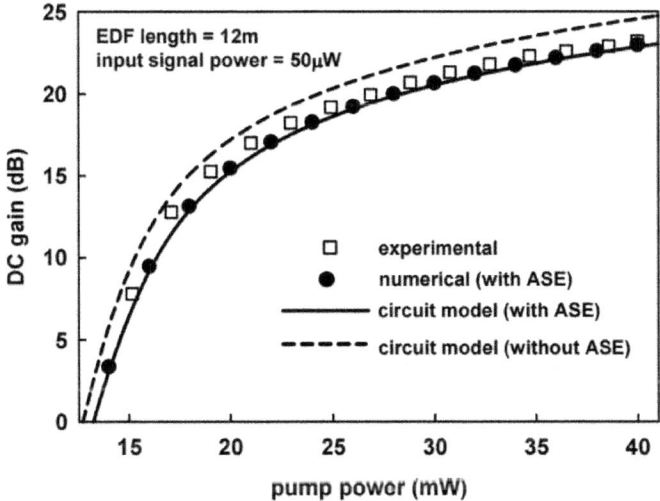

Figure 2: Measured and simulated dc gain versus pump power for EDFA

Because of the ASE influences on EDFA, the dc gain without ASE is higher than the numerical and circuit-model simulations with ASE. Similar result was also reported by (Novak & Moesle, 2002). In Fig. 2, it is observed that the simulation without ASE results in a dc gain of 2dB lager than the experimental data when the pump power is around 25mW. A good agreement between the measured data and the simulation using this circuit model is obtained. The result of this circuit-model simulation is also in very good agreement with the numerical computation.

Generally, if the EDF is subdivided into more segments, the numerical computation results would be more accurate (Yu & Fan, 1999). However, the circuit model of EDFA without ASE should not be subdivided (Bononi et al., 1997), but the EDF should be subdivided for including ASE in this model. The simulated dc gains of the EDFA are shown in Fig. 3(a), using different number of EDF segments. The dc gain deviation is about 0.2dB between the numerical result using 1200 EDF segments and the circuit-model simulation using 3 EDF segments. Using 5 EDF segments, this simulation can be very close to the numerical result. In Fig. 3(b), the total forward and backward ASE power versus pump power can be shown. It is observed that the forward ASE power is lower than the backward one, since a forward pump EDFA is considered here.

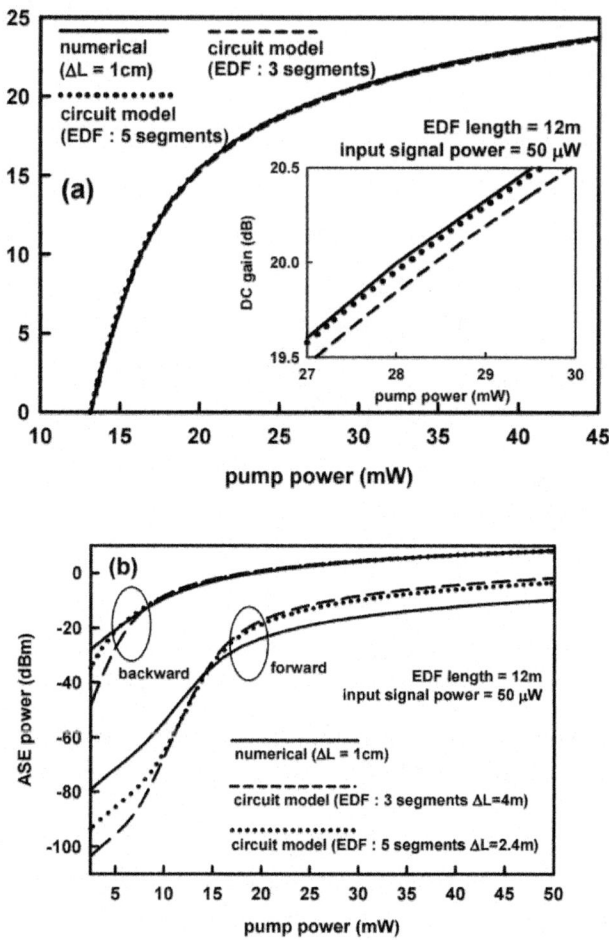

Figure 3: Static characteristics of EDFA for the different number of subdivided EDF segments (a) DC gain versus pump power (b) ASE power versus pump power.

The ASE power of circuit-model simulation is slightly higher than the numerical computation. This makes the dc gain of this circuit-model simulation slightly lower, as expected. When the pump power is around 13mW, the dc gain is around 0dB and the ASE power of circuit-model simulation is in excellent agreement with the numerical computation. The forward ASE power from this circuit model is overestimated when the pump power is above 13mW. This overestimation is believed to be due to this approximation of constant $P_{a,1}^{\pm}(z,t)$ 1 in Eq. (6) and the fact that the forward ASE grows with EDF fiber length in a longer fiber under some given pump power. It is noted that $P_{se,1}^{\pm} = 2\Gamma_1 \sigma_1^e \Delta v_{a,1} \bar{N}_2(t)/A$ A appears in Eq. (7). The use of larger se, $P_{se,1}^{\pm}$ in this circuit-model simulation can lead to the overestimation of forward ASE, especially in the first and last EDF segments. An improvement in the overestimation is observed using 5-segment EDF instead of 3-segment one, but the model with 5-segment EDF would be more complex. Therefore, the tradeoff between the calculation accuracy and the circuit model complexity should be cautiously considered. Below 13 mW pump power, an underestimation of forward ASE can be found in Fig. 3(b). It is also believed to come from this approximation of constant $P_{a,l}^{\pm}(z,t)$ 1 in Eq. (6) and the attenuation of ASE along EDF fiber. It is noted that the smaller $P_{se,1}^{\pm}$ appears in Eq. (7). A similar conclusion can be reached for the backward ASE. However, the agreement between the backward ASE of this circuit model and the numerical computation is better owing to considering a forward pumped EDFA.

ASE Spectrum of EDFA

An ASE spectrum can also be simulated through this circuit model. It is considered that there are not any input signal beams, the pump power is 55mW, the span of ASE spectrum is from 1520nm to 1560nm, and the cross sections in Fig. 2 of (Desurvire & Simpson, 1989) are used in this simulation. In Figs. 4(a) and 4(b), the numerical results of forward and backward ASE spectra with 40 ASE slots (1nm wavelength spacing) are shown as the filled circle keys. Using this circuit model with 1, 3, and 5 ASE slots, the simulation results are shown as the dotted, dash, and solid curves respectively. The stepped ASE spectrum with 5 ASE slots is coarsely similar to the numerical calculation. Although the stepped ASE spectra using 1 and 3 ASE slots are unlike with the numerical calculation, each total ASE power of those spectra is very similar. If the ASE is subdivided into more slots in this circuit-model simulation, the ASE spectrum would be more accurate, but the model would also become more complex. However, the rough spectra of circuit-model simulations are still worthwhile in the estimation of EDFA's characteristics. In Fig. 4(a), it can also be obtained that the forward ASE power of this circuit model simulation is slightly higher than that of the numerical computation, as the result in Fig. 3(b).

Frequency Response of EDFA

Using this circuit model with ASE, the frequency response of EDFA is also analyzed. The 10% signal power modulation index is used in these ac analyses. When the input pump power is 40mW, the ac gain and phase responses are shown in Figs. 5(a) and 5(b), as a function of frequency. The numerical computation is in a good agreement with this circuitmodel simulation and the ac gain response shows a high-pass characteristic (Freeman & Conradi, 1993; Liu et al., 1995; Novak & Gieske, 2002; Novak & Moesle, 2002). When the modulation frequency is below 100Hz, the maximum ac gain deviation can be near 10dB between the simulation results with and without ASE. The peak phase of simulation

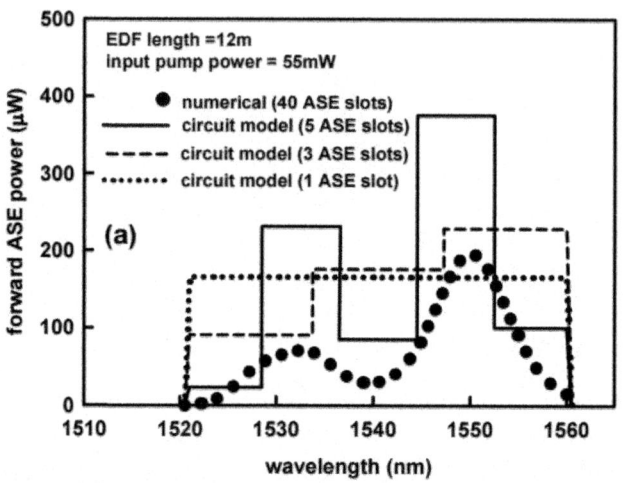

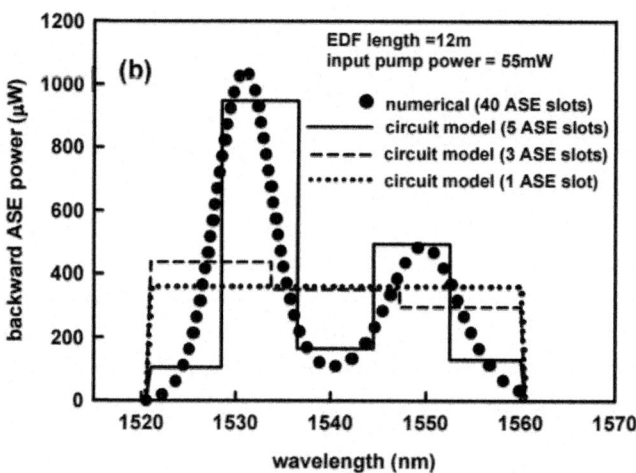

Figure 4: ASE spectra of EDFA (a) Forward ASE spectrum (b) Backward ASE spectrum.

without ASE is about 70 degree at 100Hz. With ASE it is about 30 degree at 230Hz. The deviation of peak phase between the results with and without ASE is quite large. The peak phase position shifts to higher frequency when the simulation includes ASE. The influence of ASE on frequency response of EDFA is significant in low modulation frequency region (below 100Hz). The dc gain of the numerical computation is slightly higher than that of the circuit-model simulation in Fig. 3(a), so the ac gain of the numerical computation is also slightly higher than that of this simulation in low modulation frequency region. After the numerical computation of transient analysis from Eqs. (1)-(3), the ac gain and phase can be estimated or calculated as functions of frequency. This computing process is more complex. However, using the circuit model, the frequency response of EDFA can be obtained easily and rapidly by the frequency-sweep analysis command in a SPICE simulator.

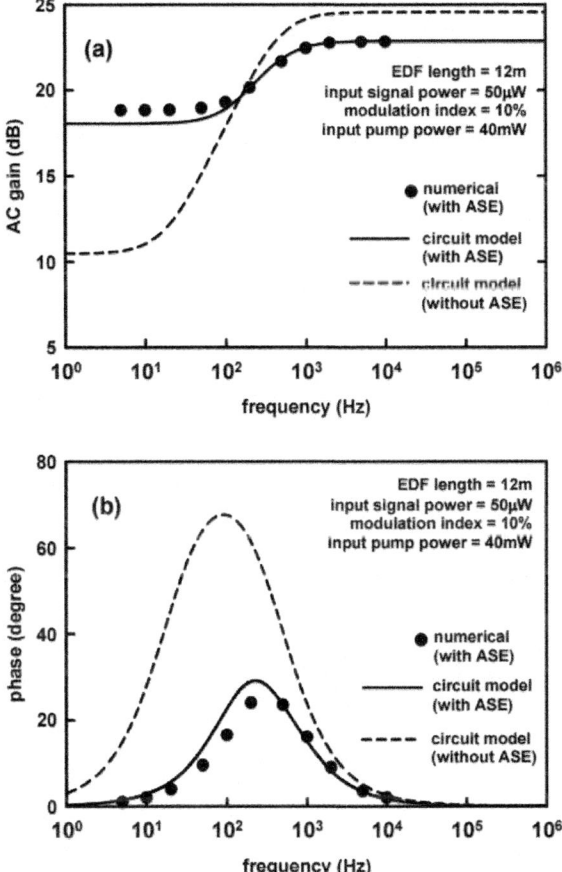

Figure 5: Frequency responses of EDFA (a) AC gain response (b) Phase response.

Transient Analysis of EDFA

Using this circuit model with ASE, the transient response of EDFA dynamics can also be obtained readily through the aid of a SPICE simulator. An eight-channel EDFA system with 35mW pump power at 980nm is considered and seven out of eight channels are added and dropped. These seven channels are represented by a 1558nm signal, while the surviving channel by a 1530nm signal, as shown in Fig. 6(a). (This system can also be regarded as a two-channel system.) $\sigma_s^a = 8.5 \times 10^{-25} m^2$ and $\sigma_s^e = 8.1 \times 10^{-25} m^2$ at 1530nm are used in this simulation. Each channel has 10μW power input to the EDFA. In Fig. 6(b), it is shown that the 1558nm output signal has a large power excursion (Giles et al., 1989; Ko et al., 1994; Wu & Lowery, 1998), when the input signals of seven channels are added simultaneously. According to the above-mentioned results, the dc gain is lower and the ac gain deviation between high and low frequency is smaller, because of the ASE influence on EDFA. Therefore, the power excursion of the transient response is lower in this simulation with ASE, as shown by the features in Fig. 6(b). A good agreement between the numerical computation and this circuit-model simulation is obtained.

When the 1558nm signal is added or dropped, the output power of 1530nm surviving channel can decrease or increase due to the effect of cross talk, as shown in Fig. 6(c). Because of the ASE influence on EDFA, the 1530nm output power of the simulation without ASE is higher, and that of the numerical computation is also slightly higher than that of the circuit model simulation with ASE. A larger difference between the result of the numerical computation and this circuit-model simulation can be observed when the 1558nm signal is dropped.

However, the agreement between this circuit model and the numerical computation is better when the 1558nm signal is added. To include the forward and backward ASE in a transient analysis, the iteration process in the two-point boundary-value problem gets complex. The numerical simulation requires a large amount of computation time and dynamic data storage, and the convergence problem is not easy to deal with. However, using the circuit model, the transient response of EDFA can be obtained conveniently through the aid of a SPICE simulator. Setting a given tolerance criteria and the number of iterations in a SPICE simulator, the convergence problem can be solved easily and the computer time can also be reduced (Avant!, 2001).

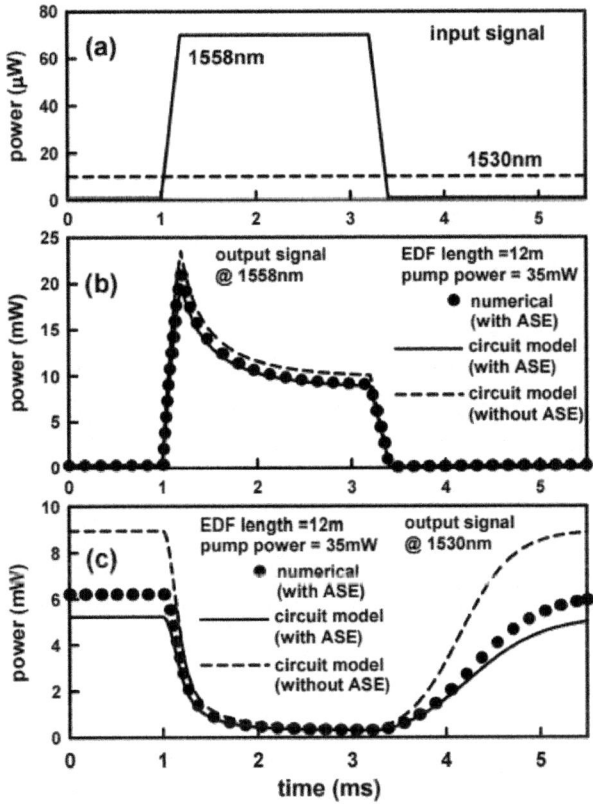

Figure 6: Transient responses of a two-channel EDFA system (a) Input signals (b) Output signal at 1558nm (c) Output signal at 1530nm.

UNIFIED CIRCUIT MODEL FOR SEMICONDUCTOR OPTICAL AMPLIFIERS AND LASER DIODES

There are two primary types of SOAs: traveling wave (TW) SOAs and Fabry-Perot (FP) SOAs. The principle of TW-SOA and FP-SOA is identical, i.e. intrinsic stimulated light amplification. The difference between TW-SOA and FP-SOA is reflectivity of cavity facets. The internal reflectivity of FP-SOA is higher than TW-SOA. Actually, an FP-SOA can be regarded as a FP LD that is biased below the threshold current. The active layer of an SOA has a positive medium gain but not large enough for laser emission. Equivalent circuit models have been separately reported for LDs (Lu et al., 1995; Rossi et al., 1998; Tsou & Pulfrey, 1997) and SOAs (Chen et al., 2000; Chu & Ghafouri-Shiraz, 1994; Sharaiha & Guegan, 2000). However, the principles of SOAs and LDs are extremely similar. In this section, a unified equivalent circuit model is presented for SOAs and LDs.

Circuit model of SOA and LD

Schematic diagrams of a TW-SOA and an FP-SOA are shown in Fig.7 (a) and (b), respectively. TW-SOAs are of a very low internal reflectivity and the incident light is amplified in single pass. FP-SOAs are of a higher reflectivity and incident light can be bounced back and forth within the cavity, resulting in resonance amplification. A basic LD structure is similar to an FP-SOA, but it doesn't need any incident light. Assume that the nonradiative recombination and carrier leakage rate can be neglected; the wavelength dependence of group velocity (v_g) can be ignored; any transport time for carriers to reach the active region is not considered. The rate equation for carrier density N, and the continuity equations for signal photon N± p and ASE photon $N^\pm_p$ and ASE photon $N^\pm_{sp}$ direction can be written as (Coldren & Corzine, 1995)

$$\frac{\partial N}{\partial t'} = \frac{\eta_i I}{qV} - R_{sp}(N) - v_g \left[g_m (N_p^+ + N_p^-) + \langle g_m \rangle (N_{sp}^+ + N_{sp}^-) \right] \quad (8)$$

$$\frac{1}{v_g} \frac{\partial N_p^\pm}{\partial t'} \pm \frac{\partial N_p^\pm}{\partial z'} = \Gamma g_m N_p^\pm - \alpha_i N_p^\pm \quad (9)$$

$$\frac{1}{v_g} \frac{\partial N_{sp}^\pm}{\partial t'} \pm \frac{\partial N_{sp}^\pm}{\partial z'} = \Gamma \langle g_m \rangle N_{sp}^\pm - \langle \alpha_i \rangle N_{sp}^\pm + \Gamma \frac{\beta_{sp}}{v_g} R_{sp} \quad (10)$$

where η_i is the internal quantum efficiency, I is the injected current, V is the volume of the active region, Γ is the confinement factor, $R_{sp}(N)$ is the spontaneous emission rate, $g_m (\approx a(N-N_0))$ is the material gain, a is the differential gain, N0 is the transparency carrier density, αi is the internal loss, and β_{sp} is the spontaneous emission factor. $\langle g_m \rangle$ and $\langle \alpha_i \rangle$ are the spectral average material gain and internal loss over ASE spectrum, respectively. The equations can be simplified by transforming to a retarded-time frame moving with velocity $v_g, t = t' - z/v_g,$ and $z = z'$. Eqs. (9) and (10) can be written as

Equivalent Circuit Models for Optical Amplifiers 235

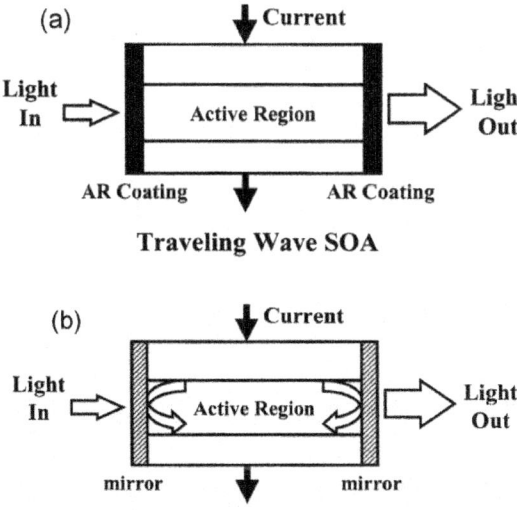

Figure 7: Schematic diagrams of (a) TW-SOA and (b) FP-SOA.

$$\pm \frac{\partial N_p^\pm}{\partial z} = \Gamma g_m N_p^\pm - \alpha_i N_p^\pm \qquad (11)$$

$$\pm \frac{\partial N_{sp}^\pm}{\partial z} = \Gamma \langle g_m \rangle N_{sp}^\pm - \langle \alpha_i \rangle N_{sp}^\pm + \Gamma \frac{\beta_{sp}}{v_g} R_{sp} \qquad (12)$$

Eqs. (11) and (12) are integrated over z from 0 to the length of active region L. The photon $N_p^\pm$ and $N_{sp}^\pm$ can be transformed into the signal power $\pm$ P and the ASE average power s $P_{sp}^\pm$, respectively. Then, by Eqs. (8), (11), and (12), these equations are shown as

$$\frac{\partial N}{\partial t} \approx \frac{\eta_i I}{qV} - R_{sp} - \frac{\Gamma g_m}{h v_s A_s g_s} \left(\frac{\partial P_s^+}{\partial z} - \frac{\partial P_s^-}{\partial z} \right) - \frac{\Gamma \langle g_m \rangle}{h v_s A_s \langle g_s \rangle} \left(\frac{\partial P_{sp}^+}{\partial z} - \frac{\partial P_{sp}^-}{\partial z} \right) \qquad (13)$$

$$P_s^\pm (z = \frac{L \pm L}{2}) = P_s^\pm (z = \frac{L \mp L}{2}) \exp\left(\int_0^L g_s dz \right) \qquad (14)$$

$$P_{sp}^\pm (z = \frac{L \pm L}{2}) = P_{sp}^\pm (z = \frac{L \mp L}{2}) \exp\left(\int_0^L \langle g_s \rangle dz + \beta_{sp} h v_{sp} A_s \int_0^L \frac{R_{sp}}{P_{sp}^\pm} dz \right) \qquad (15)$$

where $g_s = (\Gamma g_m - \alpha_i)$, h is the Plank's constant, vs is the signal frequency, vsp is the ASE central frequency, As is the cross-section area of the active region in z-direction. Assume that the distribution of carrier density N approximates a constant in z-direction and $P_{sp}^\pm (z) \approx P_{sp}^\pm (z = \frac{L \mp L}{2}) \exp(\langle g_s \rangle z)$ is integrated over z from 0 to L. These equations are shown as

$$q\frac{\partial N_T}{\partial t} \approx \eta_i I - qVR_{sp} - \frac{q\Gamma g_m}{hv_s g_s}\left[\left(P_s^{+out} + P_s^{-out}\right) - \left(P_s^{+in} + P_s^{-in}\right)\right]$$
$$-\frac{q\Gamma\langle g_m\rangle}{hv_{sp}\langle g_s\rangle}\left[\left(P_{sp}^{+out} + P_{sp}^{-out}\right) - \left(P_{sp}^{+in} + P_{sp}^{-in}\right)\right] \quad (16)$$

$$P_s^{\pm out} \approx P_s^{\pm in} G_s \quad (17)$$

$$P_{sp}^{\pm out} \approx \left(P_{sp}^{\pm in} + \beta_{sp} hv_{sp} A_s R_{sp}\frac{\langle G_s\rangle - 1}{\langle g_s\rangle\langle G_s\rangle}\right)\langle G_s\rangle = \left(P_{sp}^{\pm in} + P_{sp0}\right)\langle G_s\rangle \quad (18)$$

where N_T is the total carriers in the active region $G_s = \exp(g_s L)$, $P_{s(sp)}^{\pm in} = P_{s(sp)}^{\pm}(z = \frac{L \mp L}{2})$, and $P_{s(sp)}^{\pm out} = P_{s(sp)}^{\pm}(z = \frac{L \pm L}{2})$

The principles of FP-SOA, TW-SOA, and LD are identical, but their boundary conditions are different. The boundary conditions can be considered: 1) FP-SOA: at z = 0, $P_s^{+in} = \left(\sqrt{(1-R)P_{signal}^{in}} + \sqrt{RP_s^{-out}}\right)^2$, where P_{signal}^{in} is the incident signal light power, and $P_{sp}^{+in} = RP_{sp}^{-out}$; at z = L, $P_{s(sp)}^{-in} = RP_{s(sp)}^{+out}$, and $P_{signal}^{out} = (1-R)P_s^{+out}$, where P_{signal}^{out} is the amplified signal output power. 2) TW-SOA: the reflectivity of facets R = 0, $P_s^{-in(out)} = 0$, $P_s^{+in} = P_{signal}^{in}$, and $P_{signal}^{out} = P_s^{+out}$. 3) Laser: no incident signal, $P_s^{\pm in(out)} = 0$, $P_{sp}^{\pm in} = RP_{sp}^{\mp out}$, and $P_{signal}^{\pm out} = (1-R)P_{sp}^{\pm out}$, where $P_{signal}^{\pm out}$ is laser output power in ±z-direction. A unified equivalent circuit model of SOA and LD is developed for the Eqs. (16)-(18), as shown in Fig. 8, where $V_{Nt} = qVR_{sp}$, $E_{Nt}(V_{Nt}) = N_T(R_{sp})$, $P_{s,total}^{in(out)} = g_k\left(P_s^{+in(out)} + P_s^{-in(out)}\right)$, $P_{sp,total}^{in(out)} = \langle g_k\rangle\left(P_{sp}^{+in(out)} + P_{sp}^{-in(out)}\right)$, and $g_k = q\Gamma g_m/(hv_s g_s)$. Lossless transmission lines are employed with time delay $\Delta = L/v_g$ from light beams propagating. Using this circuit model and suitable boundary conditions, the performances of SOA and LD can be analyzed and simulated.

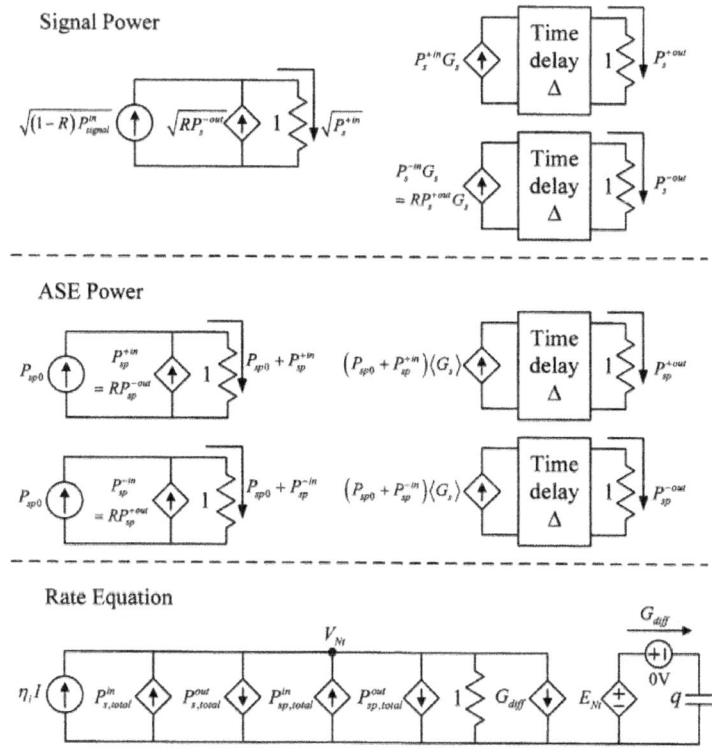

Figure 8: A unified equivalent circuit model of SOA and LD.

Model Validation and Analysis of SOA

To demonstrate the validity of this model, the gain against input signal power is simulated for SOAs. First, it is considered the TW-SOA having $R_{sp} = N/\tau_{sp} = (NT/V)/\tau sp$, τsp is the spontaneous carrier lifetime. With τ_{sp} = 4ns, ηi = 1, R = 0, Γ = 0.5, a = $5\times10^{-16}cm^2$, N0 = $10^{18}cm^{-3}$, L = 200μm, As = 0.3μm², V = 60μm³, h_{vs} = 0.8eV, υ_g = 0.75×10^8m/s, αi = 0cm-1, and β_{sp} = 0 (Adams et al., 1985), the results of the TW-SOA without ASE are shown in Fig. 9. It can be shown as expected that the gain of TW-SOAs becomes higher when the higher current injects or the lower signal light power inputs to the SOA. In Fig. 9, with the 3mA injection current and the -60dBm to -20dBm input signal power, the gain of TW-SOAs is fixed about 5.4dB. The influence of input signal power on gain becomes obvious when the higher current injects. These simulations are shown as the solid curves. The results of Ref. (Adams et al., 1985) are shown as the circle keys. A good agreement between this simulation and Ref. (Adams et al., 1985) is observed.

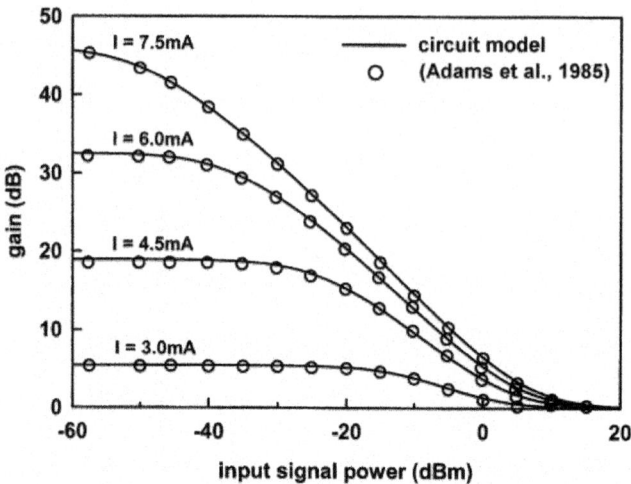

Figure 9: Gain against input signal power for TW-SOA

Next, it is considered the FP-SOA having $R_{sp} = BN^2 = B(N_T/V)^2$, is the bimolecular recombination coefficient.

The same parameters are used in the first example except R = 0.01, B = 10^{-10}cm^3/s, α_i =25cm^{-1}, β_{sp} = 10^{-4} (Adams et al., 1985), and the threshold current Ith ≈3.93mA can be obtained. FP-SOAs must be biased below the threshold current. In Fig. 4, the 0.99Ith, 0.95Ith and 0.9Ith injection currents are used, and the results of the FP-SOA without and with ASE are shown in Fig. 10(a) and (b), respectively. Similar to the TW-SOA, the gain of FP-SOAs becomes higher when the injection current more tends to the threshold current or the lower signal light power inputs to the SOA. In Fig. 10(b), it can be observed that the degeneration of the gain of FP-SOAs is influenced by ASE, it becomes more obvious when the injection current tends to the threshold current. These simulations (solid curves) are in good agreement with Ref. (Adams et al., 1985) (circle keys). The agreement can indicate the validity of this model for TW-SOA and FP-SOA.

In this simplified model shown above, the non-uniformity of carrier density is neglected. In fact, the carrier density is non-uniform along the SOA active region. In this simplified SOA model, the rate equation for spatially averaged values of carrier density is used, and the simulation results of gain are slightly high in comparison with a real SOA (Giuliani & D'Alessandro, 2000). However, the simplified model can be helpful for the coarse definition of SOA's parameters. More accurate results can be obtained by the method of cascading, i.e., by subdividing the SOA into many longitudinal sections and using a simplified model with uniform carrier density for each section (Giuliani & D'Alessandro, 2000). In this simulation, the SOA can be also divided into many sections, and the circuit model as Fig. 8 can be used for each section, as done in the multi-section circuit model of fiber lasers (Liu et al., 2002). However, the tradeoff between the result accuracy and the model complexity should be considered.

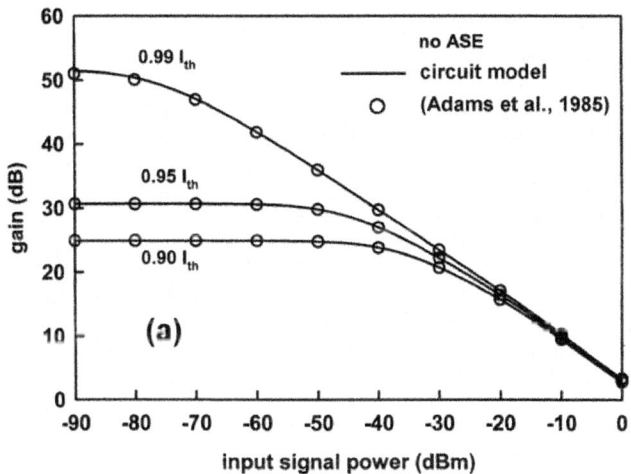

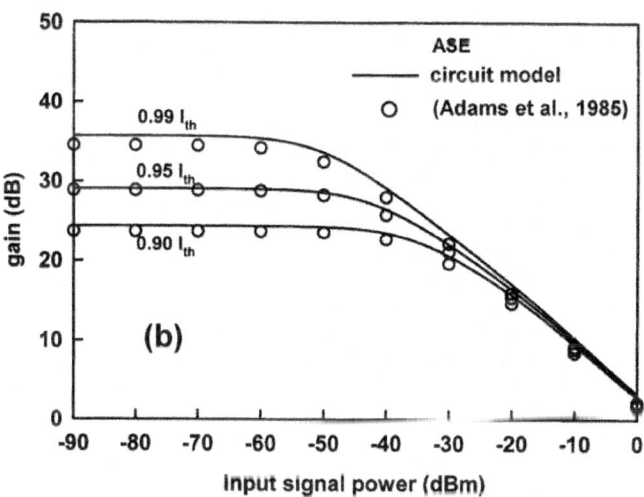

Figure 10: Gain against input signal power for FP-SOA (a) without ASE and (b) with ASE.

Static and Dynamic Analysis of Laser Diode

Then, the characteristics of LD are also analyzed by this model. The parameters used are the same as those of the above example for the FP-SOA. The light output power against injection dc current is simulated for laser diodes having different reflectivity of facets and the results are shown in Fig. 11(a). The higher reflectivity of facets is, and the lower threshold current of the laser is. The threshold current can be expressed as (Liu, 1996),

$$I_{th} = \frac{1}{\eta_i} q V R_{sp}(N_{th}) \tag{19}$$

Where $N_{th} = N_0 + \alpha_{tot}/(\Gamma a)$ and $\alpha_{tot} = \alpha_i - (\ln R)/L$ The threshold current versus reflectivity of facets is shown in Fig. 11(b). The circle keys and the solid curve represent, respectively, these simulations and the results of Eq. (19), and the results of the both methods are agreeable.

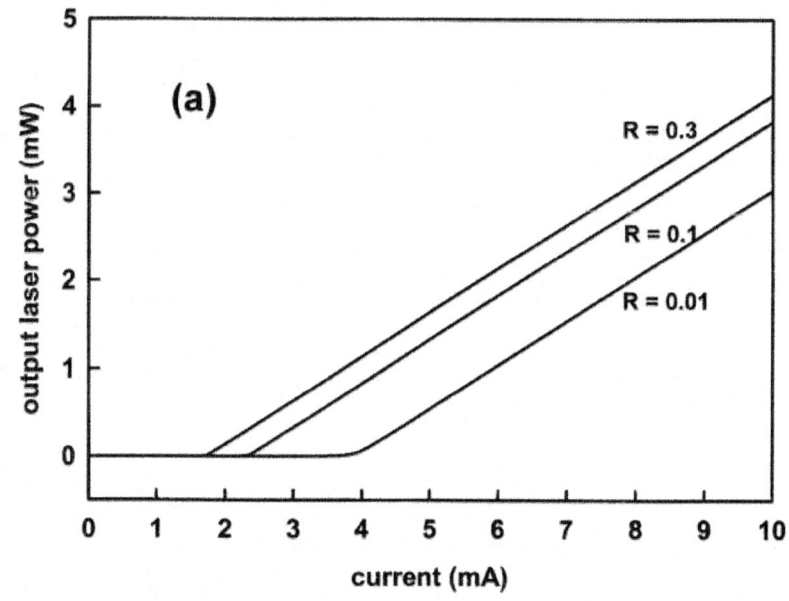

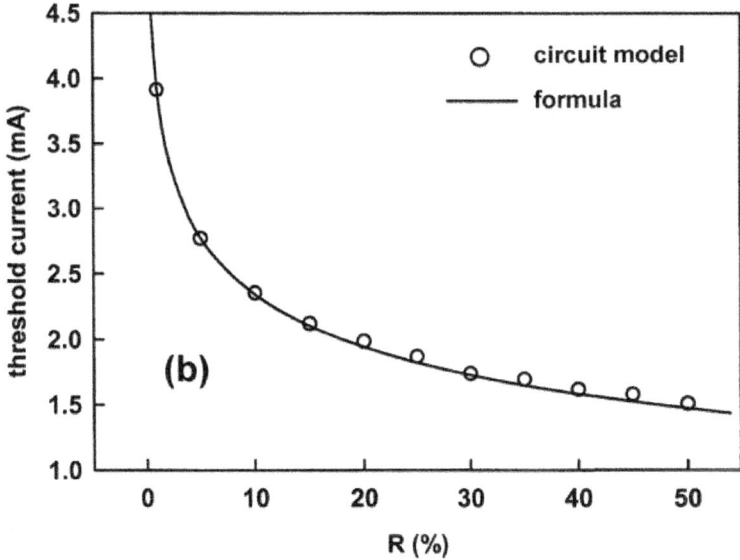

Figure 11: (a) L-I curve of LDs and (b) the threshold current vs. reflectivity of facets

In Fig. 12 (a), the frequency responses of the laser diode are shown using this circuit model. The 10% current modulation index is used in these ac analyses. The responses have different peak values at different bias currents. When the bias current becomes higher, the peak frequency becomes also higher. The peak frequency can be written as (Liu, 1996),

$$f_p = \frac{1}{2\pi}\left[\upsilon_g \alpha_{tot} \frac{R_{sp}(N_{th})}{N_{th}} \frac{I/I_{th} - 1}{1 - N_0/N_{th}}\right]^{0.5}$$

The peak frequency as a function of bias current is shown in Fig. 12(b). The circle keys exhibit these simulations and the results of Eq. (20) are represented by the solid curve. These simulations are in good agreement with the results of this formula.

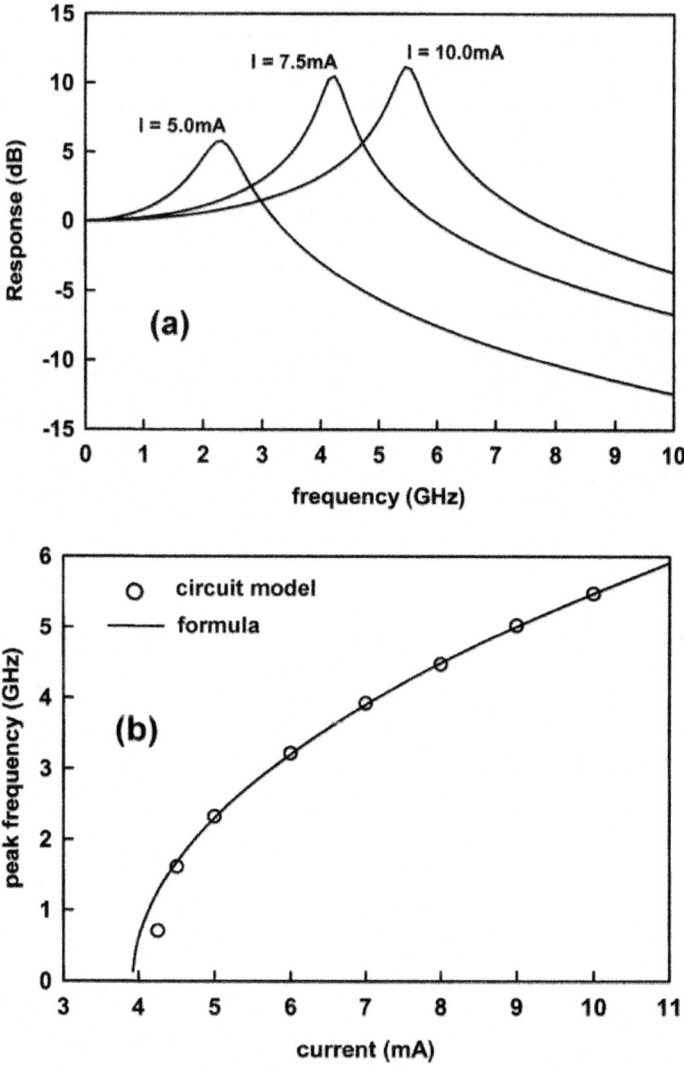

Figure 12: (a) Frequency responses of LDs and (b) the peak frequency vs. bias current

The transient responses of laser diodes are analyzed during the switching of current. The feature of relaxation oscillation and turn-on delay can be observed in this simulation, as shown in Fig. 13(a). When the turn-on current is higher, the relaxation frequency becomes higher and turn-on delay time is shorter. The turn-on delay time and the relaxation frequency can be formularized as (Liu, 1996),

$$t_d = \frac{N_{th}}{R_{sp}(N_{th})} \frac{\eta_i I/(qV)}{\eta_i I/(qV) - R_{sp}(N_{th})} \quad (21)$$

$$f_r = \frac{1}{2\pi}\sqrt{(2\pi f_p)^2 - \alpha^2} \quad (22)$$

Where $\alpha = \frac{1}{2}\frac{dR_{sp}}{dN}\bigg|_{N=N_{th}} + \frac{1}{2}\upsilon_g \alpha_{tot}(2\pi f_p)^2$ As shown in Fig. 13(b), the circle keys and the square keys represent these simulations for the turn-on delay time and the relaxation frequency, respectively. The solid curves are the results of Eqs. (21) and (22). These simulations are in agreement with the results of the formulas.

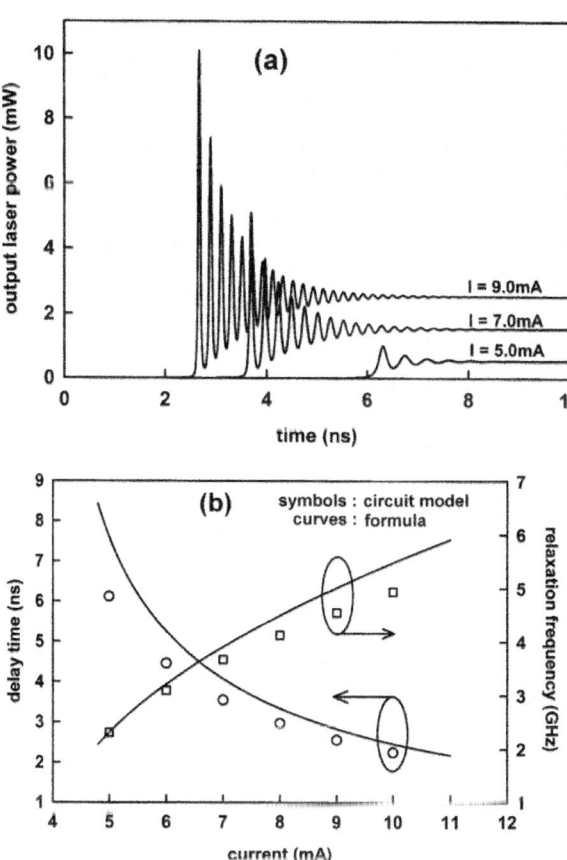

Figure 13: (a) Transient responses of LDs, and (b) the turn-on delay time and the relaxation frequency vs. bias current

CONCLUSIONS

An equivalent circuit model of EDFA including ASE is developed. A good agreement between the measured data and this simulation is obtained in the static gain analysis of EDFA. A rough ASE spectrum of EDFA can also be simulated. Furthermore, these simulation results are also in agreement with the numerical computations in the dynamic gain analysis of EDFA. The transient response analyses of a two-channel EDFA system using this circuit model have been demonstrated. Moreover, this circuit model can also be extended to simulate the multichannel, backward-pump, or bidirectional-pump EDFA systems. Besides, this approach could also be applied to develop other numerical model or Simulink model of EDFAs including ASE. A unified equivalent circuit model of semiconductor optical amplifiers and laser diodes is proposed. The model has been verified by analyzing 1) the gain against input signal power in FP-SOAs and TW-SOAs and 2) the L-I curve, small signal response, and pulse response in laser diodes. The simulation results of this model showed a good agreement with the published results. Through the aid of SPICE circuit simulator, it is convenient to implement these circuit models. These equivalent circuit models of optical amplifiers can be also extended to include other effects or devices such as modulators, fibers, multiplexers, and parasitic components. The circuit models may be of great value for integrated circuit designers requiring an equivalent circuit model for the amplifiers of the optical communication systems in order to simulate accurately the mixed photonic/electronic modules.

REFERENCES

1. Adams, M. J.; Collins, J. V. & Henning, I. D. (1985). Analysis of semiconductor laser optical amplifier. IEE Proc. J. Optoelectron., Vol. 132, No. 1, Feb. 1985, pp. 58-63, ISSN 0267- 3932
2. Araci, I. E. & Kahraman, G. (2003). Performance failure analysis of EDFA cascades in optical DWDM packet-switched networks. J. Lightwave Technol., Vol. 21, No. 5, May 2003, pp. 1156-1163, ISSN 0733-8724
3. Avant! (2001). Star-Hspice Manual Release 2001.2, Avant! Corporation
4. Barnard, C.; Myslinski, P.; Chrostowski, J. & Kavehrad, M. (1994). Analytical model for rareearth-doped fiber amplifiers and lasers. IEEE J. Quantum Electron., Vol. 30, No. 8, Aug. 1994, pp. 1817-1830, ISSN 0018-9197
5. Bononi, A.; Rusch, L. A. & Tancevski, L. (1997). Simple dynamic model of fibre amplifiers and equivalent electrical circuit. Electron. Lett., Vol. 33, No. 22, Oct. 1997, pp.1887- 1888, ISSN 0013-5194

6. Burgmeier, J.; Cords, A.; Marz, R.; Schaffer, C. & Stummer, B. (1998). A block box model of EDFAs operating in WDM systems. J. Lightwave Technol., Vol. 16, No. 7, Jul. 1998, pp. 1271-1275, ISSN 0733-8724

7. Chen, W.; Wang, A.; Zhang, Y., Liu; C. & Liu, S. (2000). Circuit model for traveling wave semiconductor laser. Solid-State Electron., Vol. 44, No. 6, Jun. 2000, pp. 1009-1012, ISSN 0038-1101

8. Chu, C. Y. J. & Ghafouri-Shiraz, H. (1994). Equivalent circuit theory of spontaneous emission power in semiconductor laser optical amplifiers. J. Lightwave Technol., Vol. 12, No. 5, May 1994, pp. 760-767, ISSN 0733-8724

9. Coldren, L. A. & Corzine, S. W. (1995). Diode Lasers and Photonic Integrated Circuit, John Wiley & Sons, ISBN 978-0471118756, New York

10. Danielsen, S. L.; Hansen, P. B. & Stubkjaer, K. E. (1998). Wavelength Conversion in Optical Packet Switching. J. Lightwave Technol., Vol. 16, No. 12, Dec. 1998, pp. 2095-2108, ISSN 0733-8724

11. Desai, N. R.; Hoang, K. V. & Sonek, G. J. (1993). Applications of PSPICE simulation software to the study of optoelectronic integrated circuits and devices. IEEE Trans. Educ., Vol. 36, NO. 4, Nov. 1993, pp. 357-362, ISSN 0018-9359

12. Desurvire, E. & Simpson, J. R. (1989). Amplification of spontaneous emission in erbiumdoped single-mode fibers. J. Lightwave Technol., Vol. 7, No. 5, May 1989, pp. 835-845, ISSN 0733-8724

13. Durhuus, T.; Mikkelsen, B.; Joergensen, C.; Danielsen, S. L. & Stubkjaer, K. E. (1996). Alloptical wavelength conversion by semiconductor optical amplifiers. J. Lightwave Technol., Vol. 14, No. 6, Jun. 1996, pp. 942-954, ISSN 0733-8724

14. Freeman, J. & Conradi, J. (1993). Gain modulation response of erbium-doped fiber amplifiers. IEEE Photon. Technol. Lett., Vol. 5, No. 2, Feb. 1993, pp. 224-226, ISSN 1041-1135

15. Giles, C. R.; Desurvire, E. & Simpson, J. R. (1989). Transient gain and cross talk in erbiumdoped fiber amplifiers. Opt. Lett., Vol. 14, No. 16, Aug. 1989, pp. 880-882, ISSN 0146-9592

16. Giullani, G. & D'Alessandro, D. (2000). Noise analysis of conventional and gain-clamped semiconductor optical amplifiers. J. Lightwave Technol., Vol. 18, No. 9, Sep. 2000, pp. 1256-1263, ISSN 0733-8724

17. Jou, J.-J.; Lai, F.-S.; Chen, B.-H. & Liu, C.-K. (2000). On-line extraction of parameters in erbium-doped fiber amplifiers. J. Chinese Ins. Eng., Vol.

23, No. 5, Sep. 2000, pp. 615- 623, ISSN 0253-3839

18. Jou, J.-J.; Liu, C.-K.; Hsiao, C.-M.; Lin, H.-H. & Lee, H.-C. (2002). Time-delay circuit model of high-speed p-i-n photodiodes. IEEE Photon. Technol. Lett., Vol. 14, No. 4, Apr. 2002, pp. 525-527, ISSN 1041-1135

19. Ko, K. Y.; Demokan, M. S. & Tam, H. Y. (1994). Transient analysis of erbium-doped fiber amplifiers. IEEE Photon. Technol. Lett., Vol. 6, No. 12, Dec. 1994, pp. 1436-1438, ISSN 1041-1135

20. Lai, F.-S.; Jou, J.-J. & Liu, C.-K. (1999). Indicator of amplified spontaneous emission in erbium doped fiber amplifiers. Elecrton. Lett., Vol. 35, No. 7, Apr. 1999, pp. 587-588, ISSN 0013-5194

21. Liu, C.-K.; Jou, J.-J. & Lai, F.-S. (1995). Second-order harmonic distortion and optimal fiber length in erbium-doped fiber amplifiers. IEEE Photon. Technol. Lett., Vol. 7, No. 12, Dec. 1995, pp. 1412-1414, ISSN 1041-1135

22. Liu, C.-K.; Jou, J.-J.; Liaw, S.-K. & Lee, H.-C. (2002). Computer-aided analysis of transients in fiber lasers and gain-clamped fiber amplifiers in ring and line configurations through a circuit simulator. Opt. Commun., Vol. 209, No. 4-6, Aug. 2002, pp. 427-436, ISSN 0030-4018

23. Liu, M. M. K. (1996). Principles and Applications of Optical Communications. Richard D. Irwin, ISBN 978-0256164152, Chicago

24. Lu, M. F.; Deng, J.-S.; Juang, C.; Jou, M. J. & Lee, B. J. (1995). Equivalent circuit model of quantum-well lasers. IEEE J. Quantum Electron., Vol. 31, No. 8, Aug. 1995, pp. 1418- 1422, ISSN 0018-9197

25. Mortazy, E. & Moravvej-Farshi, M. K. (2005). A new model for optical communication systems. Opt. Fiber Technol., Vol. 11, No. 1, Jan. 2005, pp. 69-80, ISSN 1068-5200

26. Murakami, M.; Imai, T. & Aoyama, M. (1996). A remote supervisory system based on subcarrier overmodulation for submarine optical amplifier systems. J. Lightwave Technol., Vol. 14, No. 5, May 1996, pp. 671-677, ISSN 0733-8724

27. Novak, S. & Gieske, R. (2002). Simulink model for EDFA dynamics applied to gain modulation. J. Lightwave Technol., Vol. 20, No. 6, Jun. 2002, pp. 986-992, ISSN 0733- 8724

28. Novak, S. & Moesle, A. (2002). Analytic model for gain modulation in EDFAs. J. Lightwave Technol., Vol. 20, No. 6, Jun. 2002, pp. 975-985, ISSN 0733-8724

29. O'Mahony, M. J. (1988). Semiconductor laser optical amplifiers for use in future fiber systems. J. Lightwave Technol., Vol. 6, No. 4, Apr. 1988, pp. 1556-1562, ISSN 0733- 8724

30. Pederson, B.; Dybdal, K.; Hansen, C. D.; Bjarklev, A.; Povlsen, J. H.; Vendeltorp-Pommer, H. & Larsen, C. C. (1990). Detailed theoretical and experimental investigation of highgain erbium-doped. IEEE Photon. Technol. Lett., Vol. 2, No.12, Dec. 1990, pp. 863- 865, ISSN 1041-1135

31. Rossi, G.; Paoletti, R. & Meliga, M. (1998). SPICE simulation for analysis and design of fast 1.55µm MQW laser diodes. J. Lightwave Technol., Vol. 16, No. 8, Aug. 1998, pp. 1509- 1516, ISSN 0733-8724

32. Settembre, M.; Matera, F.; Hagele, V.; Gabitov, I.; Mattheus, A. W. & Turitsyn, S. K. (1997). Cascaded optical communication systems with in-line semiconductor optical amplifiers. J. Lightwave Technol., Vol. 15, No. 6, Jun. 1997, pp. 962-967, ISSN 0733- 8724

33. Sharaiha, A. & Guegan, M. (2000). Equivalent circuit model for multi-electrode semiconductor optical amplifiers and analysis of inline photodetection in bidirectional transmissions. J. Lightwave Technol., Vol. 18, No. 5, May 2000, pp. 700- 707, ISSN 0733-8724

34. Shimizu, K.; Mizuochi, T. & Kitayama, T. (1993). Supervisory signal transmission experiments over 10000 km by modulated ASE of EDFAs. Electron. Lett., Vol. 29, No. 12, Jun. 1993, pp. 1081-1083, ISSN 0013-5194

35. Simon, J. C. (1987). GaInAsP Semiconductor laser amplifier for single-mode optical fiber communications., J. Lightwave Technol., Vol. 5, No. 9, Sep. 1987, pp. 1286-1295, ISSN 0733-8724

36. Sun, Y.; Luo, G.; Zyskind, J. L.; Saleh, A. A. M.; Srivastave, A. K. & Sulhoff, J. W. (1996). Model for gain dynamics in erbium-doped fibre amplifiers. Electron. Lett., Vol. 32, No. 16, Aug. 1996, pp. 1490-1491, ISSN 0013-5194

37. Tsou, B. P. C. & Pulfrey, D. L. (1997). A versatile SPICE model for quantum-well lasers. IEEE J. Quantum Electron., Vol. 33, No. 2, Feb. 1997, pp. 246-254, ISSN 0018-9197

38. Wu, A. W. T. & Lowery, A. J. (1998). Efficient multiwavelength dynamic model for erbiumdoped fiber amplifier. IEEE J. Quantum Electron., Vol. 34, No. 8, Aug. 1998, pp. 1325- 1331, ISSN 0018-9197

39. Yu, Q. & Fan, C. (1999). Simple dynamic model of all-optical gain-clamped erbium-doped fiber amplifiers. J. Lightwave Technol., Vol. 17, No. 7, Jul. 1999, pp. 1166-1171, ISSN 0733-8724

Chapter 13

THE SWITCHED MODE POWER AMPLIFIERS

Elisa Cipriani, Paolo Colantonio, Franco Giannini and Rocco Giofrè
University of Roma Tor Vergata Italy

INTRODUCTION

The power amplifier (PA) is a key element in transmitter systems, aimed to increase the power level of the signal at its input up to a predefined level required for the transmission purposes. The PA's features are mainly related to the absolute output power levels achievable, together with highest efficiency and linearity behaviour.

From the energetic point of view a PA acts as a device converting supplied dc power (P_{dc}) into microwave power (P_{out}). Therefore, it is obvious that highest efficiency levels become mandatory to reduce such dc power consumption. On the other hand, a linear behaviour is clearly necessary to avoid the corruption of the transmitted signal information. Unfortunately, efficiency and linearity are contrasting requirements, forcing the designer to a suitable trade-off. In general, the design of a PA is related to the operating frequency and application requirements, as well as to the available device technology, often resulting in an exciting challenge for PA designers, since not an unique approach is available. In fact, PAs are employed in a broad range of systems, whose differences are typically reflected back into the technologies adopted for PAs active modules realisation. Moreover, from the designer perspective, to improve PAs efficiency the active devices employed are usually driven into saturation. It implies that a PA has to be considered a non-linear system component, thus requiring dedicated non linear design methodologies to attain the highest available performance.

Nevertheless, for high frequency applications it is possible to identify two main classes of PA design methodologies: the trans-conductance based amplifiers with Harmonic Tuning terminations (HT) (Colantonio et al., 2009) or the Switching-Mode (SM) amplifiers (Grebennikov & Sokal, 2007; Krauss et al., 1980). In the former, the active device acts as a nonlinear current source controlled by the input signal (voltage or current for FET or BJT devices respectively).

A simplest schematic view of such an amplifier for FET is reported in Fig. 1a. Under this assumption, the high efficiency condition is achieved exploiting the device nonlinear behaviour through a suitable selection of both input and output harmonic terminations. More in general, the trans-conductance based amplifiers are identified also as Class A, AB, B to C considering the quiescent active device bias points, resulting in different output current conduction angles from 2p to 0 respectively.

The most famous solution of HT PA is the Class F approach (Gao, 2006; Colantonio et al, 2009), while for high frequency applications and taking into account practical limitations onthe control of harmonic impedances, several solutions have been successfully proposed (Colantonio et al., 2003).

Conversely, in the SM PA, the active device is driven by a very large input signal to act as a ON/OFF switch with the aim to maximise the conversion efficiency reducing the power dissipated in the active devices also. A schematic representation of a SM amplifier is depicted in Fig. 1b. When the active device is turned on, the voltage across its terminals is close to zero and high current is flowing through it. Therefore, in this part of the period the transistor acts as a very low resistance, ideally short circuit (switch closed) minimising the overlap between the current and voltage waveforms. In the other part of the period, the active device is turned off acting as an open circuit. Therefore, the current is theoretically zero while high voltage is present at the device terminals, once again minimising the overlap between voltage and current waveforms. If the active device shows a zero on resistance and an infinite off resistance, a 100% efficiency is theoretically achieved. The latter is of course an advantage over Class A or B, where the maximum theoretical efficiencies are 50 % and 78 % respectively. On the other hand, Class C could achieve high efficiency levels, despite a significant reduction in the maximum output power level achievable (theoretically 100% of efficiency for zero output power). Nevertheless, the HT PAs are intrinsically able to amplify the input signal with higher fidelity, since the active device is basically represented by a controlled current source (FET case) whose output current is directly related to the input voltage. Instead, in SM PAs the active device is assumed to be ideally driven in the ON and OFF states, thus exhibiting a higher nonlinear behaviour. However, this characteristic does not represent a trouble when signals with constant-envelope modulation are adopted.

On the basis of their operating principle, SM amplifiers are often considered as DC to RF converter rather than RF amplifiers.

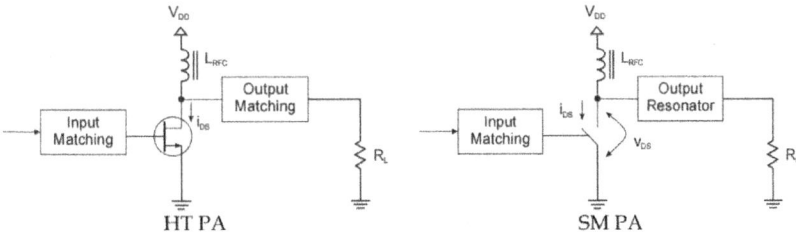

Figure 1: Simplified view of a simple single ended HT (left) and SM (right) PA.

Different SM PA classes of operation have been proposed over the years, namely Class D, S, J, F^{-1} (Cripps, 2002; Kazimierczuk, 2008), while the most famous and adopted is the Class E PA (Sokal & Sokal, 1975; Sokal, 2001) that will be described in deep detail in the following. As will be shown, these classes are based on the same operating principle while their main differences are related to their circuit implementation and current-voltage wave shaping only.

The applications of SM PAs principles have initially been limited to amplifiers at lower frequencies in the megahertz range, due to the active device and package parasitics practically limiting the operating frequencies (Kazimierczuk, 2008). They have also been applied to DC/DC power converters that also operate at lower switch frequencies (Jozwik & Kazimierczuk, 1990; Kazimierczuk & Jozwik, 1990). Recently, their principles of operation have been extended and applied to RF and microwave amplifier design, made possible by the high-performance active devices nowadays available based on silicon (Si), gallium arsenide (GaAs), silicon germanium (SiGe), silicon carbide (SiC), and gallium nitride (GaN) technologies (Lai, 2009).

SWITCHING MODE GENERIC OPERATING PRINCIPLE

The operating principle of every SM PA is based on the idea that the active device operates in saturation, thus it can be represented as a switch and either voltage or current waveforms across it are alternatively minimized to reduce overlap, so minimizing power dissipation in the device itself. If the transistor is an ideal switch, a 100% of efficiency can be achieved by the proper design of the output matching network. As reported in Fig. 1b, the output resonator can be assumed, in the simplest way, as an ideal L-C series resonator at fundamental frequency, terminated on a series load resistance (R_L). The role of the resonator is to shape the voltage and current waveforms across the switch in order to avoid power dissipation at higher harmonics. In fact, an ideal L-C series resonator shows zero impedance at resonating frequency ($\omega_0 = (LC)^{-1/2}$) and infinite impedance for every $\omega \neq \omega 0$. It follows that fundamental current only is flowing into the output load and fundamental voltage only is generated at its terminals. Consequently, 100% of efficiency is obtained (being zeroed the overlap between voltage and current waveforms over the transistor, thus being nulled the power dissipated in it) and no power is delivered at harmonic frequencies in the load, being the latter not allowed to flow into the load R_L.

In actual cases, several losses mechanisms, such as ohmic and capacitive discharge or leakage, cause an unavoidable overlapping between the voltage and current waveforms, together with power dissipation at higher harmonics, thus limiting the maximum achievable efficiency levels.

The most relevant losses in SM PA are represented by:

- parasitic capacitors, such as the device drain to source capacitance C_{ds}. The presence of such capacitance causes a low pass filter behaviour at the output of the active device, affecting the voltage wave shaping with a consequent degradation in the attainable power and efficiency levels. In fact, considering the active device as the parallel connection of a perfect switch and the parasitic capacitance C_{ds}, the higher voltage harmonics are practically shorted by C_{ds} and only few harmonics can be reasonably controlled by the loading network.

- parasitic resistance, such as the drain-to-source resistance when the transistor is conducting R_{ON} (ON state). In fact, due to the non zero resistance when the switch is closed, a relevant amount of active power will be dissipated in the transistor causing a lowering in the achievable efficiency.

- non-zero transition time, due to the presence of parasitic effects, which increase the voltage and current overlap.

- implementation losses due to the components (distributed or lumped elements) employed to realise the required input and output matching networks.

The entity of the parasitic components as well as the associated losses are strictly related to the characteristics of the active device used, especially when designing RF PA (Kazimierczuk, 2008; Lai, 2008).

THE CLASS E AMPLIFIER

Firstly presented in the early 70's in (Sokal & Sokal, 1975), the Class E power amplifier recently received more attention by microwave engineers with the growing demand of high efficient transmitters in wireless communication systems.

It has been widely adopted in constant envelope based communication systems, but represents a valid alternative if combined with envelope varying technique also, like envelope elimination and restoration or Chireix's outphasing technique (Cripps, 2002).

A complete analysis of the Class E amplifier is herein presented, making the assumption of a very idealized active device switching action. The topology considered is the most common one, firstly presented in (Sokal & Sokal, 1975), although different Class E topologies have been conceived and studied during the past (Mader et al., 1998; Grebennikov, 2003; Suetsugu & Kazimierczuk, 2005). In order to clarify Class E operation, a real device-based design is also briefly presented.

Analysis with a Generic Duty Cycle

The basic topology of a single ended Class E power amplifier is depicted in Fig. 2. The active device is schematically represented as an ideal switch and it is shunted by the capacitance C_1, which include the output equivalent capacitance of the active device also. The output network is composed by an ideal filter C_0-L_0 with a series R-L impedance.

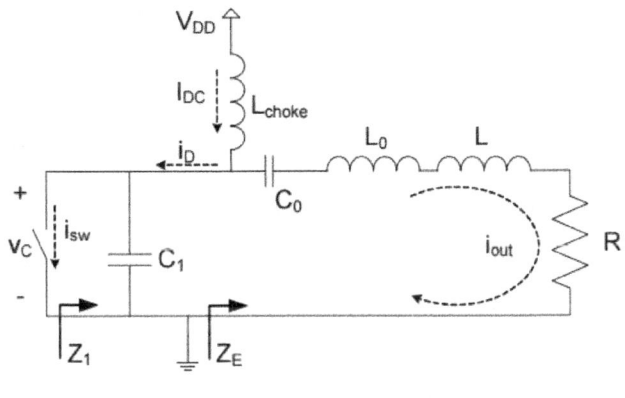

(a)

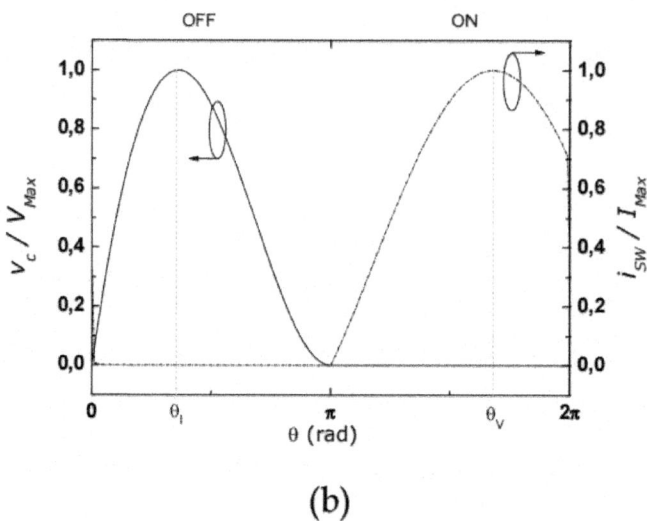

Figure 2: Basic topology of a Class E amplifier (a) corresponding ideal waveforms (b).

Such a circuit is usually analyzed in time domain, which is a straightforward but tedious process, requiring the solution of non linear differential equations. Anyway, some hypotheses can be adopted to carry out a simplified analysis useful to understand the underling operating principle.

Considering the series resonator C_0-L_0 to behave as an ideal filter, i.e. with an infinite (or high enough) Q factor, harmonics and all frequency components different from the fundamental frequency can be considered as filtered out and do not play any role in thesolution of the system. As a consequence, the current flowing into the output branch of the circuit can be assumed as a pure sinusoidal, with its own amplitude I_M and its phase f (Raab, 2001):

$$i_{out} = I_M \cdot \sin(\theta + \phi) \quad (1)$$

Where q=ω·t.

Consequently, from Kirchhoff laws the current i_D (see Fig. 2), which flows entirely through the switch during the ON period (i_{SW}) or entirely through the capacitance C_1 during the OFF period, can be written as:

$$i_D = I_{DC} - I_M \cdot \sin(\theta + \phi) \quad (2)$$

Assuming for simplicity a 50% of duty cycle (the analysis for a generic duty cycle is available in (Suetsugu & Kazimierczuk, 2007)), the current flowing into the switch i_{SW} can be expressed as:

$$i_{SW}(\theta) = \begin{cases} 0, & 0 \leq \theta \leq \pi \\ I_{DC} - I_M \cdot \sin(\theta + \phi), & \pi \leq \theta \leq 2\pi \end{cases} \quad (3)$$

And analogously the current in the capacitor C_1 becomes:

$$i_C(\theta) = \begin{cases} I_{DC} - I_M \cdot \sin(\theta + \phi), & 0 \leq \theta \leq \pi \\ 0, & \pi \leq \theta \leq 2\pi \end{cases} \quad (4)$$

While the voltage across the capacitance v_C can be easily inferred by integration of (4), resulting in the following expression:

$$v_C(\theta) = \begin{cases} \dfrac{1}{\omega \cdot C_1} \cdot \left(I_{DC}\theta + I_M \cdot \cos(\theta + \phi) - I_M \cdot \cos(\phi) \right), & 0 \leq \theta \leq \pi \\ 0, & \pi \leq \theta \leq 2\pi \end{cases} \quad (5)$$

The resulting theoretical current and voltage waveforms are depicted in Fig. 2b. It can be noted that current and voltage across the switch do not overlap, thus no power dissipation exists on the active device. The unique dissipative element in the circuit is the loading resistance R, which is active at fundamental frequency only. Then, from these assumption it follows that the DC to RF power conversion happens without losses and the theoretical efficiency is 100%.

The quantities I_{DC}, I_M and f have still to be determined as functions of maximum current and voltage allowed by the adopted active device, I_{Max} and V_{Max} respectively, and of operating angular frequency w.

For this purpose, it has to note that the capacitance C1 should be completely discharged at the switching turn on, which implies that the voltage v_C has to be null in correspondence of the instant p (see Fig. 2b):

$$v_C(\theta)\big|_{\theta=\pi} = 0 \quad (6)$$

Such condition is usually referred as Zero Voltage Switching (ZVS) condition, which implies that the capacitance C_1 should not be short circuited by the switch turn on when its voltage is still high (Sokal & Sokal, 1975).

The second condition, namely Zero Voltage Derivative Switching (ZVDS) condition, or softswitching condition, implies that the current starts to flow from zero after the switch turn on and then increases gradually, in order to prevent worsening in circuit performance due to mistuning of the waveforms (Sokal & Sokal, 1975). This condition is written as:

$$i_C(\theta)\Big|_{\theta=\pi} = \frac{d}{d\theta} v_C(\theta)\Big|_{\theta=\pi} = 0 \qquad (7)$$

Substituting (4) in the previous equations, from (6) it follows:

$$I_{DC} \cdot \pi - 2 \cdot I_M \cos(\phi) = 0 \qquad (8)$$

While from (7) it follows:

$$I_{DC} + I_M \cdot \sin(\phi) = 0 \qquad (9)$$

Thus the following relationships can be inferred:

$$\tan(\phi) = -\frac{2}{\pi} \qquad (10)$$

$$\frac{I_{DC}}{I_M} = -\sin(\phi) = \frac{2}{\pi} \cdot \cos(\phi) \qquad (11)$$

The maximum current flowing into the switch is given by:

$$I_{Max} = I_{DC} + I_M \qquad (12)$$

And it occurs in correspondence of the angle

$$\theta_I = \frac{3}{2} \cdot \pi - \phi \qquad (13)$$

Similarly, for the voltage across the switch its maximum value occurs in correspondence of the angle θ_V (see Fig. 2b), which can be inferred nulling the derivate of v_c given by (5). Thus, accounting for (11), it follows:

$$\theta_V = -2 \cdot \phi \qquad (14)$$

And

$$V_{Max} = -2 \cdot \phi \cdot \frac{I_{DC}}{\omega C_1} \qquad (15)$$

However, the value of the capacitance C_1 is still an unknown variable. It appears in the definition of the voltage waveform, and it is convenient to use voltage constraints in order to obtain its expression. In fact, its average value must be equal to the supplied DC voltage V_{DD}; thus it follows:

$$V_{DD} = \frac{1}{2\pi}\int_0^\pi v_C(\theta)d\theta \qquad (16)$$

Which solved lead to:

$$V_{DD} = \frac{1}{2\pi}\cdot\frac{1}{\omega\cdot C_1}\cdot\left(I_{DC}\cdot\frac{\pi^2}{2} + 2\cdot I_M \sin(\phi) - I_M\cdot\pi\cdot\cos(\phi)\right) \qquad (17)$$

from which the value of C1 can be finally determined remembering (11)

$$C_1 = \frac{I_{DC}}{\pi\cdot\omega\cdot V_{DD}} \qquad (18)$$

This also suggests a simple relationship between DC current and bias voltage. At this point, waveforms in Fig. 2b have been completely determined in the time domain, without recurring to the frequency domain. However, the remaining elements of the circuit, DC power, output power and output impedance have still to be determined. As stated before, all the DC power is converted to RF power and dissipated into the load resistance at fundamental frequency:

$$P_{DC} = I_{DC}\cdot V_{DD} = \frac{1}{2}\cdot I_M \cdot V_M = P_{RF} \qquad (19)$$

where V_M is the amplitude of fundamental component of the voltage across R which can be obtained by (19) and replacing (11):

$$V_M = 2\cdot\frac{V_{DD}\cdot I_{DC}}{I_M} = -2\cdot V_{DD}\cdot\sin(\phi) \qquad (20)$$

The value of the resistance R is simply obtained as the ratio between V_M and I_M:

$$R = \frac{V_M}{I_M} = 2\cdot\frac{V_{DD}}{I_{DC}}\cdot\sin(\phi)^2 \qquad (21)$$

Clearly, if a standard 50 Ohm termination is required, an impedance transformer is necessary to adapt such load to the required R value. Finally, the inductance L is computed taking into account that its reactive energy is exchanged, at every cycle, with the capacitance C_1. Thus it follows:

$$\frac{1}{2\cdot\pi}\cdot\frac{1}{\omega C_1}\int_0^\pi \left[I_{DC} - I_M \cdot \sin(\theta+\phi)\right]^2 d\theta = \frac{1}{2}\cdot\omega\cdot L\cdot I_M^2 \qquad (22)$$

where the expression in the integral represents the voltage across the capacitance C1 during the OFF period. The value for the inductance L is therefore given by:

$$L = \frac{1}{\omega\cdot C_1}\cdot\left(\frac{\pi}{2} - \frac{4}{\pi}\cdot\cos(\phi)^2\right) \qquad (23)$$

Alternatively, R and L can be found by calculation off in-phase and quadrature voltage components, as elsewhere reported (Mader et al., 1998; Cripps, 1999). The series impedance R-L can be put together in order to obtain a more compact and useful expression for the output branch impedance (Mader et al., 1998) normalized to the shunt capacitance C_1:

$$Z_E = \frac{0.28}{\omega C_1}\cdot e^{j49°} \qquad (24)$$

With reference to Fig. 2, the impedance Z_1 seen by the ideal switch is obtained by the shunt connection of the capacitance C_1 and Z_E and is herein given in its simplified formulation (Colantonio et al., 2005):

$$Z_1 = \frac{0.35}{\omega C_1}\cdot e^{j36°} \qquad (25)$$

Remaining reactive components, L_0 and C_0, are easily calculated by means of:

$$\omega^2 = \frac{1}{L_0\cdot C_0} \qquad (26)$$

Provided a high enough Q factor, the values of L_0 and C_0 are non uniquely defined and any pair of resonant element can be used.

The analysis performed here was intended for the most common case of 50% duty cycle (i.e. p conduction angle). In this case the relations are greatly simplified thanks to the properties of trigonometric functions. However, Class E approach is possible for any value of duty cycle: a detailed analysis can be found in (Suetsugu & Kazimierczuk, 2007; Colantonio et al., 2009) where all electrical properties and component values are evaluated as a function of duty cycle. It can be demonstrated that under ideal assumption the maximum output power does not occur in correspondence of a 0.5 duty cycle, but for a slightly higher value (0.511). Anyway, in terms of output power capability, this increment is extremely low (about 1‰) and a standard 0.5 duty cycle could be assumed in the design, unless differently required.

A Class E Design Example

In order to illustrate the application of the relations obtained in the analysis, a simple Class E design example is described, based on an actual active device, specifically a GaAs pHEMT.

The device exhibits a breakdown voltage of about 25 V and a maximum output current of 400 mA. From S-parameter simulation, an output capacitance of 0.35pF results at 2.5 GHz, the selected operating frequency. Considering this capacitance as the minimum value for the shunt capacitance C_1, the network elements can be easily calculated through the previous relationships.

From (20) and taking into account the maximum voltage, the bias voltage is set to V_{DD}=6V. Hence, from the inversion of (18), the DC component of drain current is determined, resulting in I_{DC}=105 mA.

At this point, using (21) and (23) or, alternatively, equation (24), the values of output matching network are R=33W and L=1.67 nH. If considering a standard output impedance of 50W, a transforming stage is necessary.

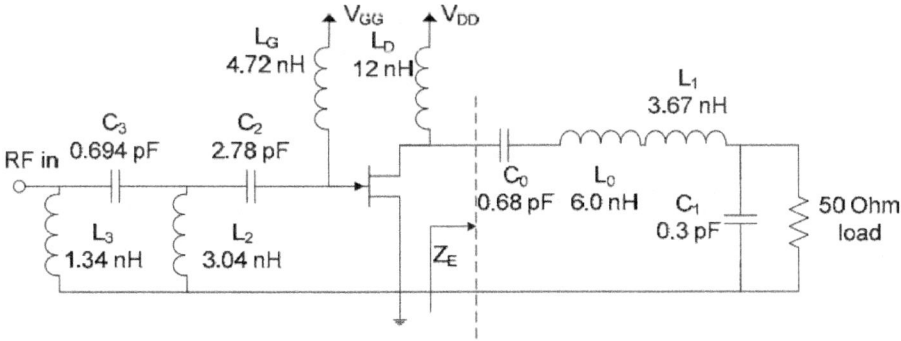

Figure 3: Schematic of a 2.5GHz GaAs HEMT Class E amplifier.

Standing the value of optimum load, the impedance matching can be easily accomplished by a single L-C cell. A series inductance - parallel capacitance configuration has been chosen. Lumped elements for the filtering output network have then determined, selecting an inductance L0=6nH and a resulting capacitance C0=0.68pF. The complete amplifier schematic is depicted in Fig.

3, while the simulated output power, gain and efficiency versus input power are shown in Fig. 4.

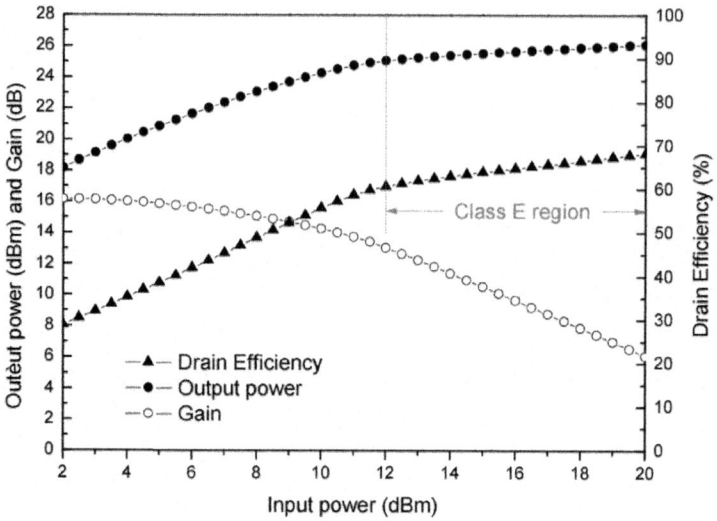

Figure 4: Simulated performance of the 2.5GHz Class E amplifier.

It is worth to notice that, under a continuous wave excitation, Class E behavior is achieved only at a certain level of compression, i.e. when the large input sinusoidal waveform implies a "square-shaping" effect on the output current, due to active device physical limits, thusapproaching a switching behavior. The output current and voltage waveforms and load line are reported in Fig. 5, showing a good agreement with the theoretical expected behavior (compare with ideal waveforms depicted in Fig. 2b).

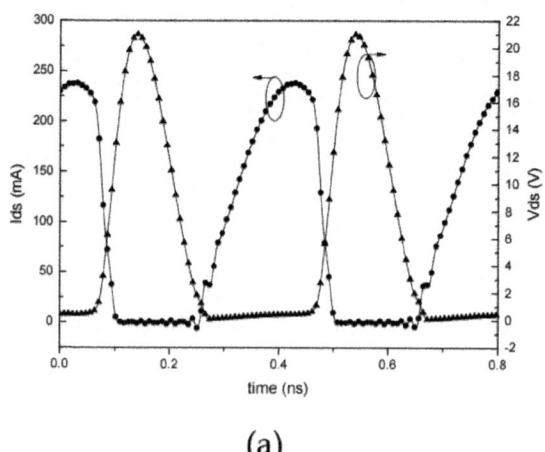

(a)

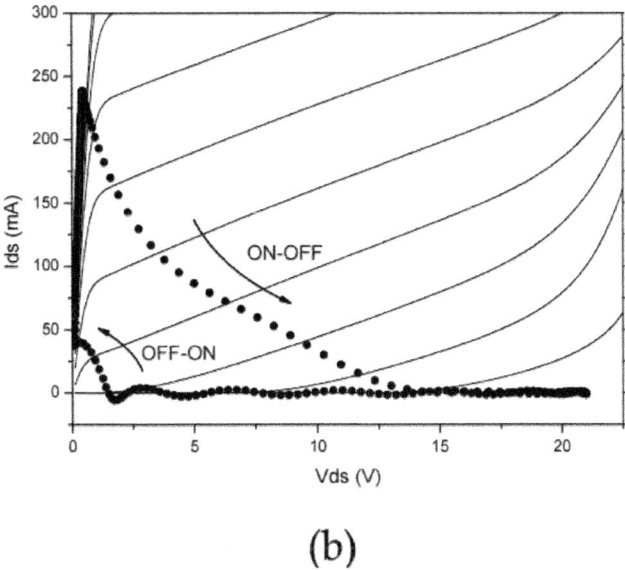

(b)

Figure 5: Output current and voltage waveforms (a) and load line (b) of the 2.5GHz Class E amplifier.

Drawbacks

As already outlined, Class E power amplifiers have some practical limitations, mainly due to their maximum operating frequency. Such limitations are partially related to the cut-off frequency of the active device, while are mainly due to the circuit topology and switching operation. In fact, as reported in (Mader et al., 1998), a Class E maximum frequency can be approximated by:

$$f_{Max} = \frac{I_{DS}}{2\pi^2 \cdot C_1 \cdot V_{DD}} \approx \frac{I_{Max}}{56.5 \cdot C_1 \cdot V_{DD}} \tag{27}$$

Practically the lower limit of C_1 is given by the active device output capacitance C_{ds}. Consequently, the value of maximum operating frequency strongly depends on the device adopted for the design, on its size and then on the maximum current it can handle. For RF and microwave devices, the maximum frequency in Class E operation is generally included between hundred of megahertz (for MOS devices) and few gigahertz (for small MESFET or pHEMT transistors).

Additionally, at microwave frequencies higher order voltage harmonic components can be considered as practically shorted by the shunt capacitance, and the Class E behavior has to be clearly approximated. In particular, the voltage wave shaping can be performed recurring to the first harmonic components only (Raab, 2001; Mader et al., 1998), while the ZVS and ZVDS conditions cannot be longer satisfied.

Truncating the ideal voltage Fourier series at the third component, the resulting waveform is reported in Fig. 6, from which it can be noted the existence of negative values. Thus it becomes mandatory to prevent such negative values of drain voltage to respect active device physical constraint and safely operations.

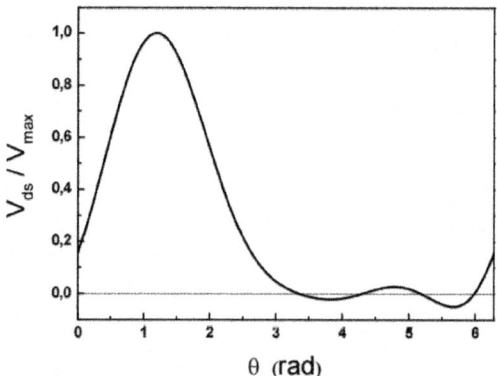

Figure 6: Three harmonics reconstructed voltage waveform.

As pointed out in (Colantonio et al., 2005), two solutions can be adopted. Obviously it is possible to increase drain bias voltage, but it would mean a non negligible increase in the DC dissipated power that in turn causes a decrease in drain efficiency levels. In addition, an increasing on peak voltage value could exceed breakdown limitations of the transistor. The other solution is based on the assumption of unaffected current harmonic components, thus optimizing the voltage fundamental component, while keeping fixed the other harmonics imposed by the network topology (i.e. the filter L_0-C_0 behavior and the device capacitance C_{ds}) (Cipriani et al., 2008). The optimization process must be implemented in a numerical form in order to reduce complexity and computing effort. The main goal is to avoid negative voltage values on drain voltage and, at the same time, maximize output power, hence efficiency. Then, for every value of frequency exceeding the maximum one, the optimum high frequency fundamental impedance, $Z_{1,HF}$, is optimized in magnitude and phase.

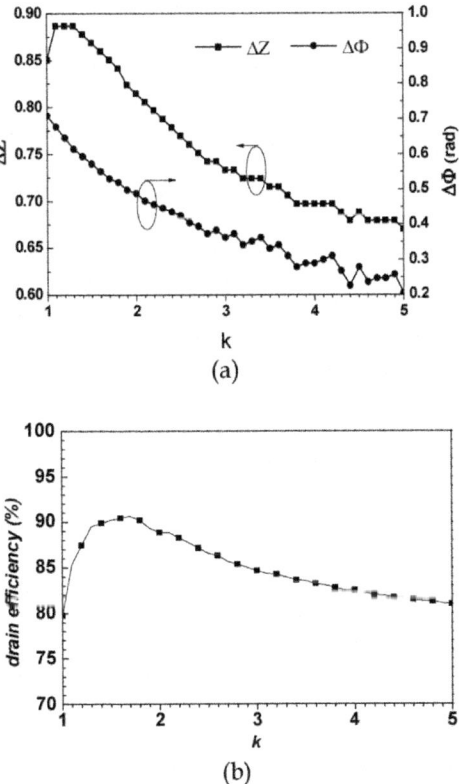

Figure 7: Class E optimum load impedance (a) and ideal efficiency (b) as a function of k.

Obtained results are expressed as normalized to ideal load impedance at maximum frequency given by (27) and depicted in Fig. 7a as a function of normalized frequency $k=f/f_{Max}$, defined as the ratio between the assumed operating frequency f and the maximum allowed one f_{Max}, defined in (27). The plot shown in Fig. 7a can be considered as a "design chart" for high frequency Class E development, once the maximum frequency is known. A quasi monotonic decrease of magnitude of fundamental impedance is observed, leading to a reduced voltage fundamental component and a reduced output power. At the same time, the phase decreases tending to almost purely resistive values. The related drain efficiency is reported in Fig. 7b, showing an increasing reduction with respect the ideal 100% value due to non ideal operating conditions.

A high frequency Class E PA example is shown in Fig. 8.

Figure 8: A 2.14 GHz LDMOS Class E PA.

The amplifier is designed using a medium power LDMOS transistor for base station application at 2.14 GHz. Once the bias point, maximum current and output capacitance of the transistor are fixed, the maximum frequency in Class E mode is directly derived. Considering a maximum current of 2.5 A, a bias voltage of 20 V and an output capacitance of 4.2 pF (estimated by S- parameters simulation), the maximum frequency in Class E results in 520 MHz, far below the frequency chosen for the design. If operating at 520 MHz, the load impedance would be $Z_1=25.1e^{j36°}$. At 2.14 GHz (4.1 times above f_{Max}), the load impedance is directly obtained by the design chart of Fig. 7a resulting in $Z_{1,HF}=17.5e^{j17°}$. Since a very simple equivalent model of the device is used, as Fig. 2 shows, this impedance is seen at the nonlinear current source terminals, so that eventual parasitic and package effects should be considered as belonging to the output load.

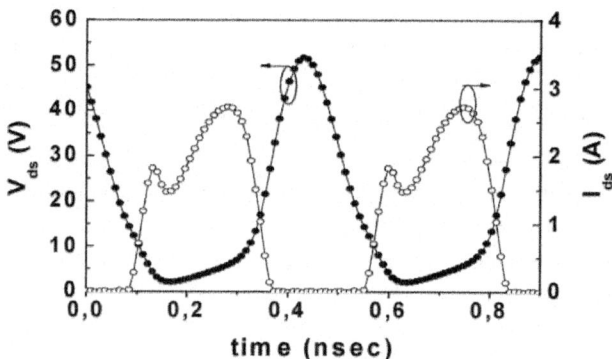

Figure 9: Simulated voltage and current drain waveform of the designed PA.

Simulated drain current and voltage waveform are depicted in Fig. 9, with reference to the internal nodes of the model. A good agreement with typical Class E waveform isremarkable, above the maximum frequency, although the perfect switching behavior cannot be satisfied.

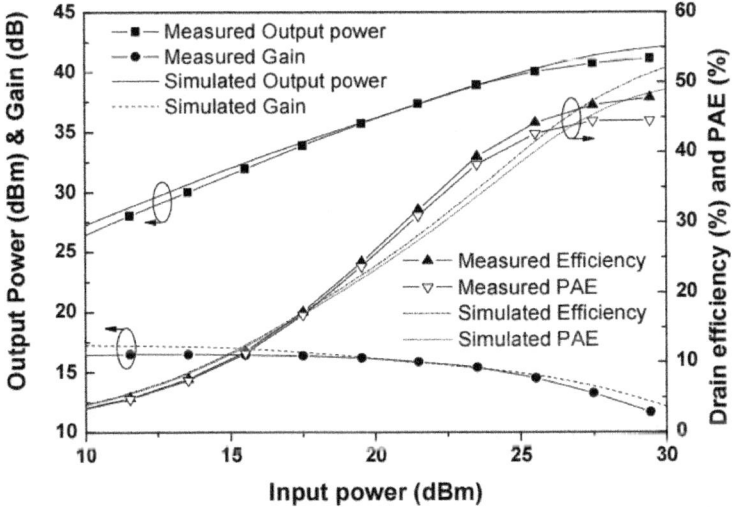

Figure 10: Simulated and measured output performance of the designed PA.

Class E Matching Network Implementation and Other Topologies

Although Class E approach has basically a fixed circuit topology, different solution can be adopted for the synthesis of output matching network. Depending on the design frequency, distributed approaches are possible: proper load conditions at fundamental have to be satisfied, according to (24), and an open circuit condition must be provided at harmonics of the fundamental frequency, usually second and third harmonics (Negra et al., 2007; Wilkinson & Everard, 2001; Xu et al., 2006).

In Fig. 11a, no lumped elements are used unless block capacitors, while the 50W matching is synthesized through a very compact and simple structure reported in Fig. 11b (Mader et al., 1998). In order to provide harmonic suppression on the load, different quarter-wave open stubs can be used at different harmonics (Negra et al., 2007), while series transmission lines and wave impedances are properly chosen to provide the correct fundamental load.

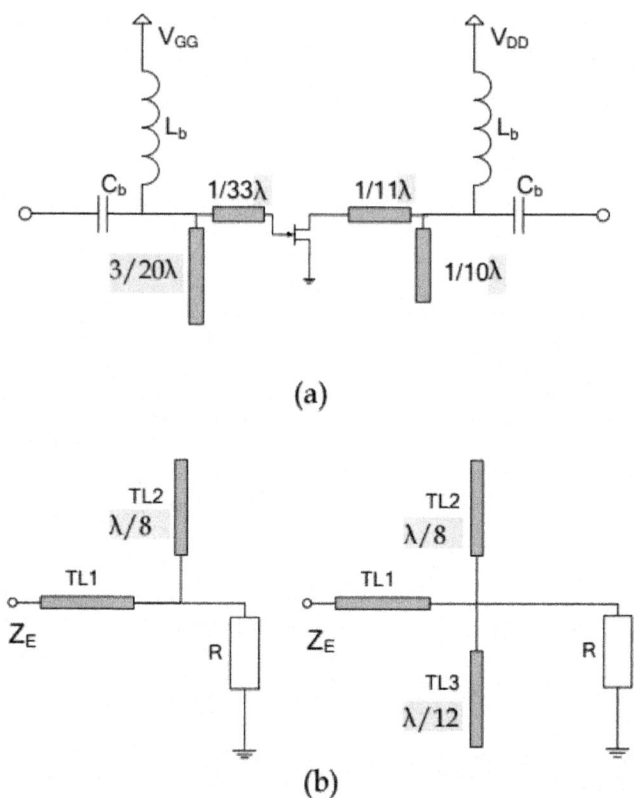

Figure 11: Some practical examples of Class E transmission lines amplifiers.

Additionally, different circuit topologies exist that can provide the same results as the classical formulation: they have been widely investigated in (Grebennikov, 2003) and are commonly referred as parallel circuit Class E and their main characteristic is the presence of a finite parallel inductor in the output network, required for the output device biasing supply, as reported in Fig. 12. As before assumed, the shunt capacitance C includes the transistor output capacitance C_{ds}.

The first circuit, in Fig. 12a, employs a very simple output matching network, which consist of a parallel inductor and a series blocking capacitance. Applying ZVS and ZVDS conditions on this circuit, and considering the transistor to behave as a perfect switch, the solution of the circuit is given by a second order non homogeneous differential equation, given by (28), which has to be solved in order to determine the value of all circuit parameters.

$$\omega^2 LC \frac{d^2 v_s(\omega t)}{d(\omega t)^2} + \frac{\omega L}{R} \frac{dv_s(\omega t)}{d(\omega t)} + v_s(\omega t) = V_{DD}$$
(28)

The values of reactive components, L and C, are:

$$C = \frac{1.025}{\omega R} \qquad L = 0.41 \frac{R}{\omega}$$
(29)

and the load resistance R is determined using the desired output power at fundamental frequency, P_1:

$$R = 1.394 \frac{V_{DD}^2}{P_1}$$
(30)

Due to the lack of any filtering action at the output, this circuit becomes not practical in applications - like telecommunications - which require harmonic suppression (Grebennikov, 2003). Moreover, a higher peak current value is obtained ($4.0 I_{DC}$ instead of $2.862\ I_{DC}$ for the classical topology) that has to be taken into account in the choice of the active device. The circuit in Fig. 12b adds a series LC filter in the output branch and it is very similar to a canonic Class E amplifier using a finite DC feed inductance, unless for the absence of the "tuning" series inductance. Providing a high Q factor for the LC series filter, the current iR flowing into the output branch can be assumed as sinusoidal: this hypothesis is used as starting point for a complete time domain analysis which is similar to what reported in paragraph 4.1. Optimum parallel capacitance C and optimum load resistance R are obtained after inferring the phase angle between in-phase and quadrature components of fundamental current:

$$\psi = \arctan\left(\frac{R}{\omega L} - \omega RC\right) = 34.244°$$
(31)

From which:

$$C = \frac{0.685}{\omega R} \qquad L = 0.732 \frac{R}{\omega}$$
(32)

Output resistance R is derived from desired fundamental output power, P_1:

$$R = 1.365 \frac{V_{DD}^2}{P_1}$$
(33)

Slightly different voltage and current peak values (Grebennikov, 2003) are obtained with respect to the traditional Class E approach:

$$V_{Max} = 3.647 \cdot V_{DD} \qquad I_{Max} = 2.647 \cdot I_{DC} \qquad (34)$$

In Fig. 12c the parallel inductance is replaced by a short-circuited short length transmission line: this solution is quite popular at microwave frequencies. In order to approximate the Class E optimum impedance at fundamental frequency, the electrical length and the characteristic impedance of the transmission line are determined starting from the optimum fundamental impedance and according to the relation (Grebennikov, 2003):

$$Z_0 \cdot \tan(\theta) = \omega L. \qquad (35)$$

The load impedance Z_E seen at device terminals should satisfy the optimum impedance at fundamental frequency, and remembering relation (31) it is rewritten as:

$$Z_E = \frac{R}{1 - j\tan\psi} \qquad (36)$$

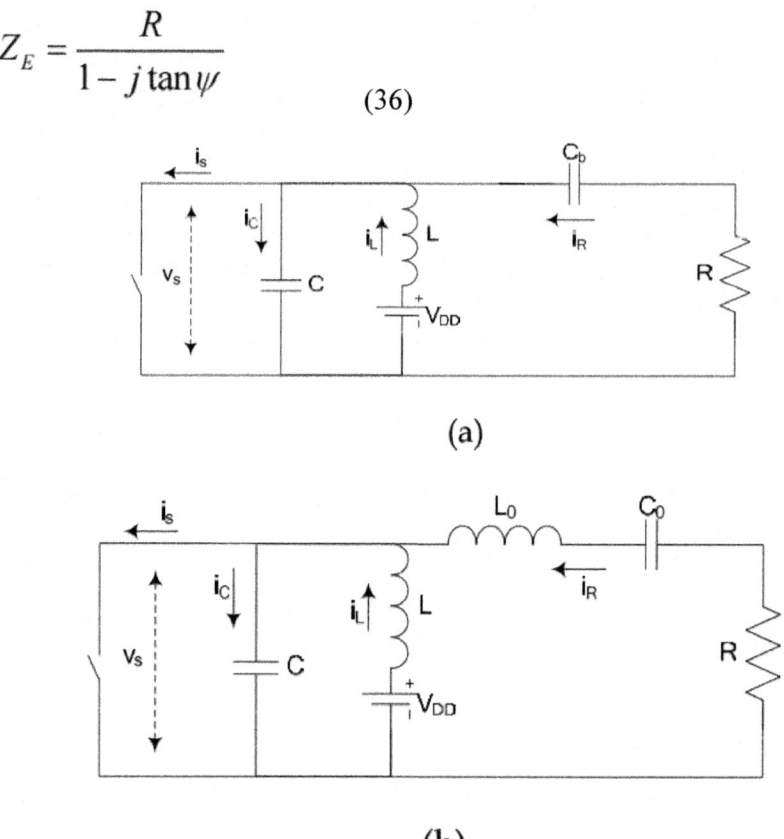

(a)

(b)

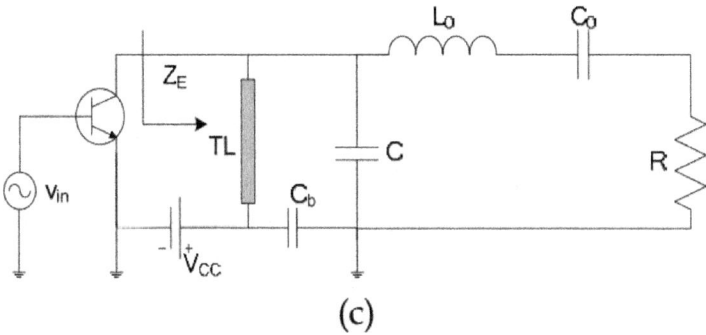

(c)

Figure 12: Parallel circuit Class E topology (a), parallel circuit Class E with output filter (b) and transmission line parallel Class E (c).

Finally, using equation (32) to determine the optimum required parallel inductance, the electrical length of the parallel transmission line can be obtained:

$$\tan \theta = 0.732 \frac{R}{Z_0} \qquad (37)$$

THE INVERSE CLASS E AMPLIFIER

The Inverse Class E amplifier, or voltage drive Class E amplifier, is commonly considered as the dual version of the Class E amplifier, in which current and voltage waveforms are interchanged, as shown in Fig. 13. It is also referred as "series-L/parallel tuned Class E", being the traditional topology defined as "parallel-C/series-tuned Class E" (Mury & Fusco, 2005; Mury & Fusco, 2007).

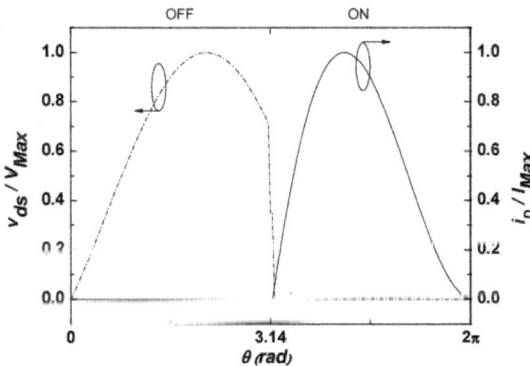

Figure 13: Ideal inverse Class E waveforms.

A former version of the Inverse Class E amplifier was reported in (Kazimierczuk, 1981): the circuit does not have shunt capacitor, while a series tuned filter and a finite DC feed inductance is considered in the output network, as depicted in Fig. 14. Although this circuit seems to be similar to those reported in Fig. 12, it implies a different behavior, due to the different characteristic of the shunting element (an inductor instead of a capacitor). When the switch is open, and provided a high enough Q factor of the series filter, the only current flowing in the circuit is the sinusoidal output current i_R, that is the inductor current i_L. The latter causes a voltage drop across the inductor, v_L, which has a cosinusoidal form. When the switch is closed, the voltage across the inductor is instantaneously constant and equal to V_{DD}. This causes a linear increase in the current i_L. The current across the switch is calculated as the difference between i_L and i_R and assumes the typical asymmetrical shape.

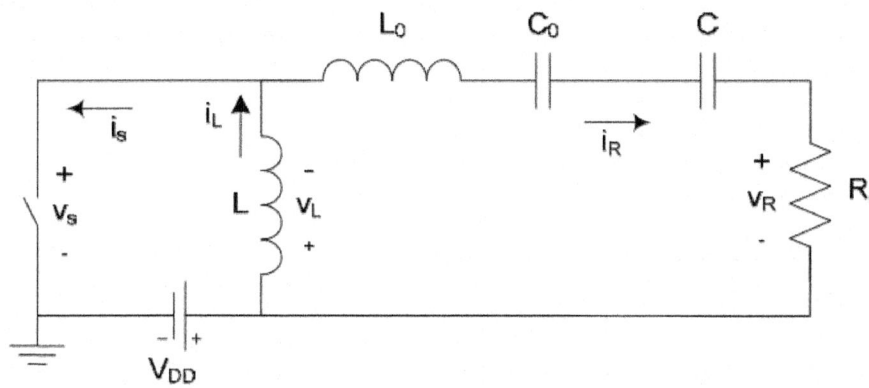

Figure 14: Inverse Class E amplifier: no-shunt-capacitor/series-tuned topology with finite inductance.

A complete analysis of the inverse Class E amplifier is reported for the first time in (Mury & Fusco, 2005; Mury & Fusco, 2007), together with a defined topology which is shown in Fig. 15 and which is substantially different from the previous version given in (Kazimierczuk, 1981). As can be seen by a comparison of Fig. 15 and the circuit depicted in Fig. 2, each component of the traditional Class E amplifier has been replaced by its dual element in a dual configuration. A DC blocking capacitance C_b is inserted in order to prevent inductance L_0 from shorting the bias voltage.

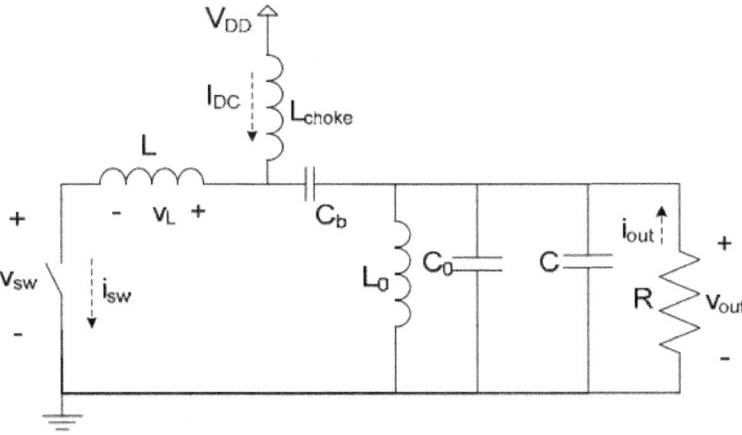

Figure 15: Inverse Class E amplifier.

Hence, the analysis of the inverse Class E amplifier can be carried out starting from the assumption of a purely sinusoidal output voltage across the output resistance R, which produces a voltage across the inductor L given by:

$$v_L(\theta) = V_{DD} - v_o(\theta) = V_{DD} \cdot (1 - a \cdot \sin(\theta + \phi)) \qquad (38)$$

This is the voltage present across the switch during the OFF time, while during the ON time the switch has no voltage across it and its current is given by integration of (38):

$$i_{SW} = \frac{1}{\omega L} \cdot \int_0^\theta v_L(\theta) d\theta \qquad (39)$$

These expressions have the same form of those reported in paragraph 4.1, unless current and voltage are interchanged: the same kind of analysis as Class E can be performed on the Inverse Class E circuit. As a consequence, the same numerical results are obtained for the dual configuration, and are summarized in Table 1, referred to a 50% duty cycle operation. As can be seen, the maximum allowable voltage for the Inverse Class E operation is much smaller than for Class E: this is an unquestionable advantage of such a circuit, because the requirement on device breakdown can be drastically relaxed.

However, it is worth to notice that in Inverse Class E amplifier the output capacitance of the active device is not taken into account and set to zero in the ideal analysis: in real world circuit, this is clearly not true. Hence, some actions have to be taken in order to compensate its presence (e.g. a shunt inductance).

Table 1: Class E and Inverse Class E comparison

Circuit component or electrical value	Class E	Inverse Class E
V_{Max}	$3.562 \cdot V_{DD}$	$2.862 \cdot V_{DD}$
I_{Max}	$2.862 \cdot I_{DC}$	$3.562 \cdot I_{DC}$
C	$\dfrac{I_{DC}}{\pi \cdot \omega \cdot V_{DD}}$	$\dfrac{0.2116}{\omega^2 \cdot L}$
R	$\dfrac{0.1836}{\omega \cdot C}$	$\dfrac{1}{0.1836 \cdot \omega \cdot L}$
L	$\dfrac{0.2116}{\omega^2 \cdot C}$	$\dfrac{V_{DD}}{\pi \cdot \omega \cdot I_{DC}}$
C_0	$\dfrac{1}{\omega \cdot Q \cdot R}$	$\dfrac{Q}{\omega \cdot R}$
L_0	$\dfrac{1}{\omega^2 \cdot C_0}$	$\dfrac{1}{\omega^2 \cdot C_0}$

CLASS F AND CLASS F-1 AMPLIFIER

Among the different switched mode amplifier design strategies, several authors include the Class F and inverse F (or Class F-1) schemes also.

Class F Theoretical Analysis

Such design strategy was introduced by Tyler at the end of fifties (Tyler, 1958) and further investigated several years later (Snider, 1967; Raab, 1996), as a simple and feasible way to improve power amplifier large-signal performances. Following the widely accepted definition, Class-F design consists in terminating the device output with open-circuit terminations at odd harmonic frequencies of the fundamental component and short-circuiting it at even harmonics. Regarding the input network, it is typically synthesized in order to guarantee maximum power transfer at the operating frequency, neglecting or at least circumventing the device input non linear generation by using short-circuit terminations.

The theoretical output voltage and current waveforms of an ideal Class F PA are depicted in Fig. 16.

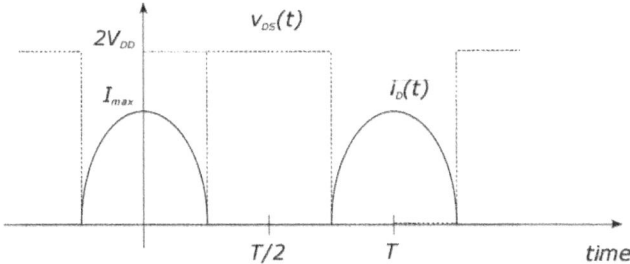

Figure 16: Ideal output voltage (v_{DS}) and current (i_D) waveforms of a Class F PA.

The current is assumed as a truncated sinusoid waves (assuming a Class B bias condition), while the voltage is squared by the proper harmonic loading conditions. The two ideal waveforms do not overlap: no output current flows for high drain/collector voltages and maximum current occurs when drain/collector voltage waveform is at its minimum. Therefore the power dissipated in the active device is nulled ($P_{diss}=0$).

Such an ideal waveforms can be easily described by their Fourier components:

$$i_D(\theta) = \sum_{n=0}^{\infty} I_n \cdot \cos(n\theta) \qquad I_n = \begin{cases} \dfrac{I_{Max}}{\pi} & n=0 \\ \dfrac{I_{Max}}{2} & n=1 \\ \dfrac{2 \cdot I_{Max}}{\pi} \dfrac{(-1)^{\frac{n}{2}+1}}{n^2-1} & n \text{ even} \\ 0 & n \text{ odd} \end{cases} \qquad (40)$$

$$v_{DS}(\theta) = \sum_{n=0}^{\infty} V_n \cdot \cos(n\theta) \qquad V_n = \begin{cases} V_{DD} & n=0 \\ -\dfrac{4 \cdot V_{DD}}{\pi} & n=1 \\ 0 & n \text{ even} \\ \dfrac{4 \cdot V_{DD}}{\pi} \dfrac{(-1)^{\frac{n+1}{2}}}{n} & n \text{ odd} \end{cases} \qquad (41)$$

where I_{Max} and V_{DD} the maximum output current and bias voltage, respectively. From the previous equations it can be noted that the current and voltage Fourier components with the same order n are alternatively zeroed, thus nulling the power delivered at harmonic frequencies also ($P_{out,nf}=0$, n>1).

The values of the ideal terminations are inferred as the ratio between the respective Fourier components V_n and I_n, i.e.:

$$Z_n = \frac{V_n}{I_n} = \begin{cases} \frac{8}{\pi} \cdot \frac{V_{DD}}{I_{Max}} & n = 1 \\ 0 & n \text{ even} \\ \infty & n \text{ odd} \end{cases} \quad (42)$$

Therefore resulting in a purely resistive terminating impedance at fundamental frequency given by $4/\pi \cdot R_{TL}$, being R_{TL} the optimum impedance of a Tuned Load scheme (i.e. short circuiting all the harmonic terminations). Conversely, short circuit condition for even harmonics and open circuit for odd ones have to be synthesized, as schematically depicted in Fig. 17.

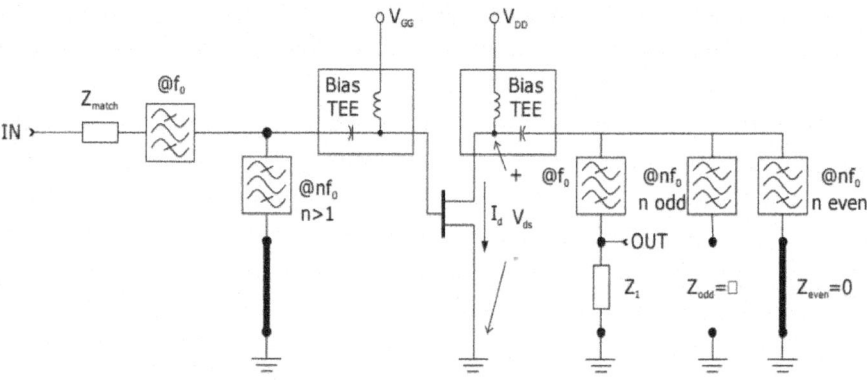

Figure 17: Ideal structure of a Class F amplifier.

The results obtained following the Class F strategy were so interesting that, before the advent of fast switching devices, such approach was widely adopted to design PAs for amplitude modulated (AM) broadcast radio transmitters (operating at LF 30-300 KHz, MF 0.3-3 MHz and HF 3-30 MHz) or for frequency modulated (FM) broadcast radio transmitters (at VHF 30-300 MHz and UHF 0.3-3 GHz) (Wood, 1992; Lu, 1992).

Nowadays, the Class F technique is generally adopted for high frequency applications in the microwave range (i.e. up to tens of Gigahertz). Examples of Class F based on GaAs devices are available at X (9.6GHz) (Colantonio et al., 2007), Ku (14.5GHz) (Ozalas, 2005) and Ka (29.5GHz) (Reece et al., 2003) bands. For high frequency applications the active devices operate in current-mode rather than in switched-mode, and the harmonic loading conditions are implemented through lumped resonating circuits.

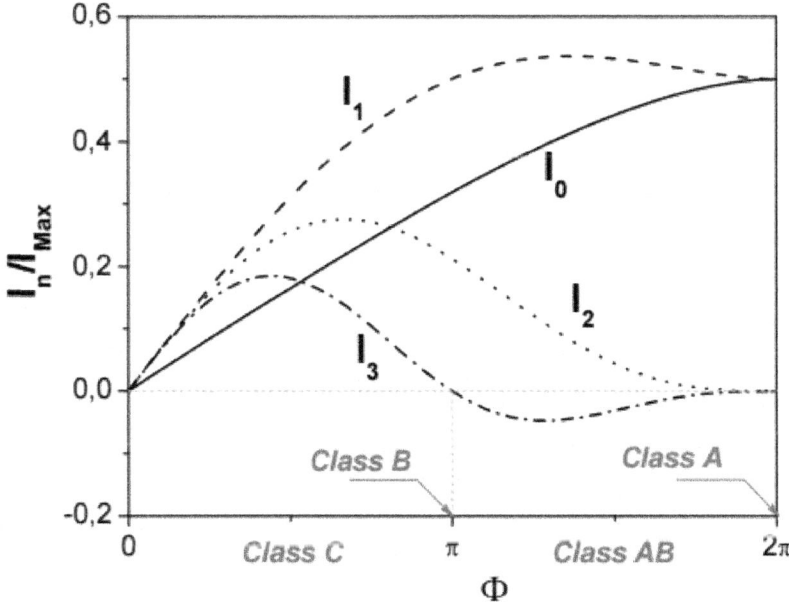

Figure 18: Fourier components I_0, I_1, I_2, I_3 normalised to the device maximum current I_{Max}, as functions of the drain CCA F.

Several research efforts were focused to clarify and to implement the harmonic terminating scheme leading to the Class F optimum behavior and its experimental validation, inferring practical design guidelines also (Duvanaud et al., 1993; Blache, 1995; Colantonio et al., 1999).

The analytical results can be easily extended to bias conditions different from Class B, still assuming for the current waveform a truncated sinusoidal wave shaping, with a conduction angle (CCA) F larger than p (Class AB), while maintaining a square voltage waveform. In this case the corresponding current harmonic components as function of the CCA depicted in Fig. 18 result, while the estimated Class F performances are depicted in Fig. 19, compared with the corresponding theoretical one for a Tuned Load condition.

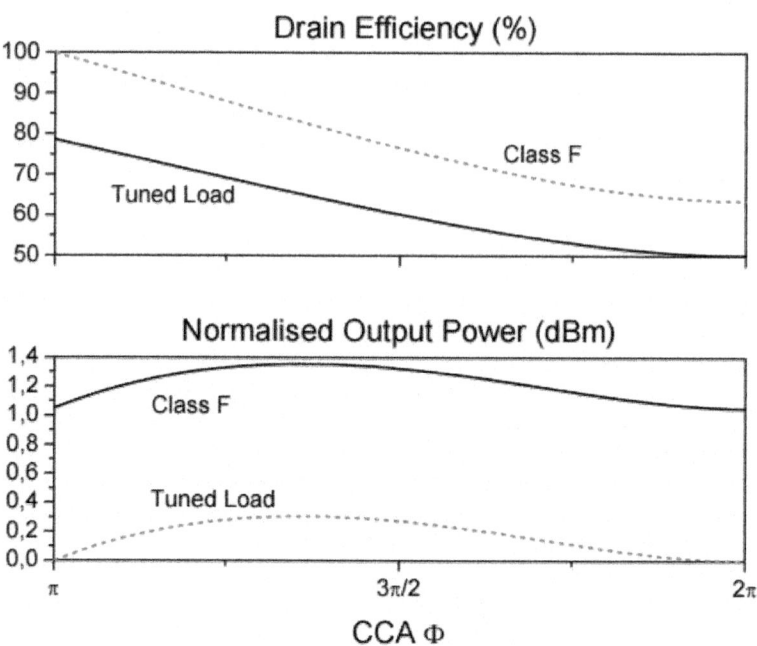

Figure 19: Ideal performance for a Class F and reference TL amplifiers as functions of the CCA F. Output power normalized to the Class A output at full swing.

Class F Practical Limitations

Referring to the ideal solution proposed to implement the Class F loading scheme, depicted in Fig. 17, it is based on the use of resonating elements ensuring low impedance values (theoretically short-circuit conditions) for even harmonic frequencies or high values for the odd harmonics (theoretically open-circuit conditions).

Clearly, for a simple and manageable mathematical formulation, the active device model has to be strongly simplified, neglecting all parasitic and reactive elements. It implies that when dealing with an actual device, the loading condition expressed by (42) has to be fulfilled across the device intrinsic output current source, as already proposed in other similar simplified approaches (Cripps, 1983). From a practical point of view, the required ideal terminations pose significant restrictions when implementing the Class technique. In fact, while it is relatively simple to realize a short-circuit termination, compensating for the device reactive elements, the realization of an open-circuit condition becomes much more cumbersome. For example, in high frequency applications the dominant capacitive behavior shown by the active device (e.g. C_{ds}) tends to short circuit the device output itself, thus practically not allowing the open-circuit loading condition for the higher-order odd harmonics.

At the same time, even if the internal C_{ds} capacitance can be effectively resonated by an external inductive element, the device output resistance (R_{ds}) cannot be removed, thus representing an upper limit for the impedance effectively synthesizable across the intrinsic current source. Therefore the realization of a true open termination is basically unfeasible. However, it has been demonstrated that lower-order harmonics are more effective in improving amplifier performance as compared to higher order ones (Raab, 2001). Therefore, accounting for the reduced number of harmonics can be effectively controlled, new terminating impedance values have been inferred not only at fundamental but also at harmonics, resulting in a different voltage harmonic ratio also (Raab, 1997; Colantonio et al., 2009). In fact, the new optimum voltage ratio becomes $|V_3/V_1|=1/6$ rather than 1/3 as in the ideal case. Simultaneously, the fundamental loading impedance becomes

$$R_F = \frac{4}{\sqrt{3}} \cdot \frac{V_{DD}}{I_{Max}}$$

(43)

resulting in an efficiency improvement of 15% only with respect to the Tuned Load theoretical case (Colantonio et al., 2009).

A further critical point is represented by the physical mechanisms generating the harmonic components of both voltage and current waveforms. If the device output only is considered, it can be described by an independent and forcing current source, whose waveform results both from the input drive level and the device physical limitations (clipping effects), being independent on the device terminating impedances. Under this assumption, the output voltage waveform is dependent on the current one, being generated by loading each harmonic current component through the respective terminations ($V_n=Z_n \times I_n$). Consequently a proper phase relationships between the output current harmonic components must be fulfilled, and in particular I_1 and I_3 must be opposite in sign. Such a condition, referring to Fig. 18 practically implies the selection of a suitable bias level close to the Class B condition, i.e. assuming a non-zero quiescent current level (deep AB bias), while leaving for instance the same harmonic terminations as derived in the ideal case. The theoretical load curve behaves as depicted in Fig. 20.

Conversely, the adoption of a Class C bias condition implies a wrong relationships among the current harmonics phase (see Fig. 18).

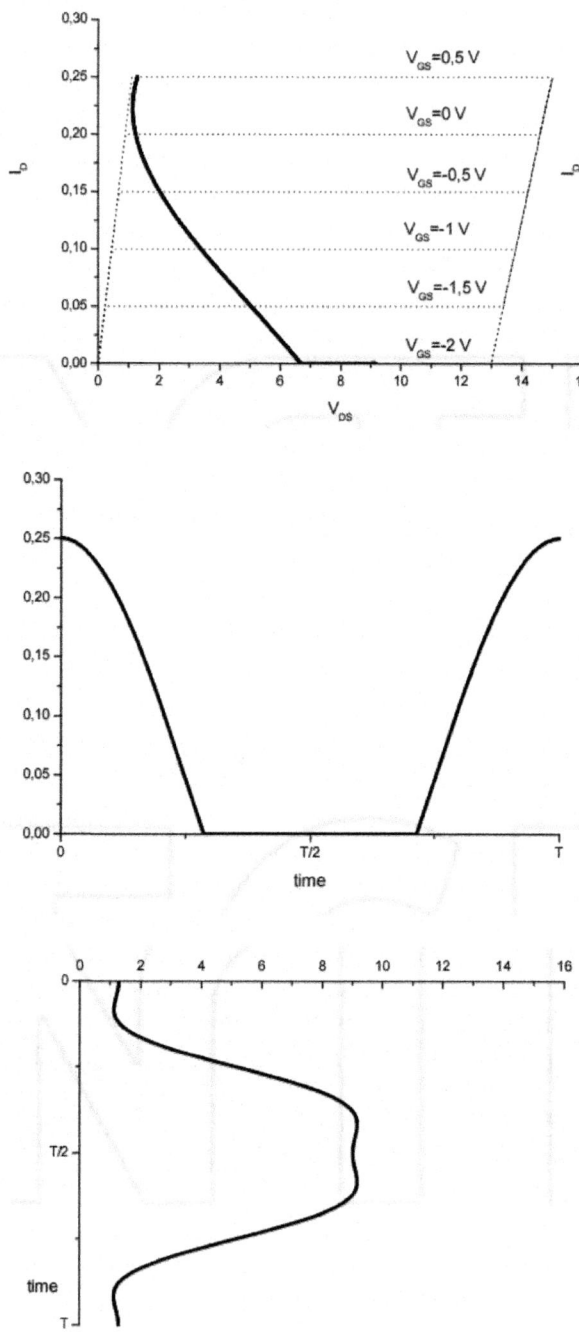

Figure 20: Theoretical load line for a Class F amplifier.

Class F Matching Networks Implementation

To design the Class F output section, resonating circuits are typically used as traps for the harmonic frequencies. Such resonators, sometimes referred as idlers, are limited to loworder harmonics, to reduce both circuit complexity and associated losses. Moreover, depending on the frequency range, their implementation can be performed either in lumped or distributed form, while accounting for also the DC bias.

For example, in Fig. 21 is depicted a possible implementation based on the use of a quarterwavelength transmission line (1/4-TL) and a parallel resonating L_0-C_0 tank to control the harmonic impedances. The optimum matching at fundamental frequency f0 is achieved by the remaining passive components L_m and C_m (Gao, 2006).

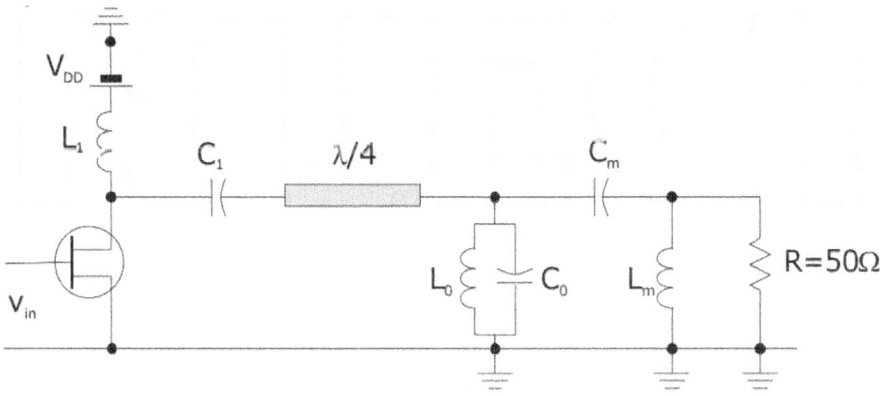

Figure 21: Example of Class F output network design.

In practical situations, to account the biasing elements and the active device output capacitance C_{ds}, other proposed solutions are schematically depicted in Fig. 22, where the design relationships to calculate the element values can be derived evaluating the impedance loading the device output current source and then imposing the short circuit condition at $2f_0$ and the open circuit one at $3f_0$. (Trask, 1999).

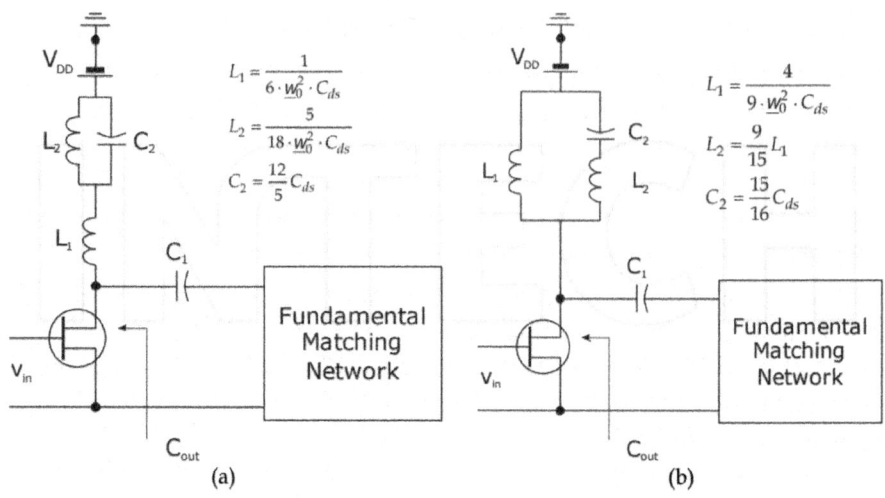

Figure 22: Practical implementations of Class F amplifier.

In a similar way, a distributed solution can be designed and implemented by using transmission lines, as for instance reported in Fig. 23 (Grebennikov, 2000; Woo et al., 2006; Negra et al., 2007).

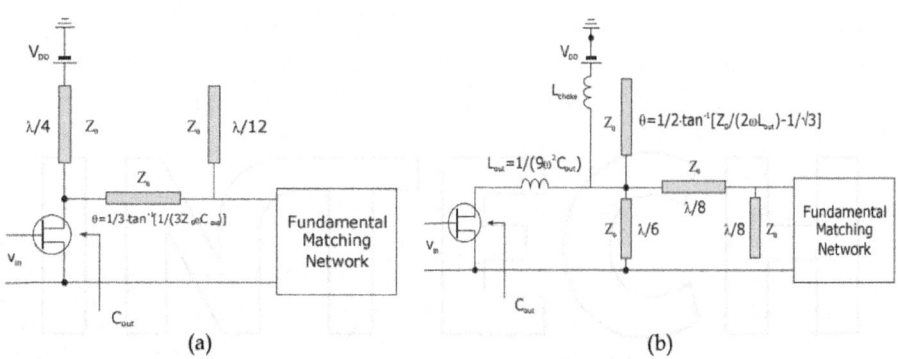

Figure 23: Distributed solutions implementing the Class F schemes.

Clearly the characteristic impedances of the TLs have to be properly selected when designing the matching network at fundamental frequency.

Class F Example

An example of hybrid (MIC) Class F PA operating at 5GHz and based on GaAs device (1mm gate periphery) is reported in Fig. 24, where both the input and output matching networks were designed on Alumina substrate by using a dixtributed approach.

Figure 24: Photo of a MIC 5GHz GaAs Class F amplifier.

The corresponding load curve and the performance as compared to a Tuned Load amplifier based on the same active device and bias point (Class B), are reported in Fig. 25 and Fig. 26, respectively.

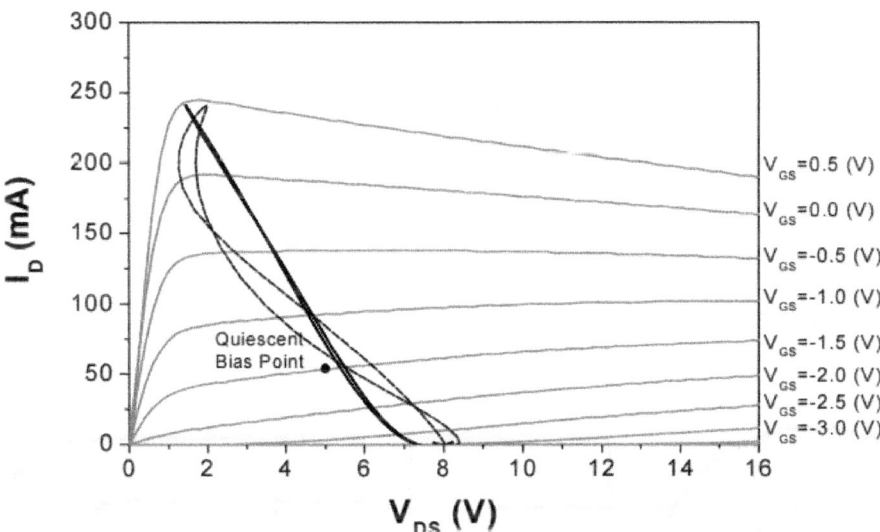

Figure 25: Load curve of 5GHz MIC Class F amplifier (dashed curve) compared with Tuned Load case (continuous curve).

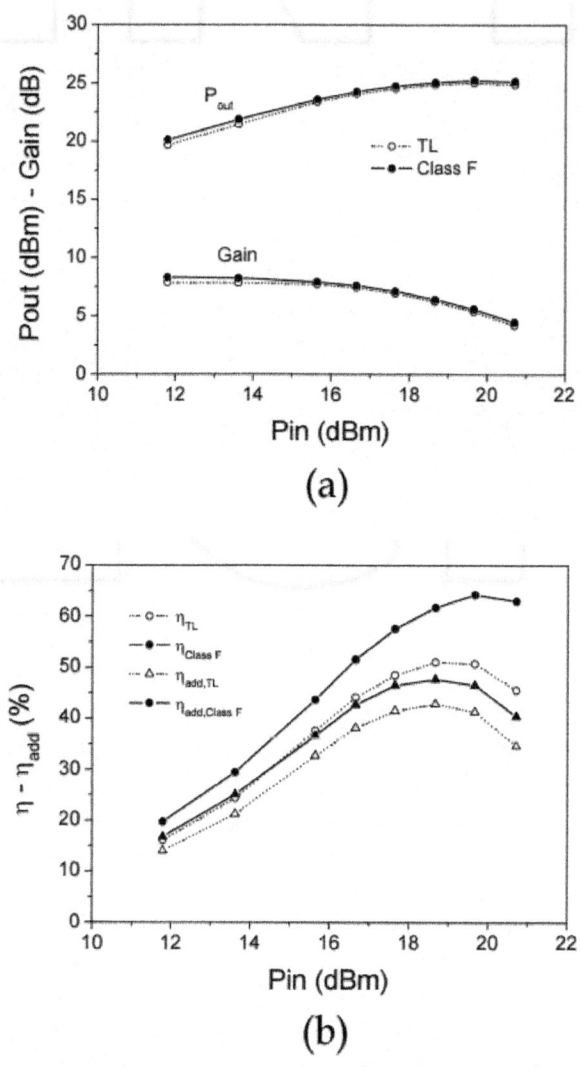

Figure 26: Output Power & Gain (a) and efficiency & power added efficiency (b) for the 5GHz MIC Class F amplifier as compared to Tuned Load amplifier.

Among the other features, it is to note that the Class F amplifier output power is higher as compared to the Tuned Load approach in the entire range of input drive. A 7-8% measured improvement (against a maximum theoretical 15% expected) is usually obtained. It is to note that at low input power levels, a higher power gain with respect to a Tuned Load is obviously expected due to the larger output load that has been assured at fundamental frequency, as given by (43). Increasing the input drive level, the use of a Class F strategy force the output voltage (and current) to approach their maximum swings value, resulting in a proper bending of the load curve (see Fig. 25), thus improving the output power at saturation level also.

The Inverse Class F Amplifier

The term "inverse" in this kind of amplifier suggests somewhat inverted in the behavior: in fact, current and voltage drain waveforms are interchanged if compared to a traditional Class F amplifier (Ingruber et al., 1998; Lepine et al., 2007; Woo et al., 2006). Inverse Class F power amplifier was introduced as a "rectangular driven Class A harmoniccontrolled amplifier", in order to combine the advantage of high efficiency of a switched mode amplifier with the high gain of Class A operation (Ingruber et al., 1998).

It employs a rectangular driving voltage, which forces the active device to operate in the ohmic region or in the interdiction region, thus justifying - more than for a Class F - the classification as a switched mode amplifier. Assuming a piecewise linear simplified model for the active device, the drain current waveform can be directly inferred, resulting in a rectangular waveform. Thus, open terminations for even harmonics and short circuit terminations for odd harmonics - except for the fundamental one - give a truncated sinusoid voltage waveform. Current and voltage waveform do not overlap, thus preventing DC power consumption on the active device; additionally, thanks to the proper harmonic terminations, no power is delivered at harmonics of fundamental frequency and 100% drain efficiency is ideally achievable.

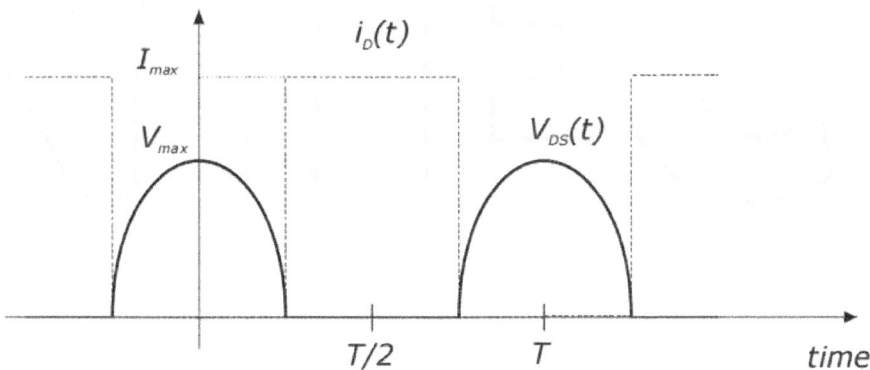

Figure 27: Ideal output voltage (v_{DS}) and current (i_D) waveforms of an inverse Class F PA.

Referring to the ideal output current and voltage waveforms of Fig. 27, DC and fundamental frequency components can be obtained using a Fourier analysis:

$$I_0 = \frac{I_{Max}}{2} \quad (44)$$

$$I_1 = \frac{1}{\pi}\int_{\frac{\pi}{2}}^{\frac{3\pi}{2}} I_{Max} \cdot \cos(\vartheta) d\vartheta = \frac{2 \cdot I_{Max}}{\pi} \quad (45)$$

$$V_0 = V_{DD} = \frac{1}{2\pi}\int_{-\frac{\pi}{2}}^{\frac{\pi}{2}} V_{Max} \cos(\vartheta) d\vartheta = \frac{V_{Max}}{\pi} \quad (46)$$

$$V_1 = \frac{1}{\pi}\int_{-\frac{\pi}{2}}^{\frac{\pi}{2}} V_{Max} \sin(\vartheta)^2 d\vartheta = \frac{V_{Max}}{2} \quad (47)$$

Then, the expression of harmonic load impedance is given by:

$$Z_n = \frac{V_n}{I_n} = \begin{cases} \frac{\pi^2}{4} \cdot \frac{V_{DD}}{I_{Max}} & n = 1 \\ \infty & n \text{ even} \\ 0 & n \text{ odd} \end{cases} \quad (47)$$

An inverse Class F amplifier requires an infinite number of resonator in order to maintain the square-shaped output current. Fig. 28 shows an immediate ideal implementation of an inverse Class F amplifier using multiple even-harmonic resonators to control the voltage and current waveforms (Grebennikov & Sokal, 2008).

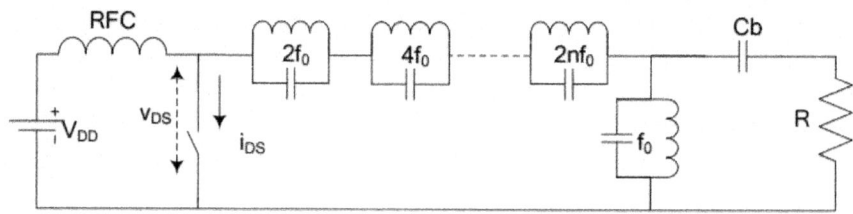

Figure 28: Inverse Class F amplifier with multiharmonic resonator circuit.

In real world design, however, a limited number of harmonics has to be taken into account. Practical output networks implementations are often based on multiharmonic quarterwavelenght stubs (Lepine et al., 2007; Woo et al., 2006). Input network design is the most critical aspect in inverse Class F design. In fact, if a continuous wave excitation is put on the gate, the required square wave driving voltage occurs only in deep saturation regime, thus affecting the gain. Ideal conditions would require a square wave driving voltage throughout the whole dynamic range of the amplifier. This can be obtained providing fundamental frequency and its second harmonic as a driving signal (Goto et al., 2001; Grebennikov & Sokal, 2008), so that an out of phase condition exist at their maximum amplitude. A network example which satisfied this condition is reported in Fig. 29.

Referring to switched mode finite harmonic operation, it is interesting to find a qualitative relation between the different classes of amplifier, as reported in (Raab, 2001). The transition from inverse Class F and Class F amplifiers is achieved by moving the second harmonic impedance (reactance) from zero to ∞, while decreasing the third harmonic impedance (reactance) from ∞ to zero, while in the intermediate condition, in which both second and third harmonic reactances have finite values, current and voltage waveforms assumes a resemblance to Class E amplifier.

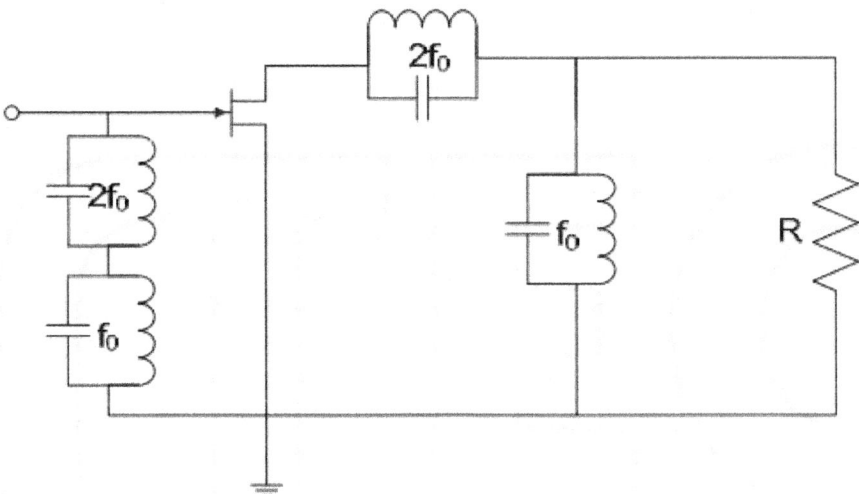

Figure 29: Bi-harmonic inverse Class F amplifier.

REFERENCES

1. Blache, F. (1995). A Novel Computerised Multiharmonic Active Load-Pull System for the Optimisation of High Efficiency Operating Classes in Power Transistors, IEEE MTT-S Int. Microwave Symp. Dig., Orlando, FL, June 1995, pp.1037-1040.
2. Cipriani, E.; Colantonio, P.; Giannini, F. & Giofré, R. (2008). Optimization of Class E Power Amplifier Design above Theoretical Maximum Frequency, Proceedings of 38th European Microwave Conference, EuMC 2008, pp. 1541 - 1544, Amsterdam, October 2008.
3. Colantonio, P.; Giannini, F.; Leuzzi, G. & Limiti, E. (1999). On the Class-F Power Amplifier Design, Intern. Journal of RF and Microwave Computer-Aided Engineering, Vol. 9, No. 2, March 1999, pp.129-149.
4. Colantonio, P.; Giannini, F.; Leuzzi, G. & Limiti, E. (2003). Theoretical Facet and Experimental Results of Harmonic Tuned PAs, Intern. Journal of RF and Microwave Computer-Aided Engineering, Vol.13, No. 6, November 2003, pp. 459-472.
5. Colantonio, P.; Giannini, F.; Giofrè, R.; Medina, M. A. Y.; Schreurs, D. & Nauwelaers, B. (2005). High frequency class E design methodologies, Proceedings of European Gallium Arsenide and Other Semiconductor Application Symposium, GAAS 2005, pp. 329 - 332, Paris, October 2005.
6. Colantonio, P.; Giannini, F.; Giofrè R. & Limiti, E. (2007). Combined Class F Monolithic PA Design, Microwave and Optical Technology Letters, Vol. 49, No. 2, February 2007, pp. 360-362.
7. Colantonio, P.; Giannini, F. & E. Limiti (2009). High Efficiency RF and Microwave Solid State Power Amplifiers, John Wiley & Sons (ISBN: 978-0-470-51300-2).
8. Cripps, S.C. (1983). A theory for the prediction of GaAs FET Load-Pull power contours, IEEE MTT-S Int. Microwave Symp. Dig., Boston, MA, May 1983, pp. 221-223.
9. Cripps, S.C. (1999). RF Power Amplifiers for Wireless Communications. Artech House, Norwood, MA (ISBN 0-89226-989-1).
10. Cripps, S. C. (2002). Advanced Techniques in RF Power Amplifier Design, Artech House, London.
11. Duvanaud, C.; Dietsche, S.; Pataut, G. & Obregon, J. (1993). High-Efficient Class F GaAs FET Amplifiers Operating with Very Low Bias Voltages for Use in Mobile Telephones at 1.75 GHz, IEEE Microwave and Guided Wave Letters, Vol.3, No. 8 , August 1993, pp.268- 270.

12. Gao, S. (2006). High efficiency class-F RF/microwave power amplifiers, IEEE Microwave Magazine, Vol. 7, No. 1, February 2006, pp. 40 – 48.
13. Goto, S.; Kunii, T.; Ohta, A.; Inoue, A.; Hosokawa, Y.; Hattori, R. & Mitsui, Y. (2001). Effect of Bias Condition and Input Harmonic Termination on High Efficiency Inverse Class-F Amplifiers, 31st European Microwave Conference, 2001. Oct. 2001, pp. 1 - 4.
14. Grebennikov, A.V. (2000). Circuit design technique for high efficiency Class F amplifiers, IEEE MTT-S Int. Microwave Symp. Dig., Boston, MA, June 2000, Vol. 2, pp. 771–774.
15. Grebennikov, A. (2003). Switched-mode RF and microwave parallel-circuit Class E power amplifiers, Intern. Journal of RF and Microwave Computer-Aided Engineering, Vol. 14, No. 1, December 2003, pp. 21-35.
16. Grebennikov, A. & Sokal, N.O. (2007). Switchmode RF power amplifiers, Newnes, Elsevier, Burlington, MA, USA.
17. Ingruber, B.; Baumgartner, J.; Smely, D.; Wachutka, M.; Magerl, G. & Petz, F.A. (1998). Rectangularly driven class-A harmonic-control amplifier, IEEE Transactions on Microwave Theory and Techniques, Vol. 46, No. 11, Part 1, November 1998, pp. 1667 - 1672.
18. Jozwik, J. J. & Kazimierczuk, M. K. (1990). Analysis and design of class-E^2 dc/dc converter, IEEE Trans. Ind. Electron., Vol. 37, No. 2, April 1990, pp. 173-183.
19. Kazimierczuk M.K. (1981). Class E tuned power amplifier with shunt inductor, IEEE Journal of Solid-State Circuits, Vol. 16, No. 1, February 1981, pp. 2 – 7.
20. Kazimierczuk, M.K. (2008). RF Power Amplifiers, Wiley, New York.
21. Kazimierczuk M. K. & Jozwik, J. (1990). Analysis and design of class E zero-currentswitching rectifier, IEEE Trans. Circuits Syst., Vol. 37, No. 8, August 1990, pp. 1000- 1009.
22. Krauss, H. L.; Bostian, C.W. & Raab, F. H. (1980). Solid State Radio Engineering, Wiley, New York.
23. Lai, J. B. (2008). Investigation into the use of high-efficiency, switched mode Class E power amplifier for high-dynamic range, pulse-mode application, PhD dissertation, University of New Mexico Albuquerque, New Mexico, December, 2008.
24. Lepine, F.; Adahl, A. & Zirath, H. (2007) L-band LDMOS power amplifiers based on an inverse class-F architecture, IEEE Transactions on Microwave Theory and Techniques, Vol. 53, No. 6, Part 2, June 2005, pp. 2007 - 2012.

25. Lu, X. (1992). An alternative approach to improving the efficiency of high power radio frequency amplifiers, IEEE Transaction on Broadcasting, Vol. 38, No. 2, June 1992, pp. 85–89.
26. Mader, T.B.; Bryerton, E.W.; Markovic, M.; Forman, M. & Popović, Z. (1998). Switched-mode high-efficiency microwave power amplifiers in a free-space power-combiner array, IEEE Transactions on Microwave Theory and Techniques, Vol. 46, No. 10, Part 1, October 1998, pp. 1391 – 1398.
27. Mury, T. & Fusco, V.F. (2005). Series-L/parallel-tuned comparison with shunt-C/seriestuned class-E power amplifier, IEE Proceedings Circuits, Devices and Systems, December 2005, pp. 709 - 717.
28. Mury, T. & Fusco, V.F. (2007). Inverse Class-E Amplifier With Transmission-Line Harmonic Suppression, IEEE Transactions on Circuits and Systems I: Regular Papers, Vol. 54, No. 7, July 2007, pp. 1555 - 1561.
29. Negra, R.; Ghannouchi, F. M. & Bächtold, W. (2007). Study and Design Optimization of Multiharmonic Transmission-Line Load Networks for Class-E and Class-F K-Band MMIC Power Amplifiers, IEEE Transactions on Microwave Theory and Techniques, Vol. 55, No. 6, Part 2, June 2007, pp. 1390 - 1397.
30. Ozalas, M.T. (2005). High efficiency class-F MIMIC power amplifiers at ku-band, 2005 IEEE Wireless and Microwave Technology Conference, WAMICON 2005, pp. 137 – 140.
31. Raab, F.H. (1996). Introduction to Class-F Power Amplifiers, RF Design, Vol.19, No. 5, May 1996, pp.79-84.
32. Raab, F.H. (1997). Class-F Power Amplifiers with Maximally Flat Waveforms, IEEE Transactions on Microwave Theory and Techniques, Vol. 45, No. 11, November 1997, pp. 2007-2012.
33. Raab, F.H. (2001). Class-E, Class-C, and Class-F power amplifiers based upon a finite number of harmonics, IEEE Transactions on Microwave Theory and Techniques, Vol. 49, No. 8, August 2001, pp. 1462 – 1468.
34. Reece, M.A.; White, C.; Penn, J.; Davis, B.; Bayne, M.Jr; Richardson, N.; Thompson, W.I.I. & Walker, L. (2003). A Ka-band class F MMIC amplifier design utilizing adaptable knowledge-based neural network modeling techniques, IEEE International Microwave Symposium Digest, June 2003, Vol. 1, pp. 615 – 618.
35. Snider, D.M. (1967). A Theoretical Analysis and Experimental Confirmation of the Optimally Loaded and Overdriven RF Power Amplifiers, IEEE Transaction on Electron Devices, Vol. 14, No. 6, June

1967, pp.851-857.
36. Sokal N. O. (2001). Class-E RF power amplifiers, QEX (published by American Radio Relay League, 225 Main St., Newington, CT 06111-1494, USA), January - February 2001, pp 9–20.
37. Sokal, N.O. & Sokal, A.D. (1975). Class E - A new class of high efficiency tuned single-ended switching power amplifiers, IEEE Journal of Solid State Circuits, Vol. SC-10, No. 3, June 1975, pp. 168-176.
38. Suetsugu, T & Kazimierczuk M.K. (2005). Voltage-clamped class E amplifier with transmission-line transformer, IEEE International Symposium on Circuits and Systems, ISCAS 2005, May 2005, Vol. 1, pp. 712 - 715.
39. Suetsugu, T. & Kazimierczuk M.K. (2007). Off-Nominal Operation of Class-E Amplifier at Any Duty Ratio, IEEE Transactions on Circuits and Systems I: Regular Papers, Vol. 54, No. 6, June 2007, pp. 1389 - 1397.
40. Trask, C. (1999). Class-F Amplifier Loading Network: A Unified Design Approach, IEEE International Microwave Symposium Digest, Vol. 1, June 1999, pp. 351 – 354.
41. Tyler, V.J. (1958). A new high efficiency high power amplifier, Marconi Rev. Vol. 21, No. 130, Fall 1958, pp. 96-109.
42. Wilkinson A. J. & Everard J. K. A. (2001). Transmission-Line Load-Network Topology for Class-E Power Amplifiers, IEEE Transactions on Microwave Theory and Techniques, Vol. 49, No. 6, June 2001, pp. 1202 - 1210.
43. Woo, Y.Y.; Yang, Y. & Kim, B. (2006). Analysis and Experiments for High-Efficiency Class-F and Inverse Class-F Power Amplifiers, IEEE Transaction on Microwave Theory and Techniques, Vol. 54, No. 5, May 2006, pp. 1969 - 1974.
44. Wood, J. (1992). The history of International broadcasting, IET History of Technology Series 19, 1992 (ISBN 0863413021)
45. Xu H.; Gao, S.; Heikman, S.; Long S. I.; Mishra, U. K. & York, R. A. (2006). A High-Efficiency Class-E GaN HEMT Power Amplifier at 1.9 GHz, IEEE Microwave And Wireless Components Letters, Vol. 16, No. 1, January 2006, pp. 22 - 24.

CITATION

CHAPTER 1
Yuanyuan Li, Wenke Lu, Changchun Zhu, Qinghong Liu, Haoxin Zhang, and Chenchao Tang, "Circuit Design of Surface Acoustic Wave Based Micro Force Sensor," Mathematical Problems in Engineering, vol. 2014, Article ID 701723, 9 pages, 2014. doi:10.1155/2014/701723.

CHAPTER 2
P. Russo, F. Yengui, G. Pillonnet, S. Taupin and N. Abouchi, "Switching Optimization for Class-G Audio Amplifiers with Two Power Supplies," Circuits and Systems, Vol. 3 No. 1, 2012, pp. 90-98. doi: 10.4236/cs.2012.31012.

CHAPTER 3
R. Stabile and K.A. Williams, Photonic Integrated Semiconductor Optical Amplifier Switch Circuits, ISBN 978-953-307-186-2.

CHAPTER 4
Peter Beshay, Joseph F. Ryan and Benton H. Calhoun, A Digital Auto-Zeroing Circuit to Reduce Offset in Sub-Threshold Sense Amplifiers, doi: 10.3390/jlpea3020159.

CHAPTER 5
Felipe Padilla, Aurora Torres, Julio Ponce, María Dolores Torres, Sylvie Ratté and Eunice Ponce-de-León (2011). Evolvable Metaheuristics on Circuit Design, Advances in Analog Circuits, Prof. Esteban Tlelo-Cuautle (Ed.), ISBN: 978-953-307-323-1, InTech, DOI: 10.5772/14688

CHAPTER 6

H. Higa and N. Nakamura, "Design of a Switched Capacitor Negative Feedback Circuit for a Very Low Level DC Current Amplifier," Circuits and Systems, Vol. 4 No. 4, 2013, pp. 356-363. doi: 10.4236/cs.2013.44048.

CHAPTER 7

Huiyong Li, Xun Li and Chen Wei, The analysis of the performance of multibeamforming in memory nonlinear power amplifier, DOI: 10.1186/1687-6180-2014-52.

CHAPTER 8

A high dynami Lei Zhang, Zhiping Yu, Xiangqing He, A high dynamic range ultralow-current-mode amplifier with pico-ampere sensitivity for biosensor applications, 10.1007/s10470-009-9342-6

CHAPTER 9

Francesco Cannone and Gianfranco Avitabile, A 12-bit track and hold amplifier for giga-sample applications, DOI 10.1007/s10470-015-0556-5.

CHAPTER 10

E. Tlelo-Cuautle, C. Sánchez-López, E. Martínez-Romero, Sheldon X.-D. Tan, Symbolic analysis of analog circuits containing voltage mirrors and current mirrors, DOI 10.1007/s10470-010-9455-y

CHAPTER 11

Ivailo Pandiev, Design and stability analysis of CFOA-based amplifier circuits using Bode criterion, DOI:10.1080/21642583.2015.1043672.

CHAPTER 12

Jau-Ji Jou and Cheng-Kuang Liu, Equivalent Circuit Models for Optical Amplifiers, ISBN: 978-953-307-186-

CHAPTER 13

Elisa Cipriani, Paolo Colantonio, Franco Giannini and Rocco Giofrè, The Switched Mode Power Amplifiers, ISBN 978-953-307-031-5.

INDEX

A
Automatic temperature 73, 74

B
Bias Temperature Instability (BTI) 78
Broadband optical signal 39

C
Computer aided design (CAD) 91
Current-feedback operational amplifiers (CFOAs) 195
Current mirror (CM 181, 182

D
DC/DC converter 27
Digital auto-zeroing (DAZ) 74, 84
Dynamic compensation 74

E
Equivalent number of bits (ENOB) 164
Erbium-doped fiber amplifiers (EDFAs) 223
Evolutionary algorithms (EAs) 89
Evolvable hardware (EHW) 90

F
Fabry-Perot (FP) 233
Four terminal floating nullor (FTFN) 182

G
Gain–bandwidth–product (GBW) 156
Genetic Algorithm (GA) 29
Genetic algorithms (GAs) 98

H
Harmonic Tuning terminations (HT) 250
High output power 135
Hybrid electronic 38

I
Input–referred–noise–current (IRNC) 156
Input–referred–offset (IRO) 158
Integrated biosensing system (IBS) 149
Integrated circuits (IC) 91
Integrated electronic information system 135
interdigital transducers (IDT) 2

L

Laser diodes (LDs) 224

M

mathematical model 1, 2
mechanical system 1, 17
Methodology 74
microelectromechanical systems (MEMS) 2
Miniaturization 116, 122, 131

N

Negative Charge Pump (NCP) 24

O

oscillator circuits 1, 6, 17

P

Pattern Search (PS) 30
Perceptual Evaluation of Audio Quality (PEAQ) 28
Pervasive digital media 37
Power amplifier (PA) 135, 249
Printed circuit board (PCB) 196
Pseudo random bit 40

S

Sample & hold (S/H) 164
Semiconductor optical amplifiers (SOAs) 38, 223
Sense amplifiers (SAs) 74
Signal integrity 38, 39
Software defined radio (SDR) 164
Software radio (SR) 163
Solid state power amplifiers (SSPAs) 136
Spontaneous emission noise 39
Spurious-free dynamic range (SFDR) 163
surface acoustic wave (SAW) 1
Switched capacitor filter (SCF) 115, 116
Switched capacitor negative feedback circuit (SCNF) 115, 116
Switched capacitor (SC) 116
Switched emitter follower (SEF) 165

T

Total Harmonic Distortion (THD) 28
Track and hold amplifiers (THAs) 164
Track-and-hold amplifier (THA) 163, 179
Traveling wave (TW) 233

V

Very large scale integration (VLSI) 90
Voltage-feedback operational amplifiers (VFOAs) 195
Voltage follower (VF) 181
Voltage mirror (VM) 181, 182

Z

Zero Voltage Derivative Switching (ZVDS) 255
Zero Voltage Switching (ZVS) 255